301.1832
R31

126092

DATE DUE			

RESEARCH IN ORGANIZATIONAL BEHAVIOR

Volume 1 • 1979

RESEARCH IN ORGANIZATIONAL BEHAVIOR

An Annual Series of Analytical Essays and Critical Reviews

Editor: BARRY M. STAW
Graduate School of Management
Northwestern University

VOLUME 1 • 1979

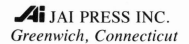 JAI PRESS INC.
Greenwich, Connecticut

Copyright © 1979 JAI PRESS INC.
165 West Putnam Avenue
Greenwich, Connecticut 06830

ISBN NUMBER: 0-89232-045-1

Manufactured in the United States of America

CONTENTS

EDITORIAL STATEMENT

This volume represents the first issue of what we hope will be a lively and productive annual series in organizational behavior. We intend the series to be a continuing forum for taking the field in new directions and for the theoretical reformation of older research issues. Essays in this annual series can provide much greater depth on important issues than could be provided in the typical journal article. This series can also provide a useful outlet for speculative theoretical work or "think pieces" that would be difficult to publish under traditional journal restrictions. Thus, if successful, the *Research in Organizational Behavior* series might well help the field get untracked from some long-term but unprofitable research trends and help it move in more fruitful directions in the future.

Featured in this first volume are invited papers from many well-known and productive scholars in the field of organizational behavior. Our approach has been to allow these authors a high degree of freedom in formulating their chapters with the only real restriction being that the essay be of interest and utility to the organizational behavior community. We believe this approach has been successful in bringing out a collection of essays which are both lively and have some chance to influence future research in the field. The essays comprising this volume range from theoretical works to critical literature reviews. We anticipate that future volumes of the *Research in Organizational Behavior* series will follow a similar format.

Barry M. Staw
Evanston, Illinois

Special note: Beginning with Volume 2 of this series, Larry L. Cummings will join *Research in Organizational Behavior* as co-editor.

RESEARCH IN ORGANIZATIONAL BEHAVIOR

Volume 1 • 1979

BEYOND OPEN SYSTEM
MODELS OF ORGANIZATION

Louis R. Pondy, UNIVERSITY OF ILLINOIS AT URBANA,
ILLINOIS
Ian I. Mitroff, UNIVERSITY OF PITTSBURGH

ABSTRACT[1]

Since the middle 1960s, the "macro" branch of organizational studies has operated with a set of assumptions commonly referred to as the open system model. No more eloquent, systematic, and widely used statement of this model exists than James D. Thompson's *Organizations in Action*. In this paper, it is argued that the open system model, as illustrated by Thompson's book, does not really satisfy the conditions of an open system. It is further argued that Thompson's model has directed our attention away from organizational dysfunctions at the macro level, and from higher mental functions of human behavior that are relevant to understanding organizations. An

Research in Organizational Behavior, Vol. 1, pp. 3–39.
Copyright © 1979 by JAI Press, Inc.
All rights of reproduction in any form reserved.
ISBN 0-89232-045-1

overarching framework of models is used to begin the development of a new set of assumptions, one that might be referred to as a cultural model of organization, the key elements of which include an emphasis on the use of language and the creation of shared meanings. In this sense, the paper attempts to invent a future for organization theory.

INTRODUCTION

Inventing a future for organization theory is the intention of this paper. Our aim is not to provide new answers to old questions, but to raise wholly new questions, ones which, more so than the old questions, address problems likely to face organizations over the next decade, and to capitalize on more advanced concepts of human behavior. Invention is always difficult, because it requires both an act of forcible escape from old assumptions grown commonplace and the creation of new and fresh ways of thinking about one's world. Escaping old assumptions is especially hard when they have served us well, when they have provided reliable guidance to our inquiry, and when they have helped us to frame questions that were answerable. Creating fresh viewpoints can be equally hard, because it seems to force us to work outside any structure whatsoever. How can one secure a new vision, a new representation of reality except by some dimly understood and unreliable leap of intuition? In fact, we believe that this fear is ungrounded and that creative insight can be achieved within a structure, provided that the structure is sufficiently open-ended. If that structure or framework also serves to help us become aware of the shortcomings of the present model, then so much the better. In this paper, we propose such a structure for organizing our thought, and use it both to critique the dominant model of organization theory and to frame new questions that will constitute a future for the field.

A few preliminaries are in order before we proceed further. The field of organizational studies is commonly segmented into "micro" and "macro" branches, the former typically, but not universally, being labeled "organizational behavior," and the latter "organization theory" (although it is by no means limited to theory). By whatever labels, the micro branch is ordinarily thought to comprise the study of individual, interpersonal, and intergroup behavior, as in the study of leadership, motivation, and job design, whereas the macro branch concerns organization-wide aspects such as structure, relations with the environment, effects of technology, and so forth. This distinction is by no means clear-cut, as will be obvious from the uncertain placement of studies of organization climate. This paper will concentrate its analysis on the macro branch, however ill-defined, and will refer to it as "organization theory," however inappropriate that terminology might seem. The domain of the paper will hopefully become clearer as we proceed. Primarily we wish to defuse any illusion that we address the state of the entire field of organization studies.

Within the subfield of organization theory there has been a remarkable unity of viewpoint and method since the mid-1960s, with a degree of coherence nearly deserving of Thomas Kuhn's label of "paradigm" (Kuhn, 1970). We shall not attempt a systematic summary of the model at this point in the paper, but it would be suggestive to say that the viewpoint has been characterized by asking how an organization could solve the problem of achieving a functional alignment of its goals, structure, technology, and environment in the presence of persistent uncertainty. And its primary method of studying this question has been to conduct systematic comparisons across organizations in hopes of discovering empirical regularities. Many scholars contributed both to the formulation of the model and empirical inquiry under its aegis. But one book stands out as the most systematic statement of the model: James D. Thompson's *Organizations in Action* (1967). Our critique of the field takes this eminent and influential treatise as its prime focus. We admit at the outset how important and useful it has been. But we also take as a working assumption that even (perhaps especially!) successful works need to have the root metaphors on which they are founded reexamined.

Within what sort of structure should the model represented by Thompson's book be reexamined? And can that same structure serve to suggest alternative approaches as well? The structure ideally should be able to incorporate the dominant perspectives within the field of organizational studies. But if it is derived from those perspectives, it is unlikely to lead us in new directions. The structure (or framework) should therefore be independent of the field, although still intimately related to the nature of its subject matter. From one point of view, the subject matter is organization (no "s"), not a collection of people and tasks, but a property of that collection having to do with orderliness and patterns. This suggests that our framework of analysis should be able to categorize pattern or organization along some dimension. But what dimension? Inasmuch as we would like to be able to extend the field in the direction of handling more complex issues, the structure should exhibit varying degrees of complexity. Suppose we put these two dimensions together—organization and complexity? Can we find or invent a framework that describes varying types of organization along a scale of complexity? If so, this might give us the handle we need both to critically examine the current state of organization theory and to extend it in new directions. In fact, such a framework does exist, and we have used it to organize our analysis and discussion. It was developed by Kenneth Boulding (1968) as part of an attempt to create a general theory of systems, and it orders systems along a hierarchy of complexity.[2] Inasmuch as theories, models, and viewpoints can be thought of as systems of ideas, it would seem to suit our purposes admirably. In the next section we describe Boulding's hierarchy, and then we shall go on to use it in inventing a future of organization theory.

BOULDING'S HIERARCHY OF COMPLEXITY: A FRAMEWORK FOR ANALYSIS

Boulding identifies nine levels of complexity. The systems in question can be either "real" systems (e.g., a cell, a chemical reaction, a tree, a bird, a man, a family). Or they can be *models* of those systems. But models are just idea-systems, so Boulding's hierarchy can be taken as a description of the complexity of *either* phenomena or models for analyzing those phenomena. This dual use of the hierarchy to describe both organizations and models of organizations will be helpful in clarifying the state and possible directions of organization theory.[3]

It should be emphasized that adjacent levels in the hierarchy differ in complexity not merely in their degree of diversity or variability, but in the appearance of wholly new system properties. For example, the difference between open-system models of level 4 and blueprinted growth models of ievel 5 is the presence of the capacity for genotypic growth and reproduction.

Boulding's levels of complexity are as follows:

Level 1: Frameworks Only static, structural properties are represented in frameworks, as in descriptions of the human anatomy, the cataloguing system used in the Library of Congress, or an organization chart of the U.S. government. The latter may be complicated, but it is not "complex" in Boulding's sense.

Level 2: Clockworks Noncontingent dynamic properties are represented in clockwork systems, as in models of a precessing gyroscope, the diffusion of innovations, or economic cycles in a laissez-faire economy.[4] The crucial difference from level 1 is that the state of the system changes over time. At any given point in time, level 2 phenomena can be described using a level 1 model.

Level 3: Control systems Control system models describe regulation of system behavior according to an externally prescribed target or criterion, as in heat-seeking missiles, thermostats, economic cycles in centrally

Figure 1. Boulding's hierarchy of system complexity.

Level 9: SYSTEMS OF UNSPECIFIED COMPLEXITY

Level 8: MULTI-CEPHALOUS SYSTEMS

Level 7: SYMBOL PROCESSING SYSTEMS

Level 6: INTERNAL IMAGE SYSTEMS

Level 5: BLUEPRINTED GROWTH SYSTEMS

Level 4: OPEN SYSTEMS

Level 3: CONTROL SYSTEMS

Level 2: CLOCKWORKS

Level 1: FRAMEWORKS

controlled economies, and the physiological process of homeostasis. The crucial difference from level 2 is the flow of information within the system between its "regulator" and its "operator," and in fact the functional differentiation between operation and regulation. For a *given* control criterion, level 3 systems behave like level 2 systems.

Level 4: Open systems Whereas a control system tends toward the equilibrium target provided to it and therefore produces uniformity, an open system maintains its internal differentiation (resists uniformity) by "sucking orderliness from its environment" (Schrödinger, 1968, p. 146). Some people have mistakenly characterized an open system as having the capacity for self-maintenance *despite* the presence of throughput from the environment, and therefore have recommended buffering the system against environmental complexity. Quite to the contrary, it is precisely the throughput of nonuniformity that preserves the differential structure of an open system. In an open system, what we might call the Law of Limited Variety operates: A system will exhibit no more variety than the variety to which it has been exposed in its environment. Examples of phenomena describable by open system models are *flames* (simple physical systems in which the transformation of oxygen and, say, methane into water, carbon dioxide, and heat maintain the system's shape, size, and color), and *cells* (biological systems involving complex chemical transformations and differentiated structures, and also the phenomenon of mitosis—duplication through cell division).

Level 5: Blueprinted growth systems   Level 5 systems do not reproduce through a process of duplication, but by producing "seeds" or "eggs" containing preprogrammed instructions for development, as in the acorn-oak system, or egg-chicken system, or other "dual level" systems. While the phenomenon of reproduction is not involved in language usage, the Chomskian distinction between the "deep-structure" and "surface-structure" of grammar seems to tap the same relationship as in acorn-and-oak (Chomsky, 1972). Both involve a rule-based generative mechanism that characterizes level 5 models. Explaining level 5 systems means discovering the generating mechanisms that produce the observed behavior. And *models* of level 5 systems will exhibit this dual level structure as well. (The intention is that at a given level there is a structural isomorphism between the model and the system. Level 5 systems do, however, have level 4, 3, 2, and 1 properties that can be described using those less complex models, so that a system and a model of that system need not be at the same level. This principle will be seen to be relevant to our later argument.)

Level 6: Internal image systems   Level 3, 4, and 5 models incorporate only primitive mechanisms for absorbing and processing information. To quote Boulding, "It is doubtful whether a tree [level 5] can distinguish much more than light from dark, long days from short, cold from hot." The essential characteristic of level 6 systems (and models of them) is a *detailed* awareness of the environment acquired through differentiated information receptors and organized into a knowledge structure or image. (Boulding argues that his hierarchy is cumulative—each level incorporates all the properties of all lower levels. However, one might argue that some sophisticated computer software systems are at level 6, yet do not exhibit the blueprinted growth of level 5, unless one wanted to describe the relationship of programming languages to machine language as "blueprinting.") Level 6 systems do not exhibit the property of *self-consciousness*. They do not know that they know. That enters at level 7. Thus, a pigeon in a Skinner box and an organization that forgot why it instituted a certain rule might be examples of level 6 systems.

Level 7: Symbol processing systems   At level 6, the system is able to process information in the form of differences in the environment. But it is unable to generalize or abstract that information into ideas, and symbols that stand for them. To do that, the system has to be conscious of itself, and this is the defining characteristic of a level 7 system. It has to be able to form the concept "*my* image of the environment," and work on it. And to work on that image, it needs a coding scheme or language. So level 7 systems are *self-conscious* language users, like individual human beings.[5] What is not so obvious is that human groups can be level 7 systems (Ackoff and Emery, 1972). The best example of what it means for a group to have an image of its environment is the process of the social construction of shared models of reality (Berger and Luckmann, 1966). That is, a group can be said to be a symbol-processing entity if its members share a common definition of reality. This is not to say that this approach is not without deep problems. For example, what does it mean for a group to be a language user as distinct from its members? Suppose the members all speak different languages. Then the group is not a language user, even though its members are, and it cannot construct a reality socially. But a group is not necessarily a language user even if its members *do* speak the same language. For a group to use language, not only must verbal interchange take place, but shared definitions of the group's situation must also be constructed.

Level 8: Multi-cephalous systems[6]   These are literally systems with several brains. Boulding's term for this level is "social organization," a collection of "individuals" (any acting unit) acting in concert. What is at issue is that the collection or assemblage of "individuals," whether they be genes, humans, human groups, or computers, creates a sense of social order, a shared culture, a history and a future, a value system—human civilization in all its richness and complexity, as an example. What distinguishes level 8 from level 7 is the elaborate shared systems of meaning (e.g., a system of law) that entire cultures, and some organizations, but no individual human beings, seem to have.[7]

Level 9:   To avoid premature closure, Boulding adds a ninth, open level to reflect the possibility that some new level of system complexity not yet imagined might emerge (see also Churchman, 1971).

Having sketched out some features of Boulding's hierarchy of complexity, let us make a bold statement that we will attempt to justify in the remainder of the paper. All human organizations are level 8 phenomena, but our conceptual models of them (with minor exceptions) are fixated at level 4, and our formal models and data collection efforts are rooted at levels 1 and 2.

Generalizing from the above conclusion, our worst fears are that the field of organization theory will take its task for the next decade to be the refinement of analysis at levels 1 through 4. Our greatest hope is that we will make an effort at moving up one or two levels in our modeling (both conceptual and formal) and begin to look at, for example, phenomena of organizational birth, death, and reproduction, the use of language, the creation of meaning, the development of organizational cultures, and other phenomena associated with the types of complexity in the upper half of Boulding's hierarchy.

EXAMINING CURRENT "OPEN SYSTEM" MODELING

Since open system models have allegedly played such a central role in organization theory in the recent past, it would be useful to sketch the present view and some of the motives for change. Empty categories in a conceptual framework of approaches tend to attract a field in their direction, but they are insufficient to divert a field entirely from a useful paradigm. We must also show why the open system model, as it has been interpreted, is too limiting.

As we have previously noted, for the last decade, thinking and research in the field of organization theory have been dominated by a point of view labeled as open system models. We have contended that most would agree that Thompson's *Organizations in Action* (1967) comes as close as any to being accepted as a paradigm statement of the "open system" perspective[8] of organization theory. Actually, Thompson intended his book to be a reconciliation of the rational or closed-system model of organizations with the natural system model, and his success in resolving this conflict for the profession probably accounts for the enthusiasm that greeted publication of the book. Despite its reconciliatory intent, the book is dominated by a natural system perspective. But within this perspective, heavy emphasis is placed on closing the system to outside influence so that rational choice can take place.

About the same time or slightly earlier, others besides Thompson (e.g., Lawrence and Lorsch, 1969; Perrow, 1967; Crozier, 1964; Burns and Stalker, 1961; Cyert and March, 1963) made important contributions to articulating the point of view that has subsequently permitted us to analyze and understand the problematic nature of uncertainty for the organization, and how uncertainty ties together technology, structure, and environment in a contingent relationship. The resulting paradigm statement generated a large amount of research and continues to do so. (For example, in the four most recent issues of *Administrative Science Quarterly* prior to completion of this paper, 35 percent of the articles reference *Organizations in Action,* even 10 years after its publication.) We have made substantial progress from where we were in 1967 in the direction pointed by Thompson. And despite Pfeffer's (1976) recent complaint that organizational behavior has been "\. . . dominated by a concern for the management of people *within* organizations," organization theory *has* researched the organization-environment interface under the guidance of open system thinking. So what is the problem?

The problem is that models not only direct attention *to* some phenomena and variables but also *away* from others. And if a model is highly successful in helping a researcher to cope with problems the model says are important, habituation will take place: The researcher will simply not "see" other problems, and he will have no basis for being receptive to competing models (Hanson, 1958). But there *are* other problems that we should be addressing, and there are competing models that we should be considering.[9] This is the motivation for our arguing that we need to go beyond open-system theory. Specifically, we offer five major reasons in support of this position:

1. By focusing on maintenance of the organization's own internal structure, open-system thoery has directed us away from ecological effects—broadly defined—of the organization's actions, to the ultimate detriment of the organization itself.

2. We should be directing our efforts to understanding massive dysfunctions at the macro level, not just explaining order and congruence. How do organizations go wrong?

3. We need to reflect in our own models conceptions of people in other fields, especially those that picture persons as having the capacities for self-awareness, for the use of language, for creative growth, and for learning from their experience.

4. Troublesome theoretical questions ignored by open-system theory are suggested by other models. For example, do organizations reproduce themselves? If so, how?

5. For the purpose of maintaining organization theory's adaptability as an inquiring system (Churchman, 1971; Mitroff, 1974), we need to discredit what we know, to change for the naked sense of change to prevent ossification of our ideas.

These five reasons for going beyond open system models of organization are closely interrelated and are, we believe, merely five different aspects of the same underlying problem with the field. Each of these motives for change are discussed in detail below. Following this, some alternative models of organization are proposed. The paper concludes with a brief examination of the implications of our position for the doing and teaching of organizational research, and the teaching of present and future managers.

MOTIVES FOR CHANGE: THE LIMITS
OF OPEN SYSTEM MODELS

The Ecology of Organizational Action

In order to understand how open system models can blind us to the
nest-fouling impact of organizations' actions on their environment, we
need to examine how open system theory has been interpreted and used
by organization theorists. Frequently, those who claim to be using an
open-system strategy are in reality using level 3 control system models.
They have failed to make the distinction, as have Haas and Drabek
(1973), between "natural" and "open" system models.

Consider Thompson(1967):

> Central to the natural-system approach is the concept of homeostasis, or self-
> stabilization, which spontaneously, or naturally, governs the necessary relationships
> among parts and activities and thereby keeps the system viable in the face of distur-
> bances stemming from the environment (p. 7).

In other words, the environment is a source of disturbance to be
adapted to, instead of the source of "information" that makes internal
organization possible. Self-*stabilization* referred to by Thompson is a
level 3 process. The equivalent level 4 process is self-*organization*. Haas
and Drabek (1973) recognize the difference between natural and open
system models, but classify Thompson (incorrectly, we believe) as an
open system theorist. What Thompson calls a closed system is equivalent
to Boulding's clockwork (level 2). Thompson made a major contribution
by formalizing organization theory at a higher level than it had been at.
But we argue that it was not at the level of open systems as understood by
Boulding and other systems theorists. There is therefore some question of
whether organization theory (as represented by Thompson's book) is
even at the open system level, to say nothing of whether it is ready to go
beyond it. So this section will have to be split into two parts: (1) the
ecological consequences of using a control system model (even though it
might be spuriously labeled as an open system model); and (2) the ecological consequences of using a true open system model. By "ecology" we
mean the structure of the organization's social, economic, and political
environment as well as of its physical and biological environment.

The ecological consequences of control system thinking   We must remember
that the aim of a control system is to produce uniformity, i.e., to decrease
variety, if it can. To the extent that the system environment is highly
varied in its texture over time, the regulator part of the system must match
the variety of the environment so that it can control that variety and
produce a uniform environment for its operator part. This is the essence
of Ashby's Law of Requisite Variety (Ashby, 1956). In Thompson's lan-

guage, this means creating the conditions necessary for rational operation at the technical core by controlling environmental uncertainty.

The ecological implication of control system thinking, both theoretical and practical, is that environments as well as organizations will become more uniform. Environments are made up of other organizations each of whom, according to this view, is following a control system strategy. Each attempts to impose uniformity on the others so that uniformity can be created "inside."[10] The result is that the entire system will grind toward a social-system-wide equilibrium. Within the context of a control system model this is a desirable state of affairs. Not so for open systems.

The ecological consequences of open system thinking The ecological consequences of open system thinking are quite different from those of control system thinking. An open system is at such a level of complexity that it can maintain that complexity *only* in the presence of throughput from a differentiated environment. If an open system insulates itself from environmental diversity and differentiation, or if it attempts actually to kill environmental diversity, then it will have only a uniform, gray soup to feed on, and eventually its own internal structure will deteriorate to the point that open system properties can no longer be maintained. If control system models are used to manage open systems, the system will be led to take precisely the *wrong* actions! The organization will attempt to insulate itself from the very diversity that it needs to survive as an open system.[11]

Suppose an open system does not attempt to buffer out variability, but exposes itself to the uncertainties of the environment. If environments are plentiful, and the system is agile, it may still extract the needed organizing information from the immediate and present environment, leave it depleted (i.e., undifferentiated) and move on to another.[12] But suppose environments are scarce. A system must then in some sense replenish its environment. It must, paradoxically, put variety back into the environment so that it can subsequently use it. But how to return variety to the environment without deorganizing the system itself?

The key to resolving this dilemma is realizing that only part of an organization's environment is given to it. Another part is enacted (Weick, 1969) by the organization itself. Some people have misunderstood Weick's concept of enactment to be identical with imagination or mental invention. But Weick means that the organization literally *does* something, and once done, that something becomes part of the environment that the system can draw on to maintain its own internal order. To put it somewhat differently, one of an organization's most crucial design decisions concerns how it attempts to design its own environment.

There is a trap here to be avoided. If the enactments are merely an expression of the system's current organization, then nothing new will be created for the system to feed on. The system can only enact what it already knows. Complex systems have an appetite for novelty (Ackoff and Emery, 1972). They need what Stafford Beer (1964) has called "completion from without." Somehow, the process of enacting an environment must escape this redundancy trap. Weick (1977) has suggested a number of strategies that apply here: *(a)* be playful, *(b)* act randomly, *(c)* doubt what you believe and believe what you doubt (i.e., discredit the existing organization). All these strategies have promise of escaping the trap.

Thinking of open systems as *needing* environmental variety sheds fresh light on the widely replicated finding that organizational complexity is positively correlated with environmental diversity. The usual explanation from contingency theory is that the organization needs to be complex in order to cope with environmental variety. Implicit in this explanation is that "surplus" complexity is possible but not necessary. The alternative explanation flowing from our analysis of open systems is that an organization is unable to maintain internal complexity *except* in the presence of environmental diversity. Surplus complexity is simply not possible from this view, but a shortage is. This might provide a basis for choosing between contingency theory and open system theory.

Hans Hoffman's view of the nature of man captures this property of open systems, especially as it relates to the necessary character of the enactment process: "The unique function of man is to live in close, creative touch with chaos, and thereby experience the birth of order" (quoted in Leavitt and Pondy, 1964, p. 58).[13]

Thus far, we have argued that organizations as open systems foul their environmental nests either by: *(a)* following a control system strategy and deliberately killing off variety in the environment; *(b)* following a short-sighted open system strategy and failing to renew the successive environments that they occupy.

Open system theory as it is currently interpreted and practiced in organization theory does not come to grips with either of these problems. Important exceptions exist (Weick, 1969; Hedberg, Nystrom, and Starbuck, 1976; Cohen and March, 1974), but they do not yet occupy center stage.

Dysfunctions in Organization Theory

One of the striking differences between organizational behavior and organization theory is that organizational behavior (the micro branch of organization studies) defines much of its research effort in terms of dysfunctions of the system. For example, there are theories of absenteeism, turnover, low productivity, industrial sabotage, work dissatisfaction, interpersonal conflict, resistance to change, and failures of communication.

Even equity theory is really a theory of *in*equities, how they are perceived and how they are resolved. But organization theory has been a theory of order.[14] We have theories of the proper match between structure and technology (Perrow, 1967), between environment and structure (Lawrence and Lorsch, 1969), between forms of involvement and forms of control (Etzioni, 1961). Thompson (1967) most eloquently speaks of administration as the "co-alignment" of goals, technology, structure, and environment, and he treats dysfunctions as neither serious nor permanent. Corrective mechanisms, true to the control system model, take care of any problems: "Dysfunctions are conceivable, but it is assumed that an offending part will adjust to produce a net positive contribution or be disengaged, or else the system will degenerate" (Thompson, 1967, pp. 6–7).

The prevailing view in organization theory offers no systematic typology of dysfunctions at the macro or systemic level. But some of the more spectacular dysfunctions have been documented in case analyses. For example, Halberstam (1972) has described the pressures for consensus decision making that operated within the Johnson White House to systematically exclude opposing points of view on our involvement in Vietnam. Janis (1972) has done the same for the decision making that led up to the Bay of Pigs invasion of 1961. Smith (1963) described a number of crises in corporate decision making. And the more recent crisis of bribery within Lockheed and the misuse of power within Watergate are familiar to the point of contempt. But organization theory as a field is currently so preoccupied with explaining order that it has not yet discovered these most interesting phenomena. (Note that it is not even necessary to argue for the study of dysfunctions on normative grounds. From a purely descriptive, nonnormative perspective, such dysfunctions are intriguing scientific happenings.)

Consider Lordstown. Much has been written analyzing how and why the workers reacted to the speed-up of the assembly line. But virtually nothing has been written explaining why General Motors made the wrong decision in the first place. Such decisions about research strategy have been termed "errors of the third kind," or E_{III}" (Mitroff and Featheringham, 1974), where E_{III} is defined as the "probability of solving the 'wrong' problem when one should have solved the 'right' one."

It's curious that in economics the situation is reversed. *Macro*economics is focused heavily on the system dysfunctions of inflation, unemployment, and recession. But *micro*economics is concerned with explaining the rationality of choice. Whether that reversal is significant we cannot tell. But in the organizational sciences, it is the macro branch that eschews the inquiry into disorder.

Like all attempts at generalization, this one suffers its exceptions. Staw and Szwajkowski's (1975) study of antitrust violations is a case in point. And Staw (1976) and Staw and Fox (1977) have studied the phenomenon of escalation experimentally. The use of power to influence the allocation of resources away from rational norms has been studied by, among others, Pfeffer and Salancik (1974) and Salancik and Pfeffer (1974). But the dominant thrust of the field has been explaining why organizations do work well, or at least how they are presently structured to perform the kinds of work they currently do.

Part of the responsibility for this functional orientation can be assigned, we believe, to Thompson's use of *organizational* rationality as a central and integrating concept. By redefining the unit of analysis as the organization *plus* the environment, we would instead be forced to define the bounds of rationality to be broader, to invoke a concept of *ecological* or *systemic* rationality (Bateson, 1972; Churchman, 1971). It is not merely the organization that adapts to the environment. The organization and its environment adapt together. Within such a model of ecological rationality, the environment's problems become also the organization's.

In a recent review of Schon's (1971) *Beyond the Stable State,* Rose Goldsen (1975) summarizes Schon's argument that institutional dysfunctions arise from a belief in the possibility of a stable state buffered against change and uncertainty:

> Change and paradox are not anomalies to be corrected, but the very nature of open systems. Learning systems accept these principles as axioms, rejecting the "myth of the stable state." Our current institutions still base themselves on that myth and it is their compulsive insistence on trying to achieve it that leads to many dysfunctions and ultimate breakdown (Goldsen, 1975, p. 464).

Goldsen, after referring to Schon's redefining of hotel chains as "total recreational systems," asks:

> Is it dysfunctional when informational breakdown in the Coca Cola Company (say) interferes with efforts to maintain sugar economies in developing nations? Is it functional when recreational systems "convert large proportions of the indigenous labor force into waiters, bellboys and cab drivers, chambermaids and prostitutes?" One man's "dysfunction" is another man's "function." (Goldsen, 1975, p. 468).

Within the paradigm represented by Thompson's seminal book, the effects described in the above quote would not be recognized. But Thompson's book was written more than a decade ago. The image of the world that it projects is a history of growth and prosperity, of munificent environments. But times have changed. It is no longer an accurate description of the world we live in. Nor is it a sensible guide to solving the problems our institutions face and have created, because it does not recognize that some of the environmental problems organizations face are of their own making.

If studying the conditions of order is incomplete, how shall we change what we do? We would argue that we need to develop a theory of error, pathology, and disequilibrium in organization (Mitroff and Turoff, 1974). And open system models as currently interpreted are of little help for that purpose.[15]

Alternative Conceptions of Man and Method

So far, we have argued that the prevailing model in organization theory *(a)* creates an artificial split between an organization and its environment, to the detriment of both, and *(b)* directs our attention away from the dysfunctional consequences of organizational action. A third reason for needing to go beyond the prevailing model of organization is that it excludes many fruitful models of human behavior. Organization theories seem to have forgotten that they are dealing with *human* organizations, not merely disembodied structures in which individuals play either the role of "in-place metering devices" (Pondy and Boje, 1976) designed to register various abstract organizational properties (e.g., complexity, formalization, etc.), or the role of passive carriers[16] of cultural values and skills. Thompson's conception of the individual is that society provides a variety of standardized models of individuals that organizations can use as inputs:

> . . . if the modern society is to be viable it must sort individuals into occupational categories; equip them with relevant aspirations, beliefs, and standards; and channel them to relevant sectors of "the" labor market (Thompson, 1967, p. 105).

Following Thompson, many macro organization theorists have downplayed man's higher capacities, including his ability to use language, his awareness of his own awareness, and his capacity to attribute meaning to events, to make sense of things. These capacities are characteristic of Boulding's level 5 through level 8. Some macro organization theorists have made language, awareness, and meaning central concepts in their theories (March and Simon, 1958; Weick, 1969; Silverman, 1971), but the dominant trend is still toward mindless conceptions of organization.[17]

There are major exceptions within the social sciences, and we can benefit from examining how they conceptualize the subject matter. Consider cultural anthropology. Geertz (1973), in *The Interpretation of Cultures*, starts out by assuming that assigning meaning to events is a central human process (see also Leach, 1972), and that the task of the anthropologist is to ferret out those meanings and the meanings that lie beneath them in multiple layers. To describe only the events is "thin" description, but to describe the layers of meaning underlying those events is "thick description." One important class of meanings is the set of beliefs about causality. To Geertz these would be problematic, requiring explanation. But to Thompson, they are given in society and organization members are simply "equipped" with them. But how do those beliefs originate and change with experience? Perrow (1967) has been influential in getting us to think of technology as well or poorly understood, as though technology could be understood without someone to do the understanding. But since technical knowledge varies from individual to individual, the degree of understanding is clearly a property of the object-observer pair, *not* of the object alone (Mitroff, 1974). Similarly, environmental uncertainty does not reside exclusively in the environment itself.[18] (We do not even deal here with the more serious problem of how an organization decides where it, i.e., the organization, leaves off and the environment begins. That boundary, too, is problematic [Weick, 1977].) We have been describing environment and technology "thinly." A thick description would probe into environment and technology as ways of classifying experience and thereby giving meaning to it.

Or, consider Harre and Secord's (1972) recent reconstruction of social psychology. They propose an "anthropomorphic model of man," in which man is treated, for scientific purposes, as if he were a human being! That is, man is endowed not only with an awareness of external events, but an awareness of his own awareness and with a capacity for language (Boulding's level 7). Most importantly, man is presumed to have generative mechanisms that produce observable behavior. The task of inquiry is to discover those mechanisms for each individual. In the prevailing organization theory paradigm, no such mechanisms are presumed. Structure is presumed to exist only at the level of empirical reality.[19] The form of explanation is therefore necessarily comparative across organizations at the same level of abstraction. But with a presumption of a "deep structure" (Chomsky, 1972) that generates the "surface structure" of observable behavior, a "theory of the individual" (Newell and Simon, 1972) makes sense, and a "science of the singular" (Hamilton, 1976) based on a case study methodology becomes rigorous science.[20] The performance programs proposed by March and Simon (1958) are such "generative mechanisms" that produce organizational behavior, and the task of inquiry is to infer the nature of those programs. It is curious that organization theory should have drawn so heavily from parts of March and Simon, but largely missed this very central point of the book.

The existence of alternative models of human behavior is insufficient by itself to cause us to desert open system models. But these alternative conceptions make us aware of phenomena that the prevailing view cannot begin to handle. In short, the higher mental capacities studied by other disciplines offer a new avenue for organization theory to explore to gain fresh insights into organizational phenomena. With isolated exceptions, those new opportunities have not been explored.

New Theoretical Questions

We have previously argued that Thompson's open system model is inadequate for dealing with important practical problems. But it is also inadequate for conceptualizing some important theoretical questions as well. No theory should be expected to cope with the full range of phenomena, of course, but neither should allegiance to a theoretical position become so strong that it prevents us from considering phenomena outside its purview.

One important class of theoretical questions addresses the phenomena of organizational birth and reproduction. Extant open system models, for the most part, are about *mature* organizations.[21] Although Thompson discusses some aspects of growth, his analysis is about continued growth of adult organizations. And it is growth whose patterns are shaped by external forces, not the blueprinted growth of Boulding's level 5. The same is true of the best known treatment of organizational growth and development (Starbuck, 1971).

Biological analogies can sometimes be carried too far, but in this case we believe it is useful to ask whether organizations "reproduce" themselves in any sense. Consider the following model.

1. The development of organizations is constrained by environmental forces, but it is directed by fundamental rules for organizing which are stored inside the organization itself. Those governing rules, or generative mechanisms, produce the observed patterns of differential functioning that make up the organization.

2. The organizing rules are stored in the brains of some, perhaps all, individuals in the organization. Those rules result from a previous process of negotiating the organizational order. Some organizing rules are also stored on *paper* (e.g., job descriptions, standard operating procedures), so that the content of the rules may transcend the tenure of any organization member.

3. When a person leaves the organization, he carries with him those organizing rules.[22] Should he be the founder of a new organization, those rules would find expression through unfolding in a new environment.

This is essentially the underlying model in a recent analysis by Kimberly (1976) of the birth of a new medical school at the University of Illinois.[23] At first glance, Pettigrew's (1976) analysis of entrepreneurship seems to tap the same phenomenon, but I believe something distinct is at work in Pettigrew's model. Whereas Kimberly is conceptualizing organizational birth as a reproductive process through the mechanism of "fathering offspring" Pettigrew seems to have a model of autonomous birth in mind. The entrepreneurs whom he has studied have formed organizations on the foundation of creative, novel myths or cultures. Those entrepreneurs do not seem to have come from any previous organizational experience.

A second important class of theoretical questions outside of Thompson's model is that dealing with higher mental capacities. We have already alluded to some of the work in the area. How people make sense of their experiences is a crucial issue for organization theory because the answer potentially overturns models of rational behavior. A phenomenological approach is that sense-making is retrospective; we can understand what we are doing only after we have done it. An action-theoretic approach argues that meanings are socially constructed, that therefore there are multiple realities. These positions have been most systematically developed within organization theory by Weick (1969) and Silverman (1971), but their influence on empirical research on organizations has been minimal. Weick's and Silverman's models are extreme points within organization theory.

It is not immediately obvious why Thompson's model precludes such theoretical questions. As we have suggested, part of the reason is that Thompson seems to have mature, already organized systems in mind. What is problematic for him is simply maintaining that organization, not creating it in the first place. A second source of blockage is the causal priority Thompson assigns to norms of rationality. How the organization comes to articulate these norms of rationality is not problematic in Thompson's model, except to say that the organization goals are negotiated within the dominant coalition. But each member of the dominant coalition is presumed to have specific interests already in mind. An alternative model reduces rationality norms to retrospective outcomes. We have not yet discussed the role that language plays in this rationalizing process, but work from other fields suggests that terms in our language affect what we see (Whorf, 1956) and even the logic we use to structure our thought (Tung-Sun, 1970; Alexander, 1967, pp. 37–39).

Change for the Sake of Change

All of the four previous reasons for going beyond open systems models have dealt with the substance and content of the theory, its shortcomings, and its neglect of new dimensions it might explore. But all that we have said about maintaining the vitality of organizations applies with equal force to the field itself. Scholars making up organization theory themselves constitute an organization. What can our analysis tell us about how the field should conduct its affairs, and whether from time to time it should change its paradigm, that is, its world view, its basic models and methods? And can a field come to believe in its models so strongly that it forgets that they are only metaphors of the phenomena being studied, not the phenomena themselves?

If a field's paradigm is too well-defined, or is believed in too strongly, then creative ideas consistent with the paradigm will gradually be selected out. If the field is to continue to be effective in working on worthwhile problems, then it must, to a certain degree, act "hypocritically," that is, in a way that it both believes and disbelieves what it knows (Weick, 1977).

We need to maintain a certain creative tension in what we take to be true (Churchman, 1971; Mitroff, 1974). Our system of scientific beliefs should be a *nearly* organized system—organized enough to provide the confidence for researching uncertain topics, but not so organized that doubt is no longer possible. The illusion of success, especially when hard-won, breeds resistance to change. Scott (1977) has voiced a similar concern:

> After searching so long for "the one best way to organize," this insight [contingency theory] was hard to come by, but having now won it, the contingency approach seems so obviously correct that we are not likely to easily give it up.

In short, we think that it is time to change for change's sake, not because we think we have the correct paradigm to replace open system models, but because we fear that some people have begun to treat contingency theory and other derivatives of open system modeling as the truth rather than as the most recent set of working assumptions. If we have begun to confuse the map with the territory, then it is time to change maps.

This concludes our litany of motives for going beyond the prevailing model of organizations. In the next section we suggest some possible alternatives.

A SKETCH OF POSSIBLE FUTURES

Bringing Empirical Work Up to Thompson's Model

We have previously argued that although the *language* of Thompson's model is at the level of open systems, the actual *content* is wedged at Boulding's level 3, the level of simple control systems; and most of the empirical research and analysis generated by the model has been at level 1, the level of static frameworks, i.e., cross-sectional comparative analysis. Therefore, one promising direction for empirical inquiry is actually to test Thompson's propositions at the proper level that reflects their dynamic rather than static content. One of the most sophisticated level 2 studies is Nystrom's (1975) analysis of the budgeting, workflow, and litigation processes within the Federal Trade Commission. Using 18 years of data from 1954 to 1971, Nystrom estimated a simultaneous, six-equation model describing funds requested and appropriated, investigations completed, formal complaints, cease and desist orders, and litigations. Two of the equations included time-lagged variables, thus making the model (as well as the data) longitudinal or dynamic. Since the endogenous time lag was only one year in length, the model could not exhibit any natural cyclical behavior, but at least some dynamic characteristics were built into the model. Nystrom's strategy of analysis is important, and serves as a prototype of level 2 analysis.

Even better is the research of Hummon, Doreian, and Teuter (1975). They have constructed a "structural control" model of organizational change that is one of the few rigorous level 3 models in the field of organizational research. A structural control model presumes the existence of equilibrium points (not necessarily stable) within a system of variables, and a set of processes that describe how the system behaves when displaced from those equilibrium points. If the system is stable, it will tend to converge on its equilibrium when displaced. Using Blau's (1970) model of structural differentiation as the content of the control model, and Meyer's (1972) data on governmental finance departments to test it, Hummon et al. demonstrated the feasibility of estimating the equilibrium points, the control processes, and therefore the stability of the system. Just as Nystrom's (1975) research serves as a paradigm for level 2 modeling, the analysis of Hummon et al. (1975) provides a paradigm case for level 3 modeling.[24]

Reformulating the Open System Model

If we are to go "beyond open system models," we must first get there in content as well as in language. This suggests a second promising direction for inquiry, now primarily at the theoretical rather than the empirical level. Before we can begin to answer questions about the behavior of open systems, we must first frame fruitful questions to ask. We believe that we have seriously misunderstood the nature of "open systems," and have confused them with "natural" or control systems, as we have argued throughout this paper. By an "open system," we seem to have meant only that the organization is influenced by the environment, or must take the environment into account, or can interact with the environment. But the interpretation advanced here has been that a high-variety environment is a *necessity* to an open system, not a problem, nor even a mere opportunity. The cognitive cycling produced by sensory deprivation provides an analog at the individual level of the phenomenon we have in mind. We are suggesting that there is a boundary between level 3 and level 4 systems across which the function of the environment undergoes a reversal. The human mind seems to be a system of sufficient complexity that it cannot continue to be a "mind" in an environment of sensory deprivation.[25] Those investigating the area of work motivation and job design have for some time realized the importance of task variety to continued satisfaction and productivity, especially for those with high growth needs (read "high system complexity"?) (Hackman and Oldham, 1975). Is it unreasonable to conjecture that organizations of sufficient complexity also need high task variety in their environments? If so, what are the implications of Thompson's strategies of buffering, smoothing, standardizing, etc.? Do they constitute a self-imposed sensory deprivation for the organization?

If an organization is to advance across the boundary between a control system and an open system, it may need to be *flooded* with variety. Otherwise the control system will have time to develop buffers against a gradually developing complexity in the environment. A dunking in a sudden lack of structure is alleged to be what brings about change in sensitivity training groups. That insight suggests that the rate at which uncertainty overwhelms an organization will be more related to the complexity of its internal structure than just the amount of environmental uncertainty that happens to exist at the time of a cross-sectional study, or the predetermined data collection periods of a longitudinal study. Since "variety floods" cannot, by definition, be anticipated, an opportunistic research strategy is forced upon us if we wish to study the level 3/level 4 metamorphosis. For example, we might wish to study organizations under conditions of natural disaster, or extreme opportunity (e.g., a small organization getting a very large influx of capital or clients). In fact, Thompson, 1967, pp. 52–54) labels organizations that arise in response to disasters "synthetic organizations," and he attributes to them many open system characteristics quite different from the buffered systems operating under norms of rationality:

> . . . headquarters of the synthetic organization . . . only occasionally emerge around previously designated officers . . . [A]uthority to coordinate the use of resources is attributed to—forced upon—the individual or group which by happenstance is at the crossroads of the two kinds of necessary information, resource availability and need . . . [W]hen normal organizations are immobilized or overtaxed by sudden disaster, the synthetic organization rapidly develops structure . . . [T]he synthetic organization emerges without the benefit of planning or blueprints, prior designations of authority, great freedom to acquire and deploy resources, since the normal institutions of authority, property, and contract are not operating (Thompson, 1967, pp. 52–53).

In short, a synthetic organization is a self-organizing open system. But our only quibble with Thompson—a major one—is that such synthetic organizing processes are not limited to natural disasters and are far more common than he suggests.

To keep our models straight, we must be careful not to endow an open system with too many properties that characterize Boulding's higher levels of system complexity. For example, we should not attribute any desire or motivation or even tendency to the system to move from level 3 to level 4, or to seek out environments rich enough in variety to maintain system complexity, or to reproduce itself by means other than mitosis-like duplication, or to have a sense of self-awareness. Those are higher-level properties. The sole property at issue in this immediate discussion has been an open system's capacity for self-organization and the important role of environmental variety in maintaining that capacity. As important as it might be to reformulate the open system model along these lines, that task does not begin to move us nearly far enough along the route toward models of a higher order of complexity. To that we turn next.

Beyond Open Systems: The Role of Language

In previous sections we have already dealt, albeit briefly, with possible research questions about organizational birth and reproduction, the concept of generative mechanisms, and with phenomenological and socially constructed realities.[26] But we have dealt only in passing with language and its relevance to organizational research in the future. It is therefore to language that we should like to direct our attention here.[27] Language plays at least four important and distinct roles in social behavior, including organizational behavior:

1. It controls our perceptions; it tends to filter out of conscious experience those events for which terms do not exist in the language.

2. It helps to define the meaning of our experiences by categorizing streams of events.

3. It influences the ease of communication; one cannot exchange ideas, information, or meanings except as the language permits.

4. It provides a channel of social influence.

Silverman has addressed the first two of these functions in his action theory of organizations:

Social reality is "pre-defined" in the very language in which we are socialized. Language provides us with categories which define as well as distinguish our experience. Language allows us to define the typical features of the social world and the typical acts of typical actors (Silverman, 1971, p. 132).

Table 1.   Information Sources used by Headquarters Executives of
Multinational Corporations

		General Management	Field of Specialization Marketing	Finance
Type of Information Source	Documentary	18%	30%	56%
	Human (face-to-face)	71%	65%	44%
	Physical inspection	11%	5%	0%
		100%	100%	100%

(From Keegan, 1974)

Language is a technology for processing both information and meanings just as production technologies process material inputs into outputs. Both types of technology constrain what inputs will be accepted and what transformations will be permitted. Languages vary in their capacity to process high-variety information. For example, the language of written communication unaided by nonverbal cues is less able to represent complex events than is the verbal plus nonverbal language of face-to-face communication. Thus we might expect face-to-face communication to be used more heavily in ill-structured fields such as "general management" than in well-structured fields such as "finance," with "marketing" falling between them.[28] Furthermore, in highly unstructured situations, even face-to-face communication may be inadequate for conveying the full meaning. We might therefore expect direct, on-site physical inspection of the phenomenon being talked about to be most common in the poorly structured areas. This is precisely what Keegan (1974) found in a study of information sources used by headquarters executives of multinational corporations, as Table 1 taken from Keegan's article shows.

Although Thompson ignores language as a variable of interest, an earlier classic in organization theory does not; in fact, March and Simon (1958, pp. 161–169) make language a central feature of their analysis of communication in organizations. Like Silverman, they recognize the importance of language in perceiving and defining reality. But they offer a thorough (and largely ignored) treatment of the effects of language on the efficiency and accuracy of communication. They define language broadly to include engineering blueprints and accounting systems as well as "natural" languages such as English. Standardized languages permit the communication of large amounts of information with minimal exchanges of symbols. On the other hand,

> . . . it is extremely difficult to communicate about intangible objects and nonstandardized objects. Hence, the heaviest burdens are placed on the communications system

by the less structured aspects of the organization's tasks, particularly by activity
directed toward the explanation of problems that are not yet well defined (March and
Simon, 1958, p. 164).

(But we should recognize the earlier point that objects become standardized by having terms in the language for referring to them. Objects are not standardized in and of themselves.)

For example, among physicists, experimental techniques and procedures probably are more ad hoc and nonstandardized than theories. Therefore, we would expect experimentalists to rely less heavily on written publications for obtaining research-relevant information from professional colleagues than theorists, and to rely more heavily on verbal, face-to-face communication than the theorists. Gaston (1972) in a study of particle physicists in the United Kingdom collected data that support our conjecture, as shown in Table 2.

With regard to the fourth function of language, the social influence function, Pondy (1977a, b) has argued that possession of a common language facilitates the exercise of social control, and that organizations can be thought of as collections of "jargon groups," within each of which specialized sublanguages grow up that set it apart from the other jargon groups in the organization. And the size and number of these jargon groups can be related to the age and size of the organizations, its technology, and the rate of turnover of personnel (Pondy, 1975). Within a scientific community, the scientific paradigm provides a language for talking about professional matters (Mitroff, 1974). When this paradigm is poorly developed, as in academic departments of sociology, political science, and English, it has been shown that the turnover of department heads is more frequent than in departments with well-developed paradigms such as mathematics and engineering, the argument being that department heads in low paradigm fields are less able to exercise social control in the resolution of professional conflicts (Salancik, Staw, and Pondy, 1976).

Not all communication operates at the level of conscious, expressed language. Some recent papers have suggested that myths, stories, and metaphors provide powerful vehicles in organizations for exchanging and

Table 2.   Forms of Communication Used By
Particle Physicists in the United Kingdom

		Experimentalists	Theorists
Form of	Verbal	66%	31%
Communication	Publications	34%	69%
		100%	100%

(From Gaston, 1972)

preserving rich sets of meaning (Boje and Rowland, 1977; Clark, 1972; Huff, 1977; Meyer and Rowan, 1977; Sproull and Weiner, 1976; Mitroff and Kilmann, 1976. This attention to the less conscious, less rational aspects of organizational language and communication provides one of the most exciting avenues for exploration open to us. It begins to approach the models characteristic of Boulding's level 8.

Let us try to place this discussion of the functions of language in organization theory in context by imagining what it must be like in an organization without the capacity for verbal language. Consider an organization of subhumans incapable of the use of language. While they could exchange signals within a finite set of messages, as is thought to characterize animal communication, they would not have the capacity for producing an infinite number of distinct sentences, as can humans. They would be incapable of reconceptualizing their relationships to each other, their technologies, or their environments. But language permits codification of those conceptualizations, and therefore sharing and social modification of them. Not only is language functional for the operation of the organization, but it is central to the evolution of organizational forms within the lifetimes of individual members. Mind need not wait for genetics to bring about change.[29] If that premise is accepted, then the fundamental structures of languages must be reflected in social organization. By "fundamental structures" we mean such characteristics as the absence of the verb "to be" in Turkish, Hopi, Hungarian, and other languages, or the use of idiographic characters in Chinese. For example, it may be easier to communicate metaphorically in Chinese than in alphabet-based languages. And the fundamental structure of language may dwarf such surface characteristics as "standardization" in their impact on organizational structure and behavior.

In summary, integrating the concept of language into formal organization theory can begin to give us a deeper understanding of perception, meaning creation, communication, and social influence. These are the four functions of language that we listed at the beginning of this section. But less obviously, it can also help us to understand the very processes by which human organizations are created and evolve. There is no better example of this organizationally creative process of language than Burton Clark's (1972) study of organizational sagas. By a "saga" Clark means a reconstruction of an organization's history that stresses its origins, its triumphs over adversity, and its tangible symbols. Clark's study of the sagas undergirding three unique colleges (Reed, Antioch, and Swarthmore) provides us with a method and a theoretical perspective that should be emulated. Through the use of language in creating and propagating a saga, an organization can become much more than just an instrumental social device; it can become a culture with a meaningful past and a meaningful future. But this image of an organization would not be possible without considering the symbolic and expressive functions of language. In short, language is a key element in moving toward a cultural metaphor of organization.

Some Implications for Doing Research

To discuss the implications for teaching and research of any theoretical position on organization theory is a tricky business. There is every likelihood that what we teach now to practitioners will create the very phenomena that we will have available to study in the future. Today's theories can enact tomorrow's facts.

To deny this likelihood is to accept the ineffectiveness of our teaching; to admit it is to reject the role of scientist in favor of one closer to that of playwright. To be quite honest, we have been unable to resolve this paradox, and it circles buzzard-like over what we have to say in this section. Our statements are either prescriptions or predictions, but we cannot tell which.

To summarize what we believe should now be obvious implications of our position for research:

1. Conceptually, the status of an organization shifts from that of an objective reality to one which is a socially constructed reality. Given such a concept of organization, to endow such concepts as technology with measurable and perceivable attributes is senseless. Instead, we need to study how participants themselves come to invoke categories such as "organization" and "technology" as a means of making sense of their experience. The resulting meanings will frequently be stored in organizational myths and metaphors to provide rationales for both membership and activity in organizations. The creation and use of myths and metaphors in organizations is a worthwhile focus for study.

2. More generally, organizations are represented as collections of "organizing rules" that generate observable behavior. While comparative analysis can document empirical regularities at the observable level, the true task of theory is to infer the generative mechanisms, or underlying models, that produce the surface behavior *in each case*. That is, to develop a theory of the individual case is a meaningful scientific activity. Determining whether collections of individuals have the same theories is a proper task for comparative analysis. What we have in mind is analogous to discovering the relationship between a given acorn and oak, and subsequently establishing it for all acorn-oak pairs. By implication, we must drop our reliance on comparative empirical analysis as the only source of scientific generalizations about organizations.[30]

3. These two conceptual hooks imply some radical methodological de-

partures as well. We suspect that questionnaire design, large sample surveys, and multivariate analysis will need to recede in importance in favor of more abstract model-building and ethnographic techniques more suitable for documenting individual cases of meaning and belief systems. This is in no sense a suggestion that we return to the purely descriptive case study. Our aim is to find out how individual organizations work, and that can best be done only one at a time. Whether a collection of individual cases work the same should be the end result of empirical inquiry, not the initial presumption as in the comparative analysis (Leach, 1961). What is at issue is what we mean by the phrase "how things work." Perhaps it would help to point out that the nature of causation changes as one ascends Boulding's hierarcy of complexity. Correlational models of causation implicit in comparative analysis are appropriate only at the levels of frameworks and clockworks, not at the level of blueprinted growth. But at level 5 and above discovering how things work means inferring the underlying model in each case.

The upshot of these three implications for research is that the concepts and methods are all being defined at a more abstract level than the level of empirical reality. Organizations are not just groups of people; they are sets of organizing rules. And "explaining" organizations is not merely establishing empirical regularities across a set of organizations; it is discovering those deeper organizing rules in each case, and only then comparing across organizations.[31] What we need is less brute-force empiricism and more of what Leach (1961, p. 5) has called "inspired guesswork."

Implications for Management Education

Thompson's view of organizations suggests that administrators should be trained in the skills of "co-aligning" environment, goals, technology, and structure in harmonious combination. And the conditions of harmony should derive from a rationality based on *organizational* well-being. (Also see Pfeffer [1976] for an excellent description of this position.) These prescriptive out-takes from Thompson's descriptive analysis have, we believe, formed the primary basis for management training in organization theory for the past decade. The position advocated in this paper has a number of contrary implications for management education:

1. By highlighting the true open system characteristics of organizations, managers can perhaps be made aware of the environmental consequences of actions taken in the narrow interests of the organization and be shown the boomerang quality of organizational rationality as the environment becomes more tightly coupled.[32] Somehow, the concept of ecology needs to be generalized and built into the conscious calculus of

administrative decision makers. The most effective—because experiential—way to do that is through large-scale, time-compressed simulations.[33] We may not be able to eliminate the motivation of self-interest, but we may be able to enlarge the manager's time perspective of rationality through such simulations.

2. By developing a typology of system dysfunctions and early warning signals, we may be able to train administrators to react adaptively when Thompson's harmony and co-alignment do not materialize according to plan. To our knowledge, nowhere do we now teach a diagnosis-and-treatment-of-macro-pathologies to managers or would-be managers.

3. We believe that the most radical implication of our position for management education derives from the view of organizations as language-using, sense-making cultures. In Thompson's view, the organization is an input-output machine, and the administrator is a technologist. In our view, the administrator's role shifts from technologist to linguist, from structural engineer to mythmaker. A key function of management according to level 8 thinking is that of helping the organization to make sense of its experiences so that it has a confident basis for future action. That is, the administrator must have a skill in creating and using metaphors. A manager's need to use metaphors skillfully suggests the conclusion that we should be teaching our institutional leaders (and organization theorists!) not only statistical analysis, but also poetry.

A FINAL NOTE

Having said what we have to say about the future of organization theory, we reflect on it as being already desperately inadequate. It is almost as though the saying of it immediately raises new problems that we must rethink at once. Given this thought and the note on which the preceding section ended, we can think of no more appropriate way to end the paper than the following, from Frederick Morgan's (1977) *Poems of the Two Worlds*.[34]

Saying

There always is another way to say it.
As when you come to a dusty hill and say,
"This is not the hill I meant to climb.
That one I've perhaps climbed already—see,
there it looms, behind me, green with trees."
And then climb as you can see the present hill.

Or when you walk through a great childhood forest
latticed with sun, carpeted in brown pine,
knowing the one you were and the one you are,
and think, "I shall not speak this forest's name
but let it densely live in what I am . . ."

The saying changes what you have to say
so that it all must be begun again
in newer reconcilings of the heart.

FOOTNOTES

1. The authors would like to acknowledge the many helpful suggestions made by David Whetten during the preparation of this paper. A number of others made helpful comments on an early draft of the paper, and we would especially like to thank George Huber, Jos Ullian, Keith Murnighan, Ray Zammuto, Karl Weick, and Gerald Salancik. Various versions of the manuscript were typed by Marsha Kopp and Norma Phegley at Illinois and Lavonne Buttyan at Pittsburgh.

2. An equally fruitful and related framework is provided by Ackoff and Emery (1972). Indeed, the similarities between Boulding's (1968) and Ackoff and Emery's framework is all the more striking since they were developed independently. Of the two, we are tempted to call the one by Ackoff and Emery the more basic, since it is the more grounded in fundamental distinctions, and in this sense, it is the more systematic. However, in this paper we mainly make use of Boulding's framework for reasons of its historical priority.

3. Ackoff and Emery (1972) employ a similar set of distinctions between "abstract" and "concrete" systems. An abstract or conceptual system is a system all of whose elements are concepts, whereas a concrete system is a system at least two of whose elements are (real) objects.

4. In Ackoff and Emery (1972), a common example of such systems is given by the class of entities called "meters," e.g., thermometers, accelerometers, etc.

5. Ackoff and Emery (1972) call examples of systems at this level "purposeful individuals." The prime example here is people, individuals who are capable of displaying "will," the autonomous creation of self-imposed goals (ends), and the ability to invent new patterns (means) of obtaining them.

6. This is our term, not Boulding's.

7. It is important to point out that Ackoff and Emery (1972) do *not* distinguish this level from that of the individual. For them, all of the concepts *necessary* to describe a purposeful individual are *sufficient* to describe a socially organized set of individuals. Note that this is *not* to say that there are no basic differences between groups and individuals. There are. Rather, this is to deny the sharp differences between the level of the individual (psychology) and that of the group (sociology) *without* thereby subsuming either science (or level) within the other (Churchman, 1968, 1971).

8. Because Thompson's point of view has been labeled as being an "open system model," we feel constrained to refer to it that way in this section, although we shall argue that relative to Boulding's definition of open system, Thompson's is not an open system approach. This labeling problem will create some unavoidable, but temporary, confusion.

9. For the necessity of perpetually considering competing models, see Feyerabend (1975),

who argues that science never outruns its need for the strongest competing models if it is to continue to advance. That is, such models are not a luxury but a dire necessity for the continued progress, i.e., development of science.

10. George Huber has pointed out to us that imposing uniformity seems to imply a proactive stance, but that control systems only react. Intentionality should properly be excised from the notion of creating uniformity. Nevertheless, the effect of control systems thinking, whether intended or not, is to create uniformity everywhere. Others have argued that we have seriously misinterpreted Thompson's emphasis on buffering the technical core. What Thompson intended, according to this line of argument, was not to homogenize the environment, but to "order its variety." It was claimed that stockpiling materials and supplies in an irregular market is a way of not only dealing with the environment, but *using* it. How stockpiling *uses* environmental variety is utterly obscure to us, and we stick by our initial interpretation that the force of all of Thompson's strategies for coping with environmental uncertainty is in the direction of producing uniformity, not only for the technical core, but outside the organization as well.

11. Ackoff and Emery (1972) put this point in a rather striking way by showing that every system must be *either* variety increasing *or* decreasing. That is, if an organization insulates itself from diversity, then this becomes equivalent to a strategy of attempting to actually reduce diversity.

12. What it means to deplete an organization of its variety is an extremely abstract concept, and perhaps an example will help to clarify what we mean. Students in MBA programs typically come from a wide variety of backgrounds ranging from engineering to the humanities. Inasmuch as students can learn from one another, this can be a major strength of the program. But the experience of going through the program together tends to make them more alike, thus reducing their original diversity. Originally, faculty members in new discipline-oriented graduate management programs also came from a wide variety of backgrounds, primarily because few programs existed to train faculty for such programs. But as programs have proliferated, they have also produced faculty members for one another according to the standards developed within the programs, thus reducing the diversity of faculty input. Some people have argued that such environments are likely never to grow uniform because they will be continually renewed by new entrants to the field and purely by chance events. We think that such assessment underestimates how closed the system can become, how much it can feed back on itself to produce the optimal inputs it thinks it needs. But in producing optimal inputs, for example, by seeking the type of students who will do best in MBA programs according to the performance criteria of grades, graduate management programs are likely to encourage uniform preparation by program aspirants. Perhaps this effect is clearer in the field of medical education where competition to gain entry to medical school creates both standardized premedical programs and selection on a very narrow base of unambiguous admissions criteria.

13. By "order" we do not mean uniformity. We mean organized complexity, departures from both uniformity and pure chaotic randomness, structured differences that have significance and meaning. And the essence of Hoffman's quote is that order, in this sense, cannot be created out of a uniform environment.

14. We must be careful here to recognize that dysfunction is defined relative to some particular value system. So interpersonal conflict might be functional for a high level of motivation or learning, and turnover might be functional for organization performance if the "right" people leave. Because macroanalysts have tended not to pay much attention to questions of value, they have consequently not paid much attention to organizational dysfunctions at the macro level. But even if one defines dysfunction more narrowly as a departure from equilibrium and from smooth operation, we think that our statement about

the focus of organization theory on the creation of order (qua stability, predictability, and uniformity) is still correct. Perhaps this was not true of the organizational sociologists of the 1940s and 1950s (e.g., Selznick, Gouldner, etc.), nor even perhaps of Crozier's (1964) analysis of power and conflict in French bureaucracies, but most macro organizational analysis is now done in schools of management rather than in departments of sociology, and this may account for the stronger functionalist orientation of the last decade. Another way to look at the micro-macro split is that the micro branch's problem-orientation is probably due to its descent from industrial psychology. Industrial psychologists tend to be practitioner-oriented and work on solving managers' applied problems of selection, turnover, training, and so forth. In contrast, organization theory (i.e., the macro branch) has a parentage in the functionalist school of sociology. Given that tradition of descriptive research and aspirations to be "scientific" in the positivist sense, it is not surprising that organization theory tends to eschew the consideration of values in inquiry and gives dysfunctions a minor role in organizational analysis.

15. Focusing on behavior away from equilibrium would also make our models more appropriate for studying (and bringing about) structural change. On the other hand, if our theories posit the existence of a correct way of organizing, changing away from that ideal coalignment makes no sense whatsoever.

16. The passive model of individual behavior implicit in the prevailing model has also been criticized by Argyris (1972). Because of his passivity, the individual is not conceived as a source of organizational change. Argyris argues, and we agree, that change, when it occurs, is conceptualized as a reaction or adjustment to change the environment or technology.

17. Some readers of an early draft have suggested that we seriously underestimated the role that language, awareness, and meaning have played in the organizational theories of Dalton, Gouldner, Goffman, and others, and that we have focused too narrowly on Thompson's book as a statement of the prevailing model. This criticism fails to recognize the fact that for the last decade Thompson's work, not that of others mentioned, has been the prime reference for research on organization, environment, technology, and structure.

18. By this we mean simply that uncertainty is a property of the object-observer pair. For a knowledgeable observer, the environment may seem quite certain, but an uninformed observer of the *same* environment may be quite baffled by what he sees or thinks may happen. Thus, it seems misguided to try to measure *the* uncertainty of the environment without specifying who the observers are and what their state of knowledge is.

19. What is at issue here is that there are different levels of reality: empirical reality or the world of actual behavior; explicit and formal assertions about the rules that govern behavior (e.g., standard operating procedures); myths, rituals, metaphors, and other implicit statements about organizational values and processes; and finally, generative mechanisms of a fundamental sort that produce behaviors of all kinds, not just organizational, and which are typically not easily accessible, for example, the general problem-solving routines of Newell and Simon (1972) or the transformational grammar of Chomsky (1972). Our argument here is that "organizational" structure (i.e., enduring patterns or models of behavior) should be studied at several levels of reality, not merely as an observed regularity at the empirical level.

20. Single case studies, especially qualitative analyses, have been out of fashion for the last 15 years. There seems to have been a presumption that one could understand phenomena at a class level, without needing to understand the individual organizations that make up the class. In fact, we argue quite the opposite, that one needs first to understand the individual case in its own terms, to build a model or theory of it. When one does compare across organizations, what is compared is not the empirical descriptions of the organiza-

tions, but the models of them. Building, and only then comparing, models of organizations is what we mean by developing theories of individuals or by creating a science of the singular case. (See Leach, 1961, pp. 1–27; Pondy and Olson, 1977.)

21. A similar criticism can be raised about most treatises on scientific method and philosophy of science; i.e., they are descriptions (prescriptions) about mature, already well-formulated theories, not about the messy process governing the birth of theories (Mitroff, 1974).

22. He also, of course, carries with him general cultural rules of organizing that are common across organizations. Peoples' personal knowledge of such organizing rules is why some administrative personnel are so valuable to pirate away from competing organizations. If all such organizing rules could be codified in a publicly accessible form, one would not need to secure the services of particular people.

23. It has been argued to us that in Kimberly's case, the founder lost control within a year and structural processes took over pushing the organization back to the norm. In fact, we have first-hand knowledge of the continuing situation that Kimberly studied, and the medical school is proceeding to engage in quite nonstandard programs such as making an institutional commitment to join M.D.-Ph.D. programs with a wide variety of academic departments on the campus. Furthermore, we make no claim that the founder will always be successful in reproducing his image of the organization. So even if the founder of the medical school had lost control, that single empirical fact would not invalidate the organization-reproduction model as a fruitful model to use in our investigations. Theories are not assertions of fact; they are guides to inquiry.

24. Richard Daft has suggested to us that the Hummon et al. model is not a level 3 model after all, because the equilibrium point of the control process is not changed. Nevertheless, it *could* be changed in principle and the system would adapt to the new equilibrium point. That is, a control target is provided for in the model, whereas by comparison in the Nystrom (1975) analysis, there is no state toward which the system is presumed to be tending; Nystrom simply has his system behaving dynamically through time. Another recent analysis that does seem more clearly to embody control system ideas within its structure is Hall's (1976) simulation model of the decline of the *Saturday Evening Post*. Hall's model is a particularly revealing one about the nature of level 3 models, because it demonstrates that control systems are not necessarily stable: The *Saturday Evening Post* after all *failed*.

25. Some care needs to be exercised here, inasmuch as some studies have shown that sensory deprivation can have short-term beneficial effects for individual behavior. For example, Suedfeld reported on an experiment in which "[P]erformance on simple tasks was seldom impaired and often improved by sensory deprivation. Complex task performance, on the other hand, was usually worse after deprivation" (Suedfeld, 1975, pp. 64–65). To carry the analogy to the organizational level, buffering against environmental uncertainty may be dysfunctional only when the technology demands innovative and nonroutine behavior. "[W]e know that sensory deprivation leads to increased daydreaming and fantasizing and to more openness and to new experiences" (Suedfeld, 1975, p. 65). However, extreme deprivation over an extended period of time can lead to a breakdown in complex cognitive processes. We see these qualifications as completely consistent with our argument that open systems need environmental variety in order to maintain their level of complexity.

26. Taken together, these research questions suggest that our underlying root metaphor of organization is shifting. If the first three levels of Boulding's hierarchy can be said to rest on a machine metaphor, and if levels 4 through 6 derive from a biological or organic metaphor, then levels 7 and 8 suggest a cultural metaphor of organizations. Some critics of our position have argued that Thompson really does operate with such a cultural metaphor of organizations. After all, he does explicate the organization's dependence on the embedding social system for belief systems, for occupational categories, and so forth, but it is a particularly

rigid conception of that relationship, and for the most part, organizations are not endowed in Thompson's model with the same kind of cultural properties and functions as the embedding social system. Thompson seems to have made the same kind of assumption as that made by Blau and Scott (1972, pp. 2–5) that the principles of "formal organization" are distinct from those of "social organization" that operate in the social system at large. But a cultural metaphor of organizations suggests that principles of social organization operate within organizations as well as outside them.

27. We define "language" here broadly to include nonverbal language (e.g., paralinguistics, kinesics, proxemics), formal languages (e.g., mathematics, computer programming), as well as natural language. Language is especially important to study not only because it is a phenomenon in its own right, but also because it provides a point of view, a linguistic metaphor, in which to study the relationship of things to the signs that stand for them. But part of our effort is simply to draw attention to language as something worth studying in organizational settings. There are some interesting language vacuums in certain parts of the social science literature. For example, Hare (1972) has recently reviewed a decade of small group research. In all of the studies of communication networks that Hare cites, not one single study so much as mentions language as a variable that was considered or investigated. And if any phenomenon would be naturally expected to include language as a variable, surely it would be communication. In any case, we attempt here to rectify that neglect of language.

28. This argument presumes that the form of a language will be functionally adapted to the setting in which it is used.

29. The argument here is that social organization among subhumans can change only by virtue of species change through genetic mechanisms. The conjecture is that in order to change social structure among a given set of living organisms, that structure must be *talked* about, verbally negotiated. Because humans can talk about their social organization, they can renegotiate it and create new forms of social organization. They need not wait for evolutionary mechanisms to change the social organization through changing the organizing rules that are genetically programmed into the organism.

30. At first glance, this proposal may seem to suggest that all macro phenomena can be reduced to micro phenomena, and that macro phenomena cannot be analyzed at the micro level of analysis—in short, that there are no emergent properties. We are suggesting nothing of the kind. We do not think it is fruitful to try to reduce all of sociology to psychology. What is being suggested instead is that generative mechanisms of a *macro* character produce observable macro level properties, and that these generative mechanisms need not be rooted in individual psyches. For example, written rules that reside in, say, filing cabinets could be such generative mechanisms that produce certain structures. Or Clark's (1972) notion of a saga as a *socially shared* story about the organization's history is a macro level generative mechanism that does not require a reductionist's retreat to individual psychology. All we are suggesting is that it would be profitable to study the relationship between organizing rules stored in the social structure (or, possibly in the minds of individuals as well) and the empirical observables that they produce. That relationship between rules and behavior need not be a perfect one, but that in no sense invalidates the study of the rules or of the rule-behavior relationship.

31. As we elaborated in note 19, there is an empirical level of reality, but there are also deeper levels of reality at which patterns reside: the levels of expressed rules, of metaphors and myths, and of inexpressible rules. All we are saying here is that organizations can be "explained," i.e., patterns can be discovered at any one of those levels of reality. Most of the current work tries to find patterns at the level of empirical reality. We are merely suggesting that we should be looking for patterns at these deeper levels of reality, too. In no sense can this proposal be construed as calling for a hiatus in macro level analysis until

cognitive psychology has told us how individuals conceptualize their worlds. But it does suggest a different strategy of macro analysis.

32. The boomerang effects referred to above include as one important subclass the so-called "tragedy of the commons," in which it is individually rational for each sheepherder to overgraze common pastures even though the collective result for all sheepherders is to destroy the grazing land.

33. The reason that time-compressed experiential learning may be necessary to expand the concept of one's self is that the long-run, indirect personal consequences of one's own actions need to be presented as contiguously and vividly as possible in order to overcome the self-environment split that is so intimate a part of our epistemology. Simply talking about it, as we are doing here, is unlikely to effect the shift.

34. Frederick Morgan, *Poems of the Two Worlds*. Urbana, Ill.: University of Illinois Press, 1977, p. 105. © 1977 by Frederick Morgan. Reprinted by permission of the author and the University of Illinois Press.

REFERENCES

1. Ackoff, R. L., and F. E. Emery (1972) *On Purposeful Systems,* Chicago: Aldine-Atherton.
2. Alexander, H. G. (1967) *Language and Thinking,* Princeton, N.J.: D. Van Nostrand.
3. Argyris, C. (1972) *The Applicability of Organizational Sociology,* Cambridge, England: Cambridge University Press.
4. Ashby, R. (1956) *An Introduction to Cybernetics,* London: Chapman & Hall.
5. Bateson, G. (1972) *Steps to an Ecology of Mind,* New York: Ballantine Books, Inc.
6. Beer, S. (1964) *Cybernetics and Management,* New York: John Wiley & Sons, Inc.
7. Berger, P. L., and T. Luckmann (1966) *The Social Construction of Reality,* Garden City, N.Y.: Doubleday & Company, Inc.
8. Blau, P. M. (1970) "A formal theory of differentiation in organizations," *American Sociological Review 35,* 201–218.
9. ——— and W. R. Scott, (1972) *Formal Organizations,* San Francisco: Chandler Publishing Company.
10. Boje, D. M., and K. M. Rowland (1977) "A dialectical approach to reification in mythmaking and other social reality constructions: the I-A-C-E model and OD," unpublished manuscript, University of Illinois, Urbana, Organizational Behavior Group.
11. Boulding, K. (1968) "General systems theory—the skeleton of science," in Walter Buckley (ed.), *Modern Systems Research for the Behavioral Scientist,* Chicago: Aldine Publishing Company.
12. Burns, T., and G. M. Stalker (1961) *The Management of Innovation,* London: Tavistock.
13. Churchman, C. W. (1971) *The Design of Inquiring Systems,* New York: Basic Books, Inc., Publishers.
14. ——— (1968) *The Systems Approach,* New York: Delacorte Press.
15. Chomsky, N. (1972) *Language and Mind* (enlarged ed.), New York: Harcourt Brace Jovanovich, Inc.
16. Clark, B. R. (1972) "The occupational saga in higher education," *Administrative Science Quarterly 17,* 178–184.
17. Cohen, M. D., and J. G. March (1974) *Leadership and Ambiguity,* New York: McGraw-Hill.

18. Crozier, M. (1964) *The Bureaucratic Phenomenon*, Chicago: University of Chicago Press.
19. Cyert, R. M., and J. G. March (1963) *A Behavioral Theory of the Firm*, Englewood Cliffs, N.J.: Prentice-Hall, Inc.
20. Etzioni, A. (1961) *A Comparative Analysis of Complex Organizations*, New York: The Free Press.
21. Feyerabend, P. K. (1975) *Against Method: Outline of an Anarchistic Theory of Knowledge*, London: Humanities Press, Inc.
22. Gaston, J. (1972) "Communication and the reward system of science: A study of national 'invisible college,'" *The Sociological Review Monograph 18*, 25–41.
23. Geertz, C. (1973) *The Interpretation of Cultures*, New York: Basic Books, Inc., Publishers.
24. Goldsen, R. K. (1975) "The technological fix: existentialist version," *Administrative Science Quarterly 20*, 464–468.
25. Haas, J. E., and T. E. Drabek (1973) *Complex Organizations: A Sociological Perspective*, New York: Macmillan, Inc.
26. Hackman, R., and G. Oldham (1975) "Development of a job diagnostic survey," *Journal of Applied Psychology 60*, 159–170.
27. Halberstam, D. (1972) *The Best and the Brightest*, New York: Random House, Inc.
28. Hall, R. I. (1976) "A system pathology of an organization: The rise and fall of the old *Saturday Evening Post*," *Administrative Science Quarterly 21*, 185–211.
29. Hamilton, D. (1976) "A science of the singular?", unpublished manuscript, CICRE, University of Illinois, School of Education, Urbana, Ill.
30. Hanson, N. R. (1958) *Patterns of Discovery*, Cambridge: Cambridge University Press.
31. Hare, A. P. (1972) "Bibliography of small group research: 1959–1969," *Sociometry 35*, 1–150.
32. Harre, H., and P. F. Secord (1973) *The Explanation of Social Behavior*, Totowa, N.J.: Littlefield, Adams & Company.
33. Hedberg, B. L. T., P. C. Nystrom, and W. H. Starbuck (1976) "Camping on seesaws: Prescriptions for a self-designing organization," *Administrative Science Quarterly 21*, 41–65.
34. Huff, A. S. (1977) "Evocative metaphors," unpublished manuscript, UCLA Graduate School of Management.
35. Hummon, N. P., P. Doreian, and K. Teuter (1975) "A structural control model of organizational change," *American Sociological Review*, 40:813–824.
36. Janis, I. (1972) *Victims of Groupthink*, Boston: Houghton Mifflin Company.
37. Keegan, W. J. (1974) "Multinational scanning: A study of information sources utilized by headquarters executives in multinational companies," *Administrative Science Quarterly 19*, 411–421.
38. Kimberly, J. R. (1976) "Contingencies in creating new institutions: An example from medical education," unpublished manuscript presented at the Joint EIASM-Dansk Management Center Research Seminar on "Entrepreneurs and the Process of Institution Building."
39. Kuhn, T. S. (1970) *The Structure of Scientific Revolutions* (2nd ed.), Chicago: University of Chicago Press.
40. Lawrence, P., and J. Lorsch, (1969) *Organization and Environment*, Cambridge: Harvard University Press.
41. Leach, E. (1972) "Anthropological aspects of language: Animal categories and verbal abuse," in Pierre Maranda (ed.), *Mythology*, London: Penguin.
42. ——— (1961) *Rethinking Anthropology*, New York: Humanities Press, Inc.

43. Leavitt, H. J., and L. R. Pondy (eds.) (1964) *Readings in Managerial Psychology* (1st ed.), Chicago: University of Chicago Press.
44. March, J. G., and H. A. Simon (1958) *Organizations*, New York: John Wiley and Sons, Inc.
45. Meyer, J., and B. Rowan (1977) "institutionalized organizations: Formal structure as myth and ceremony," *American Journal of Sociology 30*, 431–450.
46. Meyer, M. W. (1972) *Bureaucratic Structure and Authority: Coordination and Control in 254 Government Agencies*, New York: Harper & Row,
47. Mitroff, I. I. (1974) *The Subjective Side of Science: A Philosophical Inquiry into the Psychology of the Apollo Moon Scientists*, Amsterdam: Elsevier.
48. ——, and T. R. Featheringham (1974) "On systemic problem solving and the error of the third kind," *Behavioral Science 19*, 383–393.
49. ——, and R. Kilmann (1976) "On organizational stories: An approach to the design and analysis of organizations through myths and stories," in R. H. Kilmann, L. R. Pondy, and D. P. Slevin (eds.), *The Management of Organization Design: Strategies and Implementation*, New York: American Elsevier Publishing Co. Inc.
50. ——, and M. Turoff (1974) "On measuring the conceptual errors in large scale social experiments: The future as decision," *Journal of Technological Forecasting and Social Change 6*, 389–402.
51. Morgan, F. (1972) *Poems of the Two Worlds*, Urbana, Ill.: University of Illinois Press.
52. Newell, A., and H. A. Simon (1972) *Human Problem Solving*. Englewood Cliffs, N.J.: Prentice-Hall, Inc.
53. Nystrom, P. C. (1975) "Input-output processes of the Federal Trade Commission," *Administrative Science Quarterly, 20,* 104–113.
54. Perrow, C. (1967) "A framework for the comparative analysis of organizations," *American Sociological Review 32,* 194–208.
55. Pettigrew, A. M. (1976) "The creation of organizational cultures," unpublished manuscript, London Graduate School of Business Studies.
56. Pfeffer, J. (1976) "Beyond management and the worker: The institutional function of management," *Academy of Management Review 1,2,* 36–46.
57. Pfeffer, J., and C. Salancik, (1974) "Organizational decision making as a political process: The case of a university budget," *Administrative Science Quarterly 19,* 131–151.
58. Pondy, L. R. (1977a), "The other hand clapping: An information processing approach to organization power" in Tove Hammer and Samuel Bacharach, (eds.), *Reward Systems and Power Distribution,* Cornell University School of Industrial and Labor Relations.
59. —— (1977b) "Leadership is a language game," in M. McCall and M. Lombardo (eds.), *Leadership: Where Else Can We Go?,* Durham, N.C.: Duke University Press.
60. ——. (1975) "A minimum communication cost model of organizations: Derivation of Blau's laws of structural differentiation," unpublished manuscript, University of Illinois, Organizational Behavior Group, Urbana, Ill.
61. ——, and D. M. Boje, (1976) "Bringing mind back in : Paradigm development as a frontier problem in organization theory," unpublished manuscript, University of Illinois, Organizational Behavior Group, Urbana, Ill.
62. ——, and M. L. Olson, (1977) "Theories of extreme cases," unpublished manuscript, University of Illinois, Organizational Behavior Group, Urbana, Ill.
63. Salancik, G., and J. Pfeffer, (1974) "The bases and uses of power in organizational decision making: The case of a university," *Administrative Science Quarterly 19,* 353–373.
64. ——, Staw, B., and L. Pondy, (1976) "Administrative turnover as a response to unmanaged organizational interdependence: The department head as a scapegoat," un-

published manuscript, University of Illinois, Organizational Behavior Group, Urbana, Ill.

65. Schon, D. A. (1971) *Beyond the Stable State*. New York: Norton & Company, Inc.
66. Schrödinger, E. (1968) "Order, disorder, and entropy," in Walter Buckley (ed.), *Modern Systems Research for the Behavioral Scientist*, Chicago: Aldine.
67. Scott, W. R. (1977) "On the effectiveness of studies of organizational effectiveness," in P. S. Goodman and J. M. Pennings (eds), *New Perspectives on Organizational Effectiveness*. San Franscisco: Jossey-Bass Publishers.
68. Silverman, D. (1971) *The Theory of Organization*, New York: Basic Books.
69. Smith, R. A. (1971) *Corporations in Crisis*. Garden City, N.Y.: Doubleday & Company, Inc.
70. Sproull, L., and S. Weiner, (1976) "Easier 'seen' than done: the function of cognitive images in establishing a new bureaucracy," unpublished manuscript, Stanford University, Graduate School of Education.
71. Starbuck, W. H. (ed.) (1971) *Organizational Growth and Development*, Baltimore: Penguin Books Inc.
72. Staw, B. M. (1976) "Knee-deep in the big muddy: The effect of personal responsibility and decision consequences upon commitment to a previously chosen course of action," *Organizational Behavior and Human Performance 16*, 27–44.
73. ———, and F. V. Fox, "Escalation: The determinants of commitment to a previously chosen course of action," *Human Relations 30*, 431–450.
74. ———, and E. Szwajkowski, (1975) "The scarcity-munificence component of organizational environments and the commission of illegal acts," *Administrative Science Quarterly 20*, 345–354.
75. Suedfeld, P. (1975) "The benefits of boredom: Sensory deprivation reconsidered," *American Scientist 63*, 1, 60–69.
76. Thompson, J. D. (1967) *Organizations in Action*, New York: McGraw-Hill.
77. Tung-Sun, C. (1970) "A Chinese philosopher's theory of knowledge," in G. P. Stone and H. A. Farberman (eds.), *Social Psychology Through Symbolic Interaction*, Waltham, Mass.: Xerox College Publishing Co.
78. Weick, Karl E. (1969) *The Social Psychology of Organizing*, Reading, Mass.: Addison-Wesley Publishing Co., Inc.
79. ———. (1977) "On re-punctuating the problem of organized effectiveness," in P. S. Goodman and J. M. Pennings (eds.), *New Perspectives on Organizational Effectiveness*, San Francisco: Jossey-Bass Publishers.

COGNITIVE PROCESSES IN ORGANIZATIONS[1]

Karl E. Weick, CORNELL UNIVERSITY

ABSTRACT

Cognitive descriptions of organizations are built on the dual images of organizations as bodies of thought and organizations as sets of thinking practices. Traditional organizational variables such as centralization influence cognitive processes and are themselves shaped by these processes. Viewed as bodies of thought, organizations can be described as recurrent schemata, causal textures, and sets of reference levels. Viewed as sets of thinking practices, organizations can be described in terms of dominant rules for combining cognitions, routine utterances, mixtures of habituation and reflection,

Research in Organizational Behavior, Vol. 1, pp. 41–74.
Copyright © 1979 by JAI Press, Inc.
All rights of reproduction in any form reserved.
ISBN 0-89232-045-1

nature of rehearsing, and preferences for simplification. Insufficient attention has been
paid to the possibility that, for want of a thought, the organization was lost. This essay
is designed to redress that analytic imbalance.

An organization is a body of thought thought by thinking thinkers. The
elements in that formulation are thoughts (Goodman, 1968), thinking practices (Neisser, 1963), and thinkers (Jeffmar and Jeffmar, 1975). Examination of cognitive processes in organizations involves studying all three.

In reply to the question "what is an organization," we consider organizations to be snapshots of ongoing processes, these snapshots being
selected and controlled by human consciousness and attentiveness. This
consciousness and attentiveness, in turn, can be seen as snapshots of
ongoing cognitive processes or more precisely epistemological processes
where the mind acquires knowledge about its surroundings (Bougon,
Weick, and Binkhorst, 1977). In these epistemological processes, both
knowledge and the environment are constructed by participants interac-
tively.

Given a cognitive orientation, there are distinct ways to talk about
organizations. For example, an organization can be viewed as a body of
jargon available for attachment to experience just as a person can be
viewed as a glossary of labels available for attachment to ambiguous
states of arousal (London and Nisbett, 1974; Rodin, 1977). Organizational
members vary in their attentiveness to external conditions and suggestibility just as people of varying body weights differ in their external orientation and suggestibility to food cues.

Managerial work can be viewed as managing myths, images, symbols,
and labels. The much touted "bottom line" of the organization is a
symbol, if not a myth. The manager who controls labels that are meaningful to organizational members can segment and point to portions of their
experience and label it in consequential ways so that employees take that
segment more seriously and deal with it in a more organizationally appropriate manner (Pettigrew, 1975). Because managers traffic so often in
images, the appropriate role for the manager may be evangelist rather
than accountant.

Standard concepts in general cognitive psychology (Broadbent, 1971;
Reynolds and Flagg, 1977; and Scheerer, 1954) can also be adapted to
descriptions of organizations. For example, an organization member's
knowledge can be viewed as deductions drawn from a view of the world
legitimated within that organization. An organization can be characterized
by the specific syllogisms (Jones and Gerard, 1967, pp. 159–162) that are
used within it. An organization can be characterized by the contents of the
schemata members invoke routinely and through which they size up situa-

tions. Organizations can be characterized as sets of weakly held assumptions that are subject to disconfirmation thereby producing the experience of interesting times (Davis, 1971).

To describe an organization as a body of thought is to suggest that collective rather than individual omniscience is the object of interest. That emphasis in turn suggests parallels between organizations in general and scientific communities (Betz, 1971; Ravetz, 1971) in particular. Scientific communities also represent bodies of thought (paradigms), the main difference being that scientific communities reputedly (Brush, 1974) are more detached from and more objective about the thoughts they incorporate. It is conceivable that organizations could be described as scientific communities with self-interest. Organizations have a personal stake in the thoughts they accumulate and transmit, but aside from that their knowing mechanisms resemble those found in scientific organizations.

This essay is about the place of thinking in organizations. We know that people think. We know that in any organization at any moment somebody is thinking. The question then becomes, does that thinking create, accomplish, or displace anything that goes on in that organization? A researcher's job is to spot the thinking people in an organization, see what they're thinking about, and examine how those thoughts become amplified and diffused through the organization or discover why those thoughts remain localized. Much of Mintzberg's (1973) analysis suggests that managers spend little time reflecting. They are active, they act on line, they spend most of their time communicating, they have very little time to themselves, their interruptions are frequent. If they think much at all, their thinking seems to be grooved, or under the influence of the last person they talked to, or abstract and detached from the here and now (Steinbruner, 1974). That portrait, however, is an oversimplification and we will try to show why.

OBJECTS OF ORGANIZATIONAL THOUGHT

Two quotations vividly pose the issue of what is available to be thought about in organizations.

> Defining situations as real certainly has consequences, but these may contribute very marginally to the events in progress; in some cases only a slight embarrassment flits across the scene in mild concern for those who tried to define the situation wrongly. All the world is not a stage—certainly the theater isn't entirely. (Whether you organize a theater or an aircraft factory, you need to find places for cars to park and coats to be checked, and these had better be real places, which, incidentally, had better carry real

insurance against theft). Presumably, a "definition of the situation" is almost always to be found, but those who are in the situation ordinarily do not *create* this definition, even though their society often can be said to do so; ordinarily, all they do is to assess correctly what the situation ought to be for them and then act accordingly. True, we personally negotiate aspects of all the arrangements under which we live, but often once these are negotiated, we continue on mechanically as though the matter had always been settled. So, too, there are occasions when we must wait until things are almost over before discovering what has been occurring and occasions of our own activity when we can considerably put off deciding what to claim we have been doing. But surely these are not the only principles of organization. Social life is dubious enough and ludicrous enough without having to wish it further into unreality (Goffman, 1974, pp. 1–2)

Consider the situation of trial and error.

Trial implies a problematical and alternative result: either the success of the assumption put to trial or its failure. When we ask why this is so, we hit upon the presence of some "controlling" condition or circumstance in the situation—some stable physical or social fact—whose character renders the hypothesis or suggested solution either adequate or vain, as the case may be. The instrumental idea or thought, then, has its merit in enabling us to find out or locate facts and conditions which are to be allowed for thereafter. These constitute a *control upon knowledge and action,* a system of "things." . . . The method of selection by trial and error requires that relative stability, fixity and permanence be discovered in the "control" conditions in the environment, since the genesis of truth lies in the checking off of hypotheses under this more stable control. The truth of a thought may be discovered through its successful working; but we have to consider also the failures, the errors, and indeed the whole situation in which truth and error are alike possible. . . .

I may "bring about" reality perhaps, without this external control, by "willing to believe" in something for which I have no proof or reason, in cases in which this sort of event willed—as, for example, someone else's conduct—may be conditioned upon my act of will. But nature does not take to suggestions so kindly. The will of a general may stimulate his troops and so bring to him the victory he believes in; but such an act of the general's will cannot replenish the short supply of powder or shells, on which the issue of the battle perhaps more fundamentally depends" (Baldwin, 1909, pp. 72–73).

The objects of thought in organizations can be constructions of organizational members (Delia, 1977). But as Goffman, Baldwin, and others (Goldthorpe, 1973; Frankel, 1973) remind us, somewhere in most of those constructions is a grain of truth. The seeds for those constructions exist independent of the observers even though members embellish and elaborate those grains with vigor and originality.

In this essay, we intend to make the environment just as problematic for inquiry as possible. That's where the joining of organization theory and epistemology becomes crucial. Epistemology is concerned with mind and environment relationships and both are actively constructed. To argue that the chief problem of organizational theory is to articulate the relationships between organizations and environments is to miss the point that organizations and environments both undergo considerable construction in the eyes and minds of organizational members (Blumer, 1971). The environment is a problem. One way to deal with the environment is simply to say that it's "dealable." Having imposed that construction on the environment, the actor is enabled to move his world around and make sense of it.

Returning to Goffman's concern with checkrooms and parking lots and Baldwin's about ammunition, in the beginning someone enacted those settings. The trick is to go back far enough to trace that development. The "checkroom" or "parking lot" can also become sites for muggings, arenas for shooting craps, places to pile inventory, personal turf, profit centers, and much more. The thieves that Goffman is worried about certainly impose different definitions on coat rooms than do people who blithely hang their apparel in them.

We want to emphasize that in any situation where people enact their environment there usually are grains of truth that invite elaboration. Enactment isn't a hallucination. Typically it meets the environment half way. But what happens is that the actor in the organization plays a major role in unrandomizing and giving order to the bewildering number of variables that constitute those grains. Through a combination of selective attention, activity, consensual validation, and luck, organizational actors are able to stride into streams of experience where things are mixed together in random fashion and unravel those streams sufficiently so that some kind of sensemaking is possible (Schutz, 1964). In accomplishing that unraveling, people do enact their environments as well as park their cars and hang up their coats. And while doing this enactment these same people often don't assess correctly what the situation ought to be for them because "correctness" is not a dimension that can be made relevant to the situation. Neither can they "act accordingly" when cues are unreliable and instrumentalities equivocal.

Objects of thought in organization are not just enactments, they can also be described in terms of a phrase attributed to Korzybski: The map is not the territory (Hayakawa, 1961). The intent of that assertion is to remind people that their representations of the world are just that, representations rather than the thing itself. We do not take issue with that point of view but do place a different emphasis on it. The map *is* the territory if people treat it as such. That's the thrust of any formulation which talks about a definition of the situation (Ball, 1972). Things are real if people treat them as real and that's the reason why if people mistake the map for the territory, to criticize that action is less crucial than to see how people operate having made that "mistake." Treating the map as the territory also satisfies personal needs for cognition and order (Cohen, Stotland, and Wolfe, 1955).

Maps do structure the territory sufficiently so that someone can initiate activity in that territory, activity that may introduce order. We know that people make do, improvise, and act like *bricoleurs* (Levi-Strauss, 1962, pp. 16–33), which means that a map is sufficient for most people to get a sense of the situation and wade into it. They don't need to know the territory inch by inch to do something about it or in it. So when we assert that the map is the territory we take seriously the facility people demonstrate in taking bits and pieces of information and assembling them into a workable order.

Goffman and Baldwin remind us not to be carried away by images such as "definition of the situation," "negotiation of reality," and "the will to believe." Each of those concepts implies that reality is almost wholly in the hands of the actor and he can make of it what he wants. That's clearly not so. But neither is it the case that parking lots, cloak rooms, and ammunition come clearly labeled (fists can be ammunition), are consensually validated (roofs are floors for cars or ceilings for people), or mean only one thing (parking lot as gymkhana course). Much goes on in organizations after people park their cars and hang their coats and it's those other actions that we're concerned with. Since most of those other actions do involve people, they provide more latitude for invention than Baldwin admits.

The objects of organizational thought oftentimes also are dominated by the prior beliefs among members. Frequently they operate by the maxim "I'll see it when I believe it" (e.g., Gould, 1977; Nisbett and Wilson, 1977, pp. 248–249). As Einstein said, "It seems that the human mind has first to construct forms independently before we can find them in things" (cited in Rosnow and Fine, 1976, p. 63). When we want to examine cognition in organizations, one of the best starting points is to discover the beliefs through which organizational members will examine their experiences.

The importance of beliefs is dramatically apparent in the case of placebos (Bishop, 1977). Placebos are any inactive substance or procedure that is used with a medical patient under the guise of an effective treatment. Repeatedly it has been demonstrated that because the patients believe in these inactive substances or procedures, they actually work. And the amount of symptom relief has been dramatic. How placebos work isn't understood, but that they work is evident and supports the argument that beliefs do play a major role in seeing (Gregory and Bombrich, 1973). The important question for our purpose is what's the organizational equivalent of a placebo? Which beliefs, in conjunction with which innocuous assignments, produce outcomes that are disproportionate in size to those usually associated with such assignments?

One presumption that sometimes is associated with cognitive views is that people are passive. They idly sit around either registering the environment or reflecting on past experiences. In general, these people are interesting only because of their large investment in headwork. That presumption is misleading and should be corrected. The principle object of organizational thoughts is organizational acts. Acts are the raw material for cognitive work in organizations.

One of the big gaps in current organizational theory is that we just don't know much about what happens when acting precedes thinking. That's what much of this essay is about. March and Olsen have recently commented on this oversight: "If we knew more about the normative theory of acting before you think, we could say more intelligent things about the functions of management and leadership when organizations or societies do not know what they're doing" (1976, p. 79).

In terms of my own work (Weick, 1977), enactment processes are viewed as crucial in organizational sensemaking. And as Lou Pondy shrewdly noted, those processes have *never* been labeled "enthinkment." The emphasis is on actions that provide a pretext for thinking, and not the reverse. Most of our analyses elaborate the basic sensemaking recipe, 'How can I know what I think till I see what I say." In that recipe, saying or doing precede thinking and provide the objects on which thinking will dwell. To argue that thinking is detached from action is to miss the point of that recipe.

If we talk about people enacting many of thier own pretexts for sensemaking, several problems come into focus. For example, one of the dominant themes in cognitive approaches to organizations is that people simplify their situations with a vengeance and grasp them with only modest success and thoroughness. From the standpoint of enactment, these people presumably enact and impose simplifications on the world, which means that the world is then sensed as a simple display which can be monitored by attending to a relatively small number of variables. That suggests one mechanism by which Steinbruner's (1974) cybernetic theory of organizations might operate (see also Coulam, 1977). If a person enacts a simple world, it's no great accomplishment to monitor and survive within that simple world. However, the important thing to notice is that simple enactments mingle with those variables that have been ignored, suppressed, or neglected such that the composite sensation will usually be more complicated than what the actor imposed in the first place. People aren't simply presented with their own simplifications of the world. Imposed simplifications that ignore complexities nevertheless interact with those complexities and present a world that is more complicated than the observer.

In summary, when we talk about the objects of thought in organizations we emphasize that though there be grains of truth in the displays confronted by members, grains that become elaborated, additional objects of thought include maps of the organization that are treated as if they are territories, beliefs through which people see the organization, and acts that provide the pretext and raw material for sensemaking. Organizational variables become important as they affect these grains, maps, beliefs, and actions. Whenever one can hypothesize that an organizational variable such as size or formalization or centralization has a demonstrable effect on one or more of these objects of thought, cognitive theory and organizational theory have been joined.

But the causal arrow also goes the other way. Maps, beliefs, and thoughts that summarize actions, themselves constrain contacts, communication, and commands. These constraints constitute and shape organizational processes that result in structures (Berlo, 1977).

ORGANIZATIONS AS A BODY OF THOUGHT

Schemata as Bodies of Thought

The concept of schema (Axelrod, 1973) is a major tool when one wishes to think cognitively about organizations. Schemas have been given a variety of definitions and we will review quickly some qualities of the concept so that its value for the study of organizations can become apparent.

Originally the concept of schema was developed as a way to understand how well-adapted, coordinated movements were possible. When coordinated movement takes place, each successive step is made as if it were under the control of the preceding movements in the series. This implies that the position reached by the moving limb in the last preceding stage somehow is recorded and still functions even though it has concluded. The problem then became to explain how past movements retain current influence. The idea of schema was developed to suggest a standard against which all subsequent changes of posture are measured before they enter consciousness.

Bartlett (1932) used the idea of schema most extensively and the nuances he added are crucial for organizational analysis. Bartlett objected to the static quality implied by the statement that a schema is a standard against which actions are compared. He wanted to emphasize that the standard was actively doing something all the time (not just serving as a static comparison) and that it was developing from moment to moment. Bartlett toyed with the possibility of calling this developing and developed framework an "organized setting" (1932, p. 201), but abandoned that in favor of talking about a schema which he described as

"an active organization of past reactions, or of past experiences, which must always be supposed to be operating in any well adapted organic response. That is, whenever there is any order or regularity of behavior, a particular response is possibly only because it is related to other similar responses which have been serially organized, yet which operate, not simply as individual members coming one after another, but as a unitary mass. Determination by schemata is the most fundamental of all the ways in which we can be influenced by reactions and experiences which occurred sometime in the past. All incoming impulses of a certain kind, or mode, go together to build up an active, organized setting" (p. 201).

The important thing for Bartlett was that schemata are repeatedly built up on the spot. Literal recall and literal replaying of past responses never happen. Instead the individual uses the past as a point of departure and then reassembles those prior experiences together with new inputs and develops all of this in an active ongoing fashion. People build a version of the past just as they build a tennis stroke afresh, depending on the preceding balance of postures and momentary needs of the game. Every time a stroke is built or a memory is constructed it has some originality. As Bartlett notes, "In a world of constantly changing environment, literal recall is extraordinarily unimportant" (p. 24).

More recently Stotland and Canon (1972) have used schema theory as a way to understand social psychological research. Concerning schema, they observe that

"persons generate relatively abstract and generalizable rules, called schemas, regarding certain regularities in the relationships among events. Once established they serve as a guide to behavior and as a framework which influences the manner in which relevant new information will be assimilated. Since a schema is an abstraction, a general statement detailing the perceived regular co-occurrence of some categorized events, it tends to be relatively permanent and impervious to change even if a few exceptions to it are noted" (1972, p. 67).

As an illustration of how schemata are used, Stotland and Canon make the following observation:

"It is possible to quite consciously invoke or activate a schema as an assist in dealing with some concrete situation which is being faced. For instance, a person upon finding himself in a new work situation thrown together with a number of unfamiliar people might rather self-consciously turn to his schemata relevant to work, meeting new persons, effective behavior in novel surroundings. These might include such schemata as: persons who talk a great deal are usually leaders, women are easier to get to know than men, an anxiety in a novel situation is lessened by finding some old, familiar elements in it. Self-motivated arousal of such general rules of a relationship would perhaps be useful in suggesting successful ways of coping with a situation. On the other hand, it would be highly likely that schemas of this sort would be operative even though they were not intentionally activated by the person, as they can also be brought into play by external factors" (p. 68).

Thus, a schema is an abridged, generalized, corrigible organization of experience that serves as an initial frame of reference for action and perception. A schema is the belief in the phrase, I'll see it when I believe it. Schemata constrain seeing and one way in which they may do this has been described by Neisser (1976).

Figure 1.   Neisser's perceptual cycle.

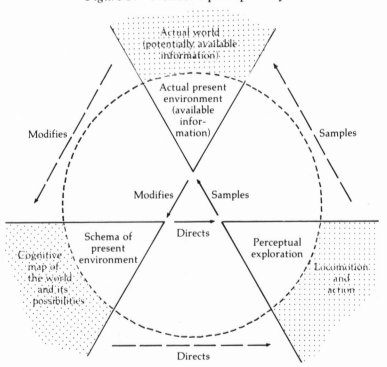

Neisser describes schemata as active, information-seeking structures that accept information and direct action. "The schema accepts information as it becomes available at sensory surfaces and is changed by that information; it directs movements and exploratory activities that make more information available, by which it is further modified" (p. 54). Neisser notes that schemata are analogous to things like formats in computer programming language, plans for finding out about objects and events, and genotypes that offer possibilities for development along certain general lines. Neisser posits a perceptual cycle to illustrate how schemata operate and this graphic provides a useful medium to describe how organizations affect their own cognition.

The perceptual cycle is continuous. A schema directs the exploration of objects, this exploration samples portions of an object, and these samples may modify the schema, which then directs further exploration and sampling, which then further modifies the schema, and this goes on continuously. Notice that the three components of Neisser's perceptual cycle, schema, exploration, and object, correspond rather neatly to the components of thinking, seeing, and saying in the sensemaking recipe, how can I know what I think till I see what I say. On the basis of what a person thinks, he sees different things in his saying and what he sees in this saying then modifies what he thinks and will then single out for closer attention in subsequent saying.

Organization theory can be coupled to schema theory by simply asking the question, given an organization, where in the perceptual cycle does that organization have an effect and how does that effect occur. The organization could directly affect the content of the perceptual cycle and the objects sampled. Less obvious is the fact that organizations could also affect perceptual processing. Imagine, for example, that the arrows associated with the activity of directing, sampling, and modifying are of various widths suggesting tighter and looser coupling among the stages. An organization that has a vision of the world that is highly resistant to change and that runs its life accordingly would have a very strong influence leading from schema to perceptual exploration with relatively looser couplings between exploration and sampling and sampling and modification. In other words, an organization with a world view that is highly resistant to change has a thick, solid arrow at the base of the triangle but relatively weak arrows between any other pair of steps.

Other properties of organizations could be coordinated with patterns of tight and loose coupling in the perceptual cycle. The activity of sampling, for example, takes time. The amount of time available for people to sample objects can have a major effect on the conclusions that are available to modify a schema. Organizations with distinctive competence (Emery and

Trist, 1969) seldom sample their environments and sometimes modify their schemata too late to deal with "sudden" competitors.

While Neisser's formulation describes the directing, sampling, and modification that goes on in just one head, it's important to realize that these activities are dispersed among many people in organizations. Boundary people, for example, are in an ideal position to sample (Aldrich and Herker, 1977, but their sampling seldom gets communicated vividly or quickly to those people who could modify schemata that bind the interpretations imposed by other people in the organization. The dispersal of the various stages of the perceptual cycle throughout an organization serves as another way to describe what an organization is like and to predict how well it will know the world it enacts.

Examples of schemata in organizations are abundant. They may exist as cognitive maps that members infer from their organizational experience (Axelrod, 1976; Bougon, Weick, and Binkhorst, 1977; Stagner, 1977).

The most conspicuous example is the standard operating procedure (Allison, 1971). A standard operating procedure is a schema that structures dealings with an environment. A standard operating procedure is a frame of reference that constrains exploration and often unfolds like a self-fulfilling prophecy (Martin, 1977). SOPs direct attention toward restricted aspects of an object which, when sampled, seemingly justify routine application of the procedure.

Janis's (1972) description of groupthink has overtones of schema theory. The phenomenon of groupthink is important because it demonstrates some of the dysfunctional consequences when people are dominated by a single schema and this domination becomes self-reinforcing. Having become true believers of a specific schema, group members direct their attention toward an environment and sample it in such a way that the true belief becomes self-validating and the group becomes even more fervent in its attachment to the schema. What is underestimated is the degree to which the direction and sampling are becoming increasingly narrow under the influence of growing consensus and enthusiasm for the restricted set of beliefs. As Janis demonstrates, this spiral frequently is associated with serious misjudgments of situations.

Notice that any idea that restricts exploration and sampling by the very nature of that restriction will come to be seen as increasingly plausible. If a person has an idea and looks for "relevant" data, there's enough complexity and ambiguity in the world that support is usually found and the idea is usually judged more plausible. One of the prominent characteristics of schemata is that they are refractory to disproof (Ross, 1977, p. 205).

Neisser would argue that schema are not that vulnerable to distortion and that by and large they pick up real checkrooms, parking lots, and ammunition. It is our contention that most "objects" in organizations consist of communications, meanings, images, and interpretations, all of which offer considerable latitude for definition and self-validation.

Notice that when we coordinated Neisser's perceptual cycle with the sensemaking recipe, the *object* in that coordination was raw talk. It's certainly obvious that saying is subject to numerous interpretations, which means it can appear to support quite divergent schemata. This seeming universal support does not arise from stupidity or malevolence on the part of the actor, but rather from a combination of an intact, reinforced schema and an equivocal object. Actors with bounded rationality presumably are more interested in confirming their schemata than in actively trying to disprove them. Even though people may build up schema anew each time they apply them, they have to start this build-up with something. And it's that something, that assumption, that retrieved portion of the past which can rather swiftly become elaborated into a schema which is like a previous schema and which has a controlling effect on what people perceive.

Thus, diagnosis of organizational schemata is a powerful means for researchers to understand much of what goes on in organizations, how its members arrive at the conclusions they do, and why they persist in conclusions that seem dated. When it is remarked that an organization is a body of thought, that can be restated as, an organization is a body of schemata that direct the exploration of objects. This directed exploration samples features that typically affirm and strengthen schemata, which means they become even more binding as recipes that organizational members apply.

Causal Textures as Bodies of Thought

Environments vary in the ease and accuracy with which cause-effect or means-ends relations can be perceived and enacted in them. This property, labeled causal texture, was described vividly in 1935 by Tolman and Brunswik and later reinterpreted by Emery and Trist (1969). The Emery and Trist work, however, dropped some of the more compelling and valuable features of the original formulation. It is the purpose of this section to revive those original features, because they illustrate how cognitive concepts can be used to diagnose organizational properties and problems. The point of departure for Tolman and Brunswik's discussion of causal texture is the fact that any distant or remote event, both in time and space, is signified with varying accuracy by local representatives. Thus, the likelihood that a manuscript will be accepted for publication may be signified more or less accurately by peer reaction to the paper, composition of the editorial board, books reviewed in previous issues, etc. Both Tolman and Brunswik were interested in the coordination between local representatives and the distant objects to which they might be coupled.

Tolman was interested in the fact that certain means objects have a stronger or weaker possibility of producing certain ends or of reaching certain goals. Working late at the office may be a stronger means to the end of being promoted than graduating from an Ivy League school. Tolman noted that means objects vary in their equivocality, by which he meant that some of them are good ways to achieve goals and other ones are poor. Thus Tolman's contribution to the causal texture notion was the idea that means objects vary in the degree to which they promote goal attainment.

Brunswik's contribution to the causal texture notion was more perceptual. Brunswik argued that means objects themselves are known with greater or lesser clarity on the basis of local signs from which their existence can be inferred. Some cues are good cues that means objects exist, but other cues are less reliable. This means that a member of an organization faces the following complexities. Cues can signify the existence of a means with varying degrees of reliability and the means objects in turn have varying degrees of equivocality when examined as suitable ways to attain ends.

Tolman and Brunswik used the colorful phrase "the lasso principle" (p. 48) to signify individual tries to connect a perceived personal deprivation with some cue, indicating that some means object will produce some

Figure 2.   Tolman and Brunswik's categories of causal texture.

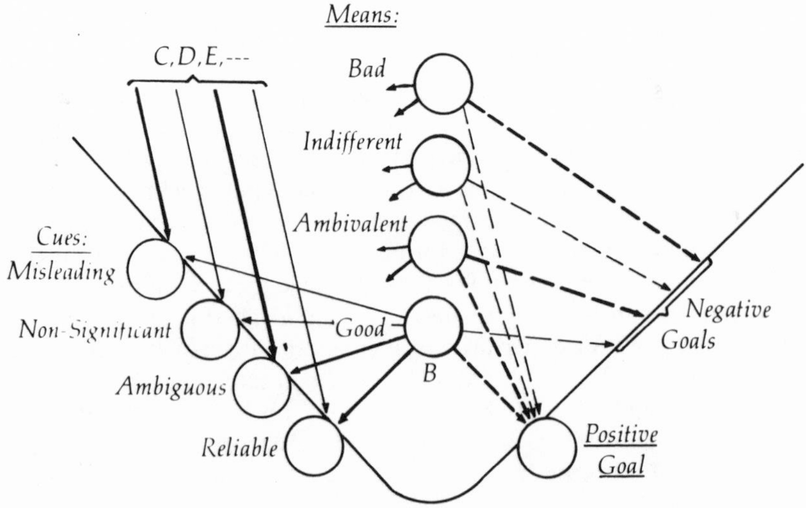

desired goal that will remove the state of deprivation. In other words, the person tries to lasso and couple together a need, a local sign, a means object, and a goal state that will satisfy the need. That seemingly simple lassoing gets complicated because of the problematic ties between means objects and goals and between signs and means objects. The nature of these complexities are depicted in the display reproduced above.

If we focus first on the right side of the diagram, we see that relative to certain goals, means objects such as peer reaction or working late can be characterized in four ways: good, ambivalent, indifferent, and bad. The differences among these four kinds of means objects are shown by the thickness of their connections to positive and negative goals. A good means object is one that leads in a relatively high percentage of cases to a positive goal and in only a small number of cases to a negative goal. An ambivalent means object leads with a relatively high probability to both positive and negative goals. An indifferent means object leads with very little probability to either a positive or negative goal. And the bad means object is one that leads with a high probability to a negative goal and with little probability to the positive goal.

These differences can be illustrated by an exhibit of maze learning. Organizational members are often described as trying to find their way through a maze, even if it is a maze they've enacted for themselves, so use of this illustration is not as ludicrous as it might seem.

Figure 3. The organization as maze.

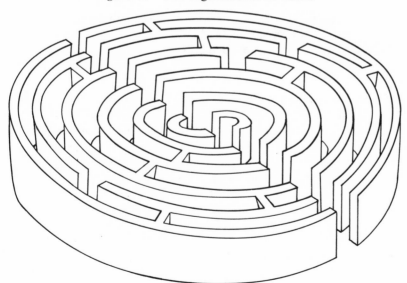

In this illustration, adapted from Tolman and Brunswik, a somewhat unusual maze is used, a maze that has multiple choices at each choice point,
making it even more correspondent with choice situations in ongoing
organizations. The maze is depicted below and is described in this way:

Figure 4.   Tolman and Brunswik's schematic maze.

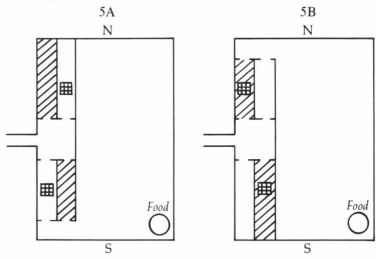

Suppose "that each choice point has four alleys instead of the usual two
issuing from it. Two of these always point south and two north. Further,
one alley in each pair is always lighted and the other dark, and one has an
electrified grill and the other no such grill. Further, in the cases of the 5A
choice-points both the two south-pointing alleys will lead on, whereas
both the two north-pointing alleys will be blinds. Also in the 5A type of
choice-point both the lighted alleys will have electric shocks and the dark
alleys will have no shocks. In the cases of the 5B choice-points, on the
other hand, everything will be just reversed; the north-pointing alleys will
lead on and the dark alleys will provide the shocks" (pp. 58–59).

To illustrate the various kinds of means objects, suppose that a maze is
constructed in which most of the choice-points are of the 5A type and
there are very few of the 5B type. In this kind of maze all of the south-
pointing dark alleys are good, because they always lead with a high degree
of frequency to the positive goal of food and they virtually never lead to
the negative goal of shock or undue exertion as in a blind alley. The
south-pointing lighted alleys will be ambivalent, because they lead with a
high degree of frequency both to food and electric shock. The north-
pointing dark alleys will be indifferent, because they lead with very little

frequency to either the positive goal or the electric shock while the north-pointing lighted alleys will be bad, because they lead with very little probability to the positive goal of food and virtually always to the negative goal of electric shock and a blind alley.

That is the learning problem that confronts a rat deprived of food just as it resembles a kind of learning problem that confronts an organization that is deprived of profit, power, visibility, share of the market, etc. If that organization wants to find a more advantageous position in its environment, it must coordinate certain means objects such as acquisitions and divestitures with more distant positive goals of better positions in an environment and avoid negative goals such as worse positions in the environment.

If we now move to the left side of Figure 2, we can see that there are four different types of cues relative to a single good means object: reliable, ambiguous, nonsignificant, or misleading. Again, the nature of these cues is represented by varying thicknesses of lines between the good means object and other means objects. For example, ambiguous cues are considered to be cues that are caused with a great frequency by both the good object and by other objects that could be potentially disruptive as far as their instrumentality to goal attainment.

Tolman and Brunswik's original description of causal texture provides a valuable means to describe environments that are both enacted and reacted to by organizations. The concepts they propose are manageable and plausible. Thus, we can assert, first, that members impose on their flow of experience the idea that that experience can be differentiated into means and ends. Second, the means they differentiate and superimpose can be good, ambivalent, indifferent, or bad. Third. we can assert that organizational members' search in their superimposed world for signs that suggest the existence of reliable paths to good outcomes.

When we embed causal textures in the realities of organizations, we begin to see why it is difficult for organizational members to learn. Because of the fluidity associated with many organizational problems (Cohen, March, and Olsen, 1972) reliable cues for good means objects are transient and rare. They're rare for several reasons. If goals are defined retrospectively rather than proactively, then means-ends coupling will occur after the fact and be loose. In a world of competitors, the value of cues can change swiftly as people try to mislead one another. Reliable cues are scarce in a world where organizations tinker with their own perceptual displays and try systematically to make it more difficult for other organizations to predict what they will do.

Tolman and Brunswik do not neglect unstable causal textures and the problems this creates for lassoing (p. 68). A dark alley which is good in some laboratories can be bad in others. They note that organisms typically bring to any new environment a set of already prepared hypotheses (read Schemata) and these hypotheses are based on the average environments they've encountered in the past. Tolman and Brunswik argue that when people are in a new situation it's crucial that they be able to discover unique features which differentiate this particular maze from the more general case of a normal maze. Once they've made this discovery people have to attach their new set of hypotheses specifically to these unique features of the novel situation. The advantage of this strategy is that people can then use their general hypotheses and schemata in normal environments and invoke specialized hypotheses only when special characteristics emerge.

Tolman and Brunswik conclude that "the wholly successful organism would be one which brings, innately, normal averagely "good" means-ends hypotheses and normal averagely "reliable" perceptual hypotheses; but which can immediately modify these innate hypotheses to suit the special conditions of the special environments; which can note and include in its cue system and in its means-end system the presence of the further identifying features of these special environments. But further, such an organism must also, if it is to become completely successful, be equally able at once to drop out such new hypotheses when the special features as to cues or means are no longer present" (p. 72).

When we argue later that an organization must complicate itself in order to survive, one way to talk about that complication is in terms of a proliferation of hypotheses and heightened sensitivity to the special conditions under which different hypotheses are to be invoked.

It's important to reemphasize that given the shifting nature of organizational goals, given the fact that goals and means are often arbitrarily imposed on ongoing flows of experience, and given the fact that turbulence is often associated with fuzzy cues and loosely coupled means-ends ties (Strauch, 1975), we assume that the combination "reliable-good" occurs infrequently and holds true for relatively short stretches of time and that the other fifteen combinations describe a greater proportion of the cognitive problems that confront organizational members.

We can now see how Emery and Trist borrowed too little. Their four types of environments—placid random, placid clustered, disturbed reactive, and turbulent—are characterized on the basis of the distribution of goals and noxiants and whether similar organizations are present in that environment. Explicit mention of the Tolman and Brunswik categories is made only for the placid clustered environment where it is merely noted that "clustering enables some parts to take on roles as signs of other parts or become means-objects with respect to approaching or avoiding" (1969, p. 247).

Those four types of environments could have been described in terms of which cues and which kinds of means objects were predominant. For example, a placid random environment could be an environment in which there is a high probability that cues will be bad and misleading or an environment in which the cues are indifferent and nonsignificant. Which of those two combinations is chosen as the operationalization of placid random will make sizable differences in how one defines the perceptual problem an organization faces and how one will design a sensing mechanism to solve that problem (Neisser, 1977).

If Tolman and Brunswik's concepts are arrayed in a matrix with the four categories of cues listed along the vertical axis and the four categories of means objects along the horizontal axis, sixteen worlds that organizations confront are depicted, not just the four originally described by Emery and Trist. The problems for organizational action, sense-making, perception, and cognition differ greatly among those sixteen cells.

The assertion that organizational members have bounded rationality takes on a different meaning if we consider the possibility that many situations in organizations involve strange mixtures of cues and means-objects toward which only hesitant and/or impulsive lassoing is possible. Bounded rationality viewed against that background is something other than one more example of the law of least effort. It is a sound strategy intentionally adopted to deal with recalcitrant, equivocal data.

That way of describing the cognitive problems of organizational members is not common, but it does capitalize more directly on what Tolman and Brunswik said.

One concern of cognitive organizational theorists is to do more fine-grained analyses of the perceptual problems that organizational members face. If an organization is viewed as a body of thought where a substantial portion of that body consists of causal textures with mixtures of cues and means-objects, analysis seems to be enhanced. If we take the assertion that organizations are goal oriented—an assertion toward which we have muted enthusiasm—we see that such an assertion says very little if it neglects the kinds of means objects potentially available for achieving those goals and the clues that people use to determine what means objects are available. Greater usage of the original version of causal texture may reduce the frequency of these oversights.

Reference Levels as Bodies of Thought
William Powers (1973) has written a powerful cybernetic analysis of perception entitled "Behavior: The Control of Perception." The key to his thinking is evident in the book's title. The basic idea is that individuals behave in the interest of controlling their sensory input so that this input remains consistent with some reference level for that sensation. When

sensory inputs begin to depart from a reference level, the individual begins to act and does whatever is necessary to bring the sensed input back toward the reference level. The output of any human control system, therefore, is a reference level that is binding on lower-order control systems in the sense that these lower-order systems act to keep incoming sensations consistent with those levels.

Applied to the perception of organizations, the Powers formulation has some interesting consequences. He argues that people have concepts of systems and these figments of the imagination serve as reference signals around which sensations are organized. In response to concepts concerning organization, people act so that sensations are produced that are consistent with these concepts. A person's concept of an organization, in other words, serves as a base line. Incoming sensations are compared against that base line and if these sensations do not deviate significantly from it no action is taken. If the incoming sensations do show some deviation from this principle of organization, the individual acts until the incoming sensations come back into compliance with that base line.

When a fan is loyal to a baseball team, he is reacting to something that is of a most ethereal nature. In the course of 20 years, a team may change stadiums, managers, owners, coaches, and players—what is left to be loyal to? Only the system concept that made it seem that there was a 'team' in the first place. Since that system concept exists in the . . . perceptions of the fan who was organized to perceive that way, there was no need for the team to have material existence. The fan will perceive it anyway, as long as he wishes rabidly to do so. Many are the stubborn old timers who insist on perceiving the *Brooklyn* Dodgers, who have merely been kidnapped temporarily to a strange state (Powers, 1973, p. 172).

Principles of organization, therefore, are reference signals against which sensations are adjusted. It is that sense in which organizations are figments of the imagination, but they are no more so than any other kinds of elements such as people, families, or ideology.

Ask yourself where is the Duke Ellington orchestra when you watch his son Mercer Ellington conduct it? There is something that is known as the Duke Ellington orchestra in the minds of many people and yet, in watching the present version of that orchestra, it's not easy to spot precisely where the organization is. Over the years personnel and music have changed, yet something persists that transcends any particular sets of charts or people who play them. What persists is the concept of the Ellington orchestra. The concept endures because observers mold whatever assemblage they observe into an entity that produces sensations consistent with the idea of the Duke Ellington orchestra.

Thus, an organization can be a body of thought in the sense that it consists of a set of higher-order reference signals in terms of which people organize their behavior. Organizational behavior then becomes behavior that generates sensations which confirm that the organization is in place and functioning.

If we couple Powers' formulation with Mintzberg's (1973) descriptions of managerial work, it could be argued that what Mintzberg has described are ways in which managers produce sensations that either match or mismatch their reference signals concerning what the nature of the organization is. As the sensed state of the world begins to deviate from these organizational concepts, actions are taken to generate the impression that in fact the organization still exists and is not disappearing. Most managerial actions that Mintzberg has described (e.g., figurehead, spokesman, disturbance handler) could be regarded as efforts to affirm that the organization is alive and well rather than collapsing and disappearing.

ORGANIZATION AS THINKING PRACTICES

Organizations are not just bodies of thought, they are sets of thinking practices that produce those bodies of thought. Some thinking practices have already been implied (schema thinking involves direction, sampling, and modification), but there are other thinking practices that appear in organizations and are responsible for the shape of these organizations and their effectiveness. The purpose of this section is to outline a few of these thinking practices and to suggest by example a means of inquiry that organization watchers should use more widely when they examine organizations.

We want to examine such things as the relations that members impose on unrelated events, the sentences that organizational members utter to themselves, the habits members invoke, the kinds of people they imagine during episodes of solitary activity, and the simplifications they desire.

The Relational Algorithm as a Thinking Practice

Organizational members frequently put cognitions and bits of information into different combinations. Their preferred ways of combining are a good example of thinking practices. A standard definition of creative activity is that it involves new combinations of old elements and old combinations of new elements. How an organization's members do combining, how frequently and elaborately they do it, the variety of elements they have available for recombination as well as the variety of their recombination formats, all can have an effect on the sensemaking that takes place.

Organizations thrive on combinations. Whether members are talking about divestiture, merging, product differentiation, incentive systems, performance evaluations, or how to treat subsidiaries, they are always taking elements and putting them together in some combination. Product differentiation, for example, can involve taking one product in some relation to another product. Take a product "against" a product (the product wipes out the effects of the original product), take a product "after" a product (a second product does a further step that the first product doesn't), take a product "under" a product (our product makes their product work better). The whole exercise of developing incentive systems in organizations is an effort to induce some relationship between a person and financial outcomes. And the same can be said for most other organizational problem solving.

A useful format for recombination is the relational algorithm developed by Herbert Crovitz (1967, 1970). Essentially the algorithm is nothing more than a list of forty-two relational words taken from Ogden's (1934) attempt to boil the English language down to 850 simple words. Crovitz feels that the essence of any idea is a statement in which one thing is taken in some relation to some other thing. The forty-two relational words are operations that can occur between the things and are listed below.

Relational Algorithm

about	at	for	of	round	to
across	because	from	off	still	under
after	before	if	on	so	up
against	between	in	opposite	then	when
among	but	near	or	though	where
and	by	not	out	through	while
as	down	now	over	till	with

Consider the famous Duncker problem: How do you get rid of an inoperable stomach tumor without harming the healthy tissue of the body, by using rays that can be modulated in intensity, and at a high enough intensity that they destroy the tumor? One solution is to take a ray *across* a ray such that the tumor is put at the point where the rays intersect. At this point the energy associated with each ray is summed and this means that the rays can be fairly low intensity so that individually they do not damage the surrounding tissue. If you run through the other forty-one words in the relational algorithm and insert them in the phrase, "take a ray (blank) a ray," you will discover that there are other solutions to Duncker's problem.

Obviously if a problem is described in sufficient detail, there will be a large number of potential domain words (the domain word in the preceding example is "ray") that can be used in the relational algorithm. And some rather massive judgments will be necessary to decide which words to use. But this very judgmental latitude means that informative individual differences in choices among groups in organizations should be evident. Groups with a dominant schema will typically single out the same domain words from every problem and repeatedly run through the same subset of relational words.

Organizational members undoubtedly also differ in the degree to which they duplicate a domain word when trying to solve a problem. In the stomach tumor problem, there are several possible domain words and combinations. Notice that if a domain word is duplicated, as in the case of the X-rays, then the problem becomes more manageable.

It is surprising how often duplicating a domain word and applying all forty-two relations can suggest a solution. For example, the problem of flood control in Japan has been partially solved by taking water against water. The Japanese have come up with a system in which they use plastic bags, fill them with water at the site where the flooding is occurring, and then they pile these plastic bags of water into dikes to control the flood water. In that example a domain word is repeated, the word "against" from the relational algorithm suggests the solution, and the solution occurs because one element is taken in some relation to another element.

There are lots of ways to recombine elements. One of the interesting things in browsing through any collection that is organized alphabetically is that unusual categories, strange suggestions, and novel objects get placed adjacent to one another. This serves as a kind of randomizing or recombination device, just as does leaving material scattered around an office so you keep rediscovering new combinations of those materials.

The point is that organizations frequently analyze their world using combinations. However, factors such as reinforcement, overload, and socialization undoubtedly focus members' attention on a relatively small number of relations present in the relational algorithm. These biases, in turn, seal off many possibilities for new interpretations of organizational events. Assessment of preferred relational words and preferred practices for isolating domain words could suggest how organizations build their bodies of thought and whether they have the resources to rebuild and redefine those bodies of thought.

How Organizations Talk to Themselves

How organizational members execute the sensemaking recipe, how can I know what I think till I see what I say, can affect their interpretation of who they are, what they are doing, and what it means to be effective, given those interpretations. There are several ways in which that basic recipe can be modified and our intent here is simply to illustrate a small number of them.

As it stands the recipe contains only the first person singular pronoun "I". If any one of those pronouns becomes a plural pronoun, we could be inspecting a phenomenon that has organizational relevance. "How can I know what we think until I see what I say" could be a minimal organizational act even if the collective quality of that assertion is modest. It's important to reiterate that organizations can be viewed as entities built up from interlocked behaviors between pairs of people. That interlocking is modest, yet it can occur around any of the four themes implicit in the sensemaking recipe: knowing, thinking, seeing, or saying.

An additional variation of the recipe occurs if the sequence of knowing, thinking, seeing, and saying is shuffled. A basic contention of some newer models of organization such as organized anarchies, garbage cans, and loose coupling is that rational sequences seldom unfold in a rational order. Steps are either omitted or occur at odd points. For example, the standard medical sequence, symptom→diagnosis→treatment in actual practice often becomes symptoms→treatment→diagnosis. On the basis of data about the effectiveness of treatments, physicians can make a diagnosis after the fact, but a priori diagnosis seems to have little to do with treatment effectiveness. Applied to organizations we can ask what kinds of activities in organizations are suggested by these shufflings: "How can we say what I see till I think what I know?"; "how can we see what we know till I think what I say." By altering the sequence of sensemaking activities in the sensemaking recipe, it is possible to characterize both the unique problems that confront an organization and the different styles of thinking adopted by subunits and actors within that organization.

If we look more closely at specific words in the recipe, new possibilities for diagnosing organizational thinking practices and categorizing styles of organizing become apparent. It's not apparent, for example, that organizational members would always want to know *what* they think. People in organizations also want to know how they think, when they think, why they think, and where they think. If we examine those additional questions, we begin to learn something about when people become concerned about seeing what they say.

Inclusion of the word "know" in that recipe has a ring of certainty that may be a luxury in organizations. Instead of saying how can I know what I think, we can imagine the people in organizations settle for how can I invent, intuit, glimpse, discover, get-a-sense-of, grasp what I think till I see what I say. Each of those amendments makes sensemaking more probabilistic and more consistent with the notion of satisficing.

Consider this problem: How can I know what I think because I forgot what I said? The only way the sensemaking recipe works is if people can remember or retrieve what they've said (Krippendorff, 1975). Organizations characterized by cluttered files, empty traditions, or sloppy minutes have poor memories and will have trouble resorting to seeing what they say as a means to understand their existence.

Up to a point faulty memories are functional. Albert Speer, writing about his experiences during the Third Reich, notes that one of the best interventions for improving German industry during the war were the Allied bombing raids. Speer notes that these bombing raids destroyed the filing facilities of factories and also destroyed the traditions and procedures that had been mainstays of those bureaucracies. "Speer was so enamored with the results of these bombing raids that, upon learning of the destruction of his ministry, in the Allied air raid of November 22, 1943, he commented: "'Although we have been fortunate in that large parts of the current files of the ministry have been burned and so relieved us for a time of useless ballast, we cannot really expect that such events will continually introduce the necessary fresh air into our work'" (Speer, 1971, cited in Singer and Wooton, 1976, p. 87).

Moderately faulty memories may lead to healthy improvisation when people look back over their poorly stored words and try to make sense of them. But with no memory whatsoever, virtually nothing is available for people to make sense of.

If members of an organization can know things only a posteriori, that also suggests that how things are said or accomplished will influence the amount and variety of interpretations that can be imposed on them. If organizational members are encouraged to generate complicated sentences that suggest interesting possibilities of what people have been up to, then the lives of the people doing the interpreting should be both more interesting and more adaptive to changing circumstances. If people in organizations, however, are reinforced for uttering the same simple sentences over and over, then their degrees of freedom for subsequent retrospective sensemaking will be curtailed sharply. When people inspect those bland displays, their interpretations will be impoverished, redundant, and potentially nonadaptive.

As a final point, it's probable that organizations vary dramatically in how they stop an episode of sensemaking. It's not obvious when people should stop their saying, start their seeing, conclude that they have a thought in hand, or that they have exhausted the meaning of a display. A repeated problem in organizational thinking and problem solving is that organizational members think too long and go right past the solution to a problem without realizing they've done so. How organizational members discover when to stop thinking is not obvious, it may distinguish their thinking styles and be responsible for many of the products that come from their thinking practices.

Habits as Alternatives to Thought

Most organizations seem to exist for some time on a small number of ideas, a full array of standard operating procedures, and a handful of well practiced schemata. In any situation where researchers are tempted to examine cognitive processes in organizations, close attention should be paid to the question of whether it's even necessary for thought to occur.

A particularly good essay developing this point is Warren Thorngate's article entitled "Must We Always Think Before We Act?" (1976).

Thorngate argues that people who adopt a cognitive approach seem to neglect two crucial facts. First, thought takes time and mental effort and people try to avoid cognitive strain. Second, most social interactions are redundant either in terms of the situations and people involved, or in terms of the behaviors necessary to sustain the interaction. In the case of most social interaction, familiarity is the rule. Given those costs of thinking, it makes sense to act habitually in most social encounters, because this frees attention and short-term memory for activities such as vigilance concerning events surrounding the interaction and rehearsal of future responses. Control by habit also means that response latencies will be short and responses will be smooth, thereby turning the interaction into a pleasant, self-fulfilling prophecy.

On the basis of these speculations Thorngate induces the general rule that, "If a response generated in an interaction is judged to be satisfactory, it will tend to be reproduced under subsequent, equivalent circumstances from habit rather than thought" (p. 32). Thorngate notes that much cognitive research involves situations where subjects are confronted with unfamiliar environments and are urged to be accurate rather than fast. Given these demands to deal accurately with the unfamiliar, it's not surprising that people have to think in order to manage these complexities. Extrapolating from those situations to organizational situations where people are confronted with the familiar, rewarded for speed, and where gross inaccuracies can be tolerated or rectified is precarious.

Thorngate's concluding comments about future research given these realities are relevant for organizational researchers.

"To date we have almost no empirical data relevant to assessing the role of habit in social interaction or its relative popularity as a determinant of social behavior. Field research is necessary to determine the redundancy of social interactions across people, topics and situations so that we can estimate how often habits may be invoked. We need to develop methods of distinguishing between thoughtful and thoughtless behaviors in both field and laboratory settings. Studies must be undertaken to determine how often and how quickly behavior shifts from thought to habit, and to determine the relative popularity of various cognitive processes which can be used to generate responses in unfamiliar situations. It would also be of some interest to quantify the mental effort associated with various cognitive processes or mental heuristics and to determine the trade-offs between the simplicity or "crudeness" of a heuristic and the number of maladaptive responses it produces in social interactions" (p. 34).

Investigation of the mix between habit and thinking is crucial in any analysis of cognition in organizations. It's important to notice that when people are acting out of habit their attention is focused somewhere and it's important to understand where that attention is focused and with what consequences. Virtually all discussions of thinking in organizations up to now have been concerned with instrumental thought. The concern has been with ideas that are instrumental in solving problems or giving the organization new definitions of itself.

Almost no attention has been paid to consummatory cognitive activities in organizations such as daydreaming and fantasizing (McKellar, 1957). If consummatory cognition accompanies habitual actions it could deflect or accelerate those routines and/or influence the next nonhabitual action the person performs. It does not seem productive to assume automatically that thoughtful accompaniments of habit are irrelevant to the ways in which those habits unfold. The question remains, what happens to thoughts during thoughtless episodes. What people dwell on and what this does to their activities in the next interval are potentially crucial questions when one thinks cognitively about organizations. The question of whether demands for speed undercut reliance on thought does deserve closer attention. Thorngate's automatic coupling of those two into a negative relationship remains to be demonstrated in organizations.

The Invocation of Others in Thinking Practices

Much thinking involves implicit conversations with phantom others. Who those phantom others are, the conditions under which they are invoked, and the vigor of their presumed exchanges with the thinker are all variables that differ among organizational members. The nature of the phenomenon has been described by Lofland (1976, p. 100):

All encounters involve people in immediate interaction, but not all interactants need be in separate bodies. By means of memory, consciousness, and symbolization, humans summon particular past humans (more accurately, a residue composite of one) and composite categories of persons ("them," "my family," "the government," etc.) into the forefront of consciousness, taking account of what are projected to be their belief and action when dealing with a situation. No other person need physically be present for there to be social interaction in this sense. It is *social* interaction in that the individual is taking other people into account when constructing his own action.

Moreover, people interact sheerly with themselves, or, rather, different aspects or dispositions of themselves. To the degree that people engage in internal dialogue with themselves, we may speak of self-encounters.

The audience "present" at most episodes of thinking is potentially influential on the way in which that thinking unfolds and the conclusions that come from it. Efforts to diagnose the audience that is invoked during thinking is a crucial aspect of cognitive analyses of organizations. Thinking is never a solitary activity and this means it is important in any understanding of cognition and organizations to discover precisely what people uttering what things keep that organizational thinker from being a solitary individual.

Simplification in Thinking Practices

A substantial portion of the cognitive literature pertaining to organizations can be viewed as variations on the theme that organizational members simplify and vulgarize the data to which they're exposed. Organizational members try to manage uncertainty by imposing categorical inferences rather than probabilistic judgments (Steinbruner, 1974, p. 110), they operate under the constraint of consistency to introduce cognitive economies, they manage inconsistent information by collapsing or stretching time, by wishful thinking, by inferring the impossibility of implied action, or any one of numerous other techniques which have been documented repeatedly (Slovic, Fischhoff, and Lichtenstein, 1976).

It's not evident that cognitive organizational theory is best served by yet one more documentation on the phenomenon of simplification. What we need instead is to cultivate a sensitivity to thinking practices that move beyond simplicity, thinking practices that complicate rather than simplify the world (e.g., Jacob, 1977). We need to understand how people can reverse some of the potential rigidities imposed by schemata. Not much attention has been paid to the issue of how to move beyond simplicity and reverse the tendency of organizations to encourage and operate on increasingly impoverished views of the world.

It's hard to find mechanisms that make organizations smarter, it's easy to find mechanisms that seem to erode their intelligence. Dixon's (1976) recent analysis of military incompetence suggests that there is considerable intelligence scattered throughout military organizations, but that norms of toughness and manliness make cultivation of itnelligent analyses next to impossible. Norms favoring toughness, realism, and doing, summarized in the maxim, "We're all business and mean business," favor simplicity and work against complexity.

People who have been concerned with organizational self-design (Clark, 1975; Kilmann, Pondy, and Slevin, 1976) have noted that organizations seldom have mechanisms for generating new structures that complicate their existence. Most research on groups demonstrates that groups are simplifying collectivities that vulgarize the minds that are in them. When, for example, groups accommodate to the least accomplished member, or foster polarized beliefs, or become solution centered rather than problem centered, potentially complex analyses are excluded from consideration.

Thus it seems crucial to learn more about ways to reverse simplifications. Most organizational researchers are familiar with Bartlett's (1932) serial reproduction situation in which a story is passed along from person to person, details drop out, and the final version of the story is a caricature of the original. Interestingly, nobody has ever sent the story back through in reverse order to see if it *regains* some of its original complexity and then is available for reinspection, reinterpretation, and redefinition.

If we want to make organizational members more complicated and reverse some of the effects of simplification, then somehow we have to make it possible for members to reexamine original rich displays and come away from those reexaminations with different interpretations of what they might mean. If uncertainty can be regenerated as well as absorbed, then theoretically it should be possible to recomplicate original observations that have become simplified. And if original complicated observations can be reinstated, then the organization has the opportunity to reexperience some of those original data and become more intelligent in handling them.

Investigators such as Watzlawick, Weakland, and Fisch (1974) have suggested that intentional confusion is a means to introduce complications and induce insights. Plausible as that suggestion is, it can be double edged. Confusion can also heighten arousal so much that people notice even less about current situations than they noticed before and this, in turn, leads them to rely even more heavily on previous overlearned schemata. The fact that the overlearned schemata do remove the confusion means that they are strongly reinforced right when the intention was to extinguish them. Thus, faced with confusion, the person has become more simplified and more attached to the simplifications that eased him through.

Not much attention has been paid to the ways in which simplifications can be reversed, yet it is not obvious that such reversals are impossible. Research has documented the existence of simplifications but has ignored the question, how can simplifying processes be reversed so that they generate complexities? If, for example, people in organizations were held accountable and rewarded for their success in complicating those who report to them, something other than simplifications should be observed.

CONCLUDING STATEMENT

We have tried to suggest that viewing an organization as a body of thought sustained by a set of thinkers and thinking practices reveals determinants of effectiveness that other formulations miss. Prevailing descriptions suggest that organizational members do many things by standard operating procedures and out of habit, deal with redundant situations with which they are familiar, spend time actively communicating rather than reflecting, argue that "ideas are a dime a dozen" and therefore simple to find, act on the basis of incomplete analyses, and in general seem to get along quite well without much resort to thought. It's the very obviousness of that description which leads us to suspect that we've missed something. The fact that thinking seems so trivial if not impossible under organizational conditions strikes us as all the more reason to look carefully for the places where it occurs, the ways in which it unfolds, and the organizational properties it constrains.

Analyses of cognitive processes in organizations suggest that there are numerous questions which deserve some attention:

1. Precisely what is the nature of the argument that says, if organizations thought more they'd be in better shape? Conceivably organizational members don't need to think more frequently, but rather differently, and on different occasions.

2. How does an organization discover it's ignorant? Should it even try to assess the degree of its ignorance (Schneider, 1962).

3. How do single cognitions generated by a single actor become amplified in organizations? We need detailed case studies of the ways in which single thoughts become diffused through an organization. Subtitle for such an exercise would be, "For want of a thought the organization was lost."

4. If people have bounded rationality and prefer simplification, is it possible to equip them with standard operating procedures that are smarter than they are? It seems no great trick to equip people with smart routines, but the problem is to preserve that complexity and buffer it from the simplifications of operators. Admissions decisions (Dawes, 1976) are a good example of standard operating procedures that can be smarter than their operators.

5. In any analysis of cognition and organizations we can ask, what are the provocations to cognition and who are the provokers? To look at cognition in organizations may be to look at forcing functions rather than at voluntary participation in thinking processes. Many people in organizations throw others off guard and routinely generate inconsistencies. These people can be viewed as provokers of thinking and how they function, where they function, and when they function may be important information.

6. Is cognitive organizational theory basically a tops down view of the organization? When applying a cognitive formulation to organizations, it is easy to argue that the images and thoughts of powerful people are crucial but that images, thoughts, expectations, and sensemakings of less powerful people are not. But, is it the case that lower level cognitions are not inconsequential?

A pair of aphorisms from the excellent collection by Auden and Kronenberger (1966) portray the kind of theoretical tension that potentially is associated with cognitive analyses of organizations and of the way they can enliven our efforts to say interesting things about organizations:

1. "Thinking is more interesting than knowing, but less interesting than looking" Goethe (p. 350).

2. "The world of reason is poor compared to the world of senses—until *or, but, because, when, if, and, unless* populate it with endless possibilities" Kaufmann (p. 342).

FOOTNOTE

1. I am grateful to Rosemary Burke for discussions concerning the content of this paper. This work was supported by the National Science Foundation through Grant BNS 75-09864.

REFERENCES

1. Aldrich, H., and D. Herker (1977) "Boundary spanning roles and organization structure," *The Academy of Management Review* 2, 217–230.
2. Allison, G. T. (1971) *Essence of Decision: Explaining the Cuban Missile Crisis,* Boston: Little, Brown, & Co.
3. Auden, W. H., and L. Kronenberger (1966) *The Viking Book of Aphorisms,* New York: Viking.
4. Axelrod, R. (1973) "Schema theoery: An information processing model of perception and cognition," *American Political Science Review 67,* 1248–1266.
5. Axelrod, R. (ed.). (1976) *Structure of Decision: The Cognitive Maps of Political Elites,* Princeton, N.J.: Princeton University Press.

6. Baldwin, J. M. (1909) *Darwin and the Humanities,* Baltimore: Review Publishing Co.
7. Ball, D. W. (1972) " 'The definition of situation': Some theoretical and methodological consequences of taking W. I. Thomas seriously." *Journal for the Theory of Social Behavior 2,* 61–82.
8. Bartlett, F. C. (1932) *Remembering,* Cambridge: Cambridge University Press.
9. Berlo, D. K. (1977) "Communication as process: Review and commentary," in B. D. Ruben (ed.), *Communciation Yearbook,* I, New Brunswick, N.J.: Transaction Books, pp. 11–27.
10. Betz, F. (1971) "On the management of inquiry," *Management Science 18,* B117–133.
11. Bishop, J. E. (August 1977) "Potent non-drugs," *Wall Street Journal.*
12. Blumer, H. (1971) "Social problems as collective behavior," *Social Problems 18,* 298–306.
13. Bougon, M., K. E. Weick, and D. Binkhorst (1977) "Cognition in organizations: An analysis of the Utrecht Jazz Orchestra," *Administrative Science Quarterly 22,* 606–639.
14. Broadbent, D. E. (1971) *Decision and Stress,* New York: Academic Press, Inc.
15. Brush, S. G. (1974) "Should the history of science be rated X?" *Science 183,* 1164–1172.
16. Clark, P. (1975) "Organizational design: A review of key problems," *Administration and Society 7,* 213–256.
17. Cohen, A. R., E. Sotland, and D. M. Wolfe (1955) "An experimental investigation of need for cognition," *Journal of Abnormal and Social Psychology 51,* 291–294.
18. Cohen, M. D., J. G. March, and J. P. Olsen (1972) "A garbage can model of organizational choice," *Administrative Science Quarterly 17,* 1–25.
19. Coulam, R. F. (1977) *Illusions of Choice: The F-111 and the Problem of Weapons Acquisition Reform,* Princeton, N.J.: Princeton University Press.
20. Crovitz, H. F. (1967) "The form of logical solutions," *The American Journal of Psychology 80,* 461–462.
21. ———. (1970) *Galton's Walk,* New York: Harper & Row, Publishers.
22. Davis, M. S. (1971) "That's interesting: Towards a phenomenology of sociology and a sociology of phenomenology," *Philosophy of Social Science 1,* 309–344.
23. Dawes, R. M. (1976) "Shallow psychology," in J. S. Carroll & J. W. Payne (eds.), *Cognition and Social Behavior,* Hillsdale, N.J.: Erlbaum, pp. 3–11.
24. Delia, J. G. (1977) "Constructivism and the study of human communication," *The Quarterly Journal of Speech 63,* 66–83.
25. Dixon, N. F. (1976) *On the Psychology of Military Incompetence,* New York: Basic Books, Inc., Publishers.
26. Emery, F. E., and E. L. Trist (1969) "The causal texture of organizational environments," in F. E. Emery (ed.), *Systems Thinking,* Middlesex, England: Penguin Books, Inc., pp. 241–257.
27. Frankel, C. (1973) "The nature and sources of irrationalism," *Science. 180,* 927–931.
28. Goffman, E. (1974) *Frame Analysis,* New York: Harper & Row, Publishers.
29. Goldthorpe, J. H. (1973) "A revolution in sociology?" *Sociology 6–7,* 449–462.
30. Goodman, P. S. (1968) "The measurement of an individual's organization map," *Administrative Science Quarterly 13,* 246–265.
31. Gould, S. J. (1977) "The continental drift affair," *National History LXXXVI.*
32. Gregory, R. L., and E. H. Gombrich (eds.) (1973) *Illusion in Nature and Art,* New York: Charles Scribner's Sons.
33. Hayakawa, S. I. (1961) "The word is not the thing," in P. R. Lawrence, J. C. Bailey, R. L. Katz, J. A. Seiler, C. D. Orth III, J. V. Clark, L. B. Barnes, and A. N. Turner, *Organizational Behavior and Administration,* Homewood, Ill.: Dorsey, pp. 397–400.

34. Jacob, F. (1977) "Evolution and tinkering," *Science 196*, 1161–1166.
35. Janis, I. R. (1972) *Victims of Groupthink*, Boston: Houghton Mifflin Company.
36. Jeffmar, M., and C. Jeffmar (1975) "A system approach to cognition," *General Systems 20*, 65–69.
37. Jones, E. E., and H. B. Gerard (1967) *Foundations of social psychology*, New York: John Wiley & Sons, Inc.
38. Kilmann, R. H., L. R. Pondy, and D. P. Slevin (eds.) (1976) *The Management of Organization Design* (2 vols.), New York: North-Holland.
39. Krippendorff, K. (1975) "Some principles of information storage and retrieval in society," *General Systems 20*, 15–35.
40. Levi-Strauss, C. (1962) *The Savage Mind*, Chicago: University of Chicago Press.
41. Lofland, J. (1976) *Doing Social Life*. New York: John Wiley & Sons, Inc.
42. London, H., and R. E. Nisbett (eds.) (1974) *Thought and Feeling*, Chicago: Aldine Publishing Company.
43. March, J. G., and J. P. Olsen (1976) *Ambiguity and Choice in Organizations*, Bergen, Norway: Universitetsforlaget.
44. Martin, M. (1977) "The philosophical importance of the Rosenthal effect," *Journal for the Theory of Social Behaviour 7*, 81–97.
45. McKellar, P. (1957) *Imagination and Thinking*. New York: Basic Books, Inc., Publishers.
46. Mintzberg, H. (1973) *The Nature of Managerial Work*, New York: Harper & Row.
47. Neisser, U. (1963) "The multiplicity of thought," *British Journal of Psychology 54*, 1–14.
48. ———. (1976) *Cognition and Reality*, San Francisco: W. H. Freeman and Company Publishers.
49. ———. (1977) Gibson's ecological optics: Consequences of a different stimulus description," *Journal for the Theory of Social Behaviour 7*, 17–28.
50. Nisbett, R. E., and T. D. Wilson (1977) "Telling more than we can know: Verbal reports on mental processes," *Psychological Review 84*, 231–259.
51. Pettigrew, A. M. (1975) "Towards a political theory of organizational intervention," *Human Relations 28*, 191–208.
52. Powers, W. T. (1973) *Behavior: the Control of Perception*, Chicago: Aldine Publishing Company.
53. Ravetz, J. R. (1971) *Scientific Knowledge and its Social Problems*, New York: Oxford University Press.
54. Reynolds, A. G., and P. W. Flagg (1977) *Cognitive Psychology*, Cambridge, Mass.: Winthrop Publishers, Inc.
55. Rodin, J. (1977) "Research on eating behavior and obesity: Where does it fit in personality and social psychology?" *Personality and Social Psychology Bulletin 3*, 333–355.
56. Rosnow, R. L., and G. A. Fine (1976) *Rumor and Gossip*, New York: American Elsevier Publishing Co., Inc.
57. Ross, L. (1977) "The intuitive psychologist and his shortcomings: Distortions in the attribution process," in L. Berkowitz (ed.), *Advances in Experimental Social Psychology*, Vol. 10, New York: Academic Press Inc., pp. 173–220.
58. Scheerer, M. (1954) "Cognitive theory," in G. Lindzey (ed.), *Handbook of Social Psychology* Vol. 1, Reading Mass.: Addison-Wesley Publishing Co., Inc. pp. 91–142.
59. Schneider, L. (1962) "The role of the category of ignorance in sociological theory: An exploratory statement," *American Sociological Review 27*, 492–508.
60. Schutz, A. (1964) "The stranger: An essay in social psychology," in A. Schutz, *Collected Papers*, Vol. 2. The Hague: Martinus Nijhoff. pp. 91–105.

61. Singer, E. A., and L. M. Wooton (1976) "The triumph and failure of Albert Speer's administrative genius: Implications for current management theory and practice," *The Journal of Applied Behavioral Science 12*, 79–103.
62. Slovic, P., B. Fischhoff, and S. Lichtenstein (1976) "Cognitive processes and societal risk taking," in J. S. Carroll & J. W. Payne (eds.), *Cognition and Social Behavior*, Hillsdale, N.J.: Erlbaum, pp. 165–184.
63. Speer, A. (1971) *Inside the Third Reich*, New York: Avon Books.
64. Stagner, R. (1977) "New maps of deadly territories" (review of Structure of decision by R. Axelrod (ed.), *Contemporary Psychology 22*, 547–549.
65. Steinbruner, J. D. (1974) *The Cybernetic Theory of Decision*, Princeton, N.J.: Princeton University Press.
66. Sotland, E., and L. K. Canon (1972) *Social Psychology: A Cognitive Approach*, Philadelphia: W. B. Saunders Company.
67. Strauch, R. E. (1975) "'Squishy' problems and quantitative methods," *Policy Sciences 6*, 175–184.
68. Thorngate, W. (1976) "Must we always think before we act?," *Personality and Social Psychology Bulletin 2*, 31–35.
69. Tolman, E. C., and E. Brunswik (1935) "The organism and the causal texture of the environment," *Psychological Review 42*, 43–77.
70. Watzlawick, P., J. Weakland, and R. Fisch (1974) *Change*, New York: W. W. Norton & Company, Inc.
71. Weick, K. E. (1977) "Enactment processes in organizations," in B. Staw and G. Salancik (eds.), *New Directions in Organizational Behavior*, Chicago: St. Clair pp. 267–300.

ORGANIZATIONAL LEARNING: IMPLICATIONS FOR ORGANIZATIONAL DESIGN

Robert Duncan, NORTHWESTERN UNIVERSITY

Andrew Weiss, NORTHWESTERN UNIVERSITY

ABSTRACT

Macro organizational theorists (Thompson, 1967; Terreberry, 1967) have implied the notion of organization learning, but have never identified the processes involved. There is no question that the concept of organizational learning is complex and difficult to specify. However, it remains central to our understanding of how organizations and their members behave over time. This paper develops a concept of the process within the organization by which organizational members develop knowledge about action-outcome relationships and the effect of the environment on these relationships. Or-

Research in Organizational Behavior, Vol. 1, pp. 75–123
Copyright © 1979 by JAI Press, Inc.
All rights of reproduction in any form reserved.
ISBN 0-89232-045-1

ganizational learning is defined as different from individual learning. Having developed this conceptualization of organizational learning, a "middle range" theory of organizational learning as it relates to the organizational design process is presented.

I. INTRODUCTION

A problem of major interest to organizational theorists centers on the observation that different organizations have different levels of success, even among firms which are in the same industry and which do business in the same market. This can be generalized beyond business firms. Schools which have similar student populations, hospitals which serve the same kind of community, even voluntary associations with similar intentions are differentially successful. In other words, organizations which operate in environments which are relatively the same in terms of components, demands on the organization, and other characteristics may be differentially effective.

The contingency perspective which currently is dominant in organizational theory represents an approach to understanding this problem. Under this perspective, organizations are treated as open systems which engage in exchanges with their environments. The internal structures and processes which comprise an organization are argued to reflect the chraracteristics of their environments. When cast in terms of organizational effectiveness, the central proposition in this perspective becomes: Organizational effectiveness is directly related to the degree that internal organizational structures (and processes) "fit" the characteristics of the organization's environment—that the characteristics of the organization are appropriately matched to the characteristics of its environment. As environments change, the internal structure and processes of the organization must change to maintain this fit. In other words, the organization ideally must be designed to meet the demands of its environment.

If one assumes that the contingency perspective has some validity, it is imperative to address the issue of how this fit is obtained and maintained. In fact, unless an understanding of this process of "finding a fit" is included in a thoery constructed within the contingency approach, such a theory tells us little about either how organizations operate at a descriptive level or how to design and/or improve them at a normative level. If it is assumed that such "fits" are accidental, arising out of a set of actions by organizational members that happen to work at a given point in time, we are left with a theory of organizational luck. The effective organization would be that one in which decision makers stumble on an approach. But, why then can't it stumble out of this approach? Perhaps more important, how could an organization be consistently effective over time given that changes occur in its environment?

A selection/retention concept of organizational change in an evolutionary model also falls short. If the focus is on the organization, this invests the organization with something like a self-awareness—a reification that is both misleading and dangerous. If the focus is on individuals, one still is left with the problem of how such selection can be translated into the design of the organization.

The key to this problem seems to be in how the matching of organizational and environmental characteristics is achieved deliberately, through the actions of individual decision makers. At any level of the organization—work unit, division, subunit, organization, or firm—the characteristics of the organization reflect decisions made on some basis and for some reason. The delegation of authority, the creation of rules and procedures, the structuring of work into jobs, units, and departments or divisions, and even the choice of professional or nonprofessional employees, all of which have been offered as important characteristics of the organization which are related to the environment and effectiveness, reflect such decisions.

Some insight into this problem of fit can be obtained from the strategy literature. Strategy, it has been argued, leads to organizational structure. Chandler (1962) uses the term in such a way as to suggest that strategy is a conscious position taken by dominant members of a firm, which is used to direct and organize activities which in turn generate organizational structure, at least the overall structure of the firm, e.g., divisional versus functional or centralized versus decentralized. Strategy leads to the design of the organization. Rumelt (1974) has examined the diversification of some 200 Fortune 500 firms during the period 1949–1969. He finds that under this diversification strategy, the tendency was for firms to move from functional to divisional organization designs.

The focus here, of course, is on those members of the organization referred to by Thompson as the dominant coalition (1967). The political nature of organizations has been suggested by a number of authors (e.g., Pettigrew, 1973). The dominant coalition represents those organizational members who at any point in time have the power to influence the strategies, goals, and design of the organization.

Given this perspective, the problem of the maintenance of the correspondence of organizational structures and activities can be stated in terms of the ability of organizational members, *and particularly its dominant coalition,* to consciously determine strategies and to derive organizational designs which are effective given the organization's environment. The problem, however, that this raises concerns *how* this becomes possible. It requires that some concept be advanced that can be used to explain how such conscious action can be undertaken. In other words, what is it that underlies the ability of members of the dominant coalition of an organization to identify strategies and courses of action which they believe will lead to some desired outcome given the environment they face?

The concept of organizational learning is a way of identifying the process by which the organization is designed to deal with its environment. The concept of organizational learning has been discussed by various organizational theorists. Cyert and March (1963), Terreberry (1967) and March and Olsen (1976) have essentially argued that effective organizations are those in which members have a capacity to learn to predict changes in their environments, search for alternative sources of raw materials or new markets, and have a memory in which is stored information about possible interchangeable input/output components. Organizational learning thus becomes that process in the organization through which members of the dominant coalition develop, over time, the ability to discover when organizational changes are required and what changes can be undertaken which they believe will succeed.

Unfortunately, these organizational theorists have not offered any insight into how this learning takes place, what the specific outcomes of learning are (except for changes in organizational activity), who in the organization learns, or how the dominant coalition can utilize this learning process. To be useful in understanding the processes of organizational adaptation and effectiveness, these issues must be addressed.

Argyris and Schon (1978) have provided a more documented approach to organizational learning. Their approach to organizational learning focuses on decision makers' theories of action. According to Argyris and Schon

> Organizational learning occurs when individuals, acting from their images and maps, detect a match or mismatch of outcome to expectation which confirms or disconfirms organizational theory–in-use. The learning agents must discover the sources of error—that is, they must attribute error to strategies and assumptions in existing theory-in-use. They must invent new strategies, based on new assumptions, in order to correct error (1978: p. 19).

While Argyris and Schon's is the first systematic attempt to deal with the processes of organizational learning, they do not deal directly with the issue of how one designs the organization for learning in terms of the proper environment organizational structure fit, nor do they deal with where these images and maps come from.

This paper then has two objectives. First, we will develop a concept of organizational learning and specify what the outcomes of this learning process are and how these outcomes are available to and used by the dominant coalition. Second, we will develop a middle range theory of organizational learning with respect to the organizational design process.

II. THE NATURE OF ORGANIZATIONS

The development of a concept of organizational learning which can be applied to these theoretical problems requires that such a concept be understood as an organizational process. That is, a process characteristic of organizations which has as its function the provision of some organizationally useful outcome. Thus, it must be conceptualized in terms of some specific concept of how organizations operate (and indeed what they are) and what kinds of outcomes from the organizational learning process would be useful.

We therefore begin the development of the concept of organizational learning with a concept of the organization. From this concept of the organization, an understanding of what organization learning provides will be derived. This in turn will allow the definition and discussion of the organizational learning process itself.

Defining the Organization

At the most fundamental level, formal organizations can be understood to be a group of individuals who engage in activities which transform, or support the transformation of, some set of inputs into some set of outputs. These inputs could include people, as in the case of schools or hospitals, physical materials, such as in manufacturing, or information, as in the case of insurance or accounting forms. The outcomes could include educated people, physical products, or a certified audit. Accomplishing this transformation represents the basic purpose of the organization. The reason organizations exist is to undertake some transformation and *organize* the activities of individuals with respect to it.

We are not here concerned with the goals of the organization—while these are important they need not be considered in identifying the organizational purpose. It is important to point out that whatever the goal of a given organization, it must involve some kind of organized activity which has some purpose. If the activity undertaken in the organization is not purposeful, there would be no reason to organize and no basis for doing so regardless of the goals or how that purpose is established.

The concept of organization thus must include the provision that activity within the organization be coordinated in some way. It is through coordination that activity becomes organized with respect to some purpose. Unless the activities of individuals need to be integrated in some way, there is no reason to form an organization. It is the purpose of the organization, the transformation process, that requires this coordination of individual activity.

This concept of coordinated, purposeful activity distinguishes organizations from other kinds of collective behavior. The association of individuals at some place and time does not imply that an organization exists. Crowds do not constitute an organization. Nor is purpose alone sufficient. A social club may be formed as the basis for interaction among members at a specified time and location. This could be considered to be done to transform the members in some way—to provide desired human contact and so create a change in their general life satisfaction. This could not usefully be considered an organization, however, if the interaction between members is informal or uncoordinated. However, if the membership engages in some activity to, for example, raise funds and if this activity is characterized by a coordinated effort, the members can be said to have formed an organization.

Following Silverman (1971), then, we conceive of the organization as a system of purposeful action. Individuals engage in action to achieve some specific outcome. These outcomes are related to more general outcomes which are related to the overall purpose of the organization: the transformation of some input into a specific output. These actions may include the task involved in the transformation process itself—such as the teaching of students or assembly of an automobile—or the activities to coordinate and support these tasks—administrative and decision-making activities such as organizational control, conflict management, or the hiring of personnel.

We do not wish to imply that all individual action within an organization is related to this purpose, nor do we mean to imply that the purpose of an individual in undertaking a given action is the individual's motivation. Individuals become members of organizations for a number of reasons and engage in many actions for these reasons. These actions are, however, not the issue here. Whatever their reasons for engaging in a given action, it is the specific organizational outcome—e.g., bolting on a tire or developing a new product—that is the purpose that concerns us here. That these actions are engaged in to achieve rewards, either intrinsic or extrinsic, formally in the reward system or external to it, is not important to this discussion. It is the organizationally relevant outcomes, e.g., those related to or involved in the transformation process, which are important. The fact remains that all organizational members do engage in tasks which are purposeful with regard to the transformation process even if not all actions engaged in by these members are purposeful in this sense.

Thus, the organization is conceived of as a system of actions which are each related to specific outcomes. These outcomes are related to the overall transformation process—the purpose of the organization. Thus, these actions are purposeful and represent deliberate activities. The basic unit of the organization can thus be understood as an action-outcome link, or a task, which is related to other tasks in the organization.

Organizational Effectiveness and Knowledge

Given the perspective of the organization as a system of purposeful action, we suggest that the effectiveness of the organization will reflect the ability of the organization's members, and particularly the dominant coalition, to determine particular actions which will achieve a given desired outcome. This will at the most general level involve the choice of strategic, or long-term, courses of action which are intended to achieve the long-term goals established for the organization. It will also involve the choice and specification of the actions which comprise the transformation process and the administration and support of this process.

These choices will be based on some prior knowledge about the relationships between actions and outcomes. The choice of a given action represents a decision made somewhere in the organization. If we assume, following Thompson (1967), that organizational decision makers operate under norms of rationality, the decision maker must base the decision on some understanding or belief that the action will indeed yield the desired organizational outcome. Regardless of where in the organization this decision is made, the rationality of the decision turns on the existence of such knowledge.

Given the above discussion, it is now possible to sketch a concept of organizational effectiveness. If an organization is viewed as a system of purposeful action in which the basic element is an action-outcome relationship, the *most fundamental concept of effectiveness must be the degree to which firm or organizational actions lead to the outcomes intended.* If the constituent actions undertaken within the organization are not effective in this sense, the overall transformation process cannot be achieved.

Organizational effectiveness as defined here can be broken down into three components: (1) the action-outcome relations achieve the intended outcomes as specified by the dominant coalition, (2) the ability to continue achieving organizational outcomes by specifying changes in organization action given changes in conditions which affect action-outcome relationships, and (3) organizational actions are consistent—different activities in different parts of the organization are coordinated. Each of these three components of effectiveness can be discussed in terms of organizational knowledge.

Action-outcome decions   The first component of effectiveness, the ability to achieve intended outcomes, can be understood as a function of the existence of knowledge about the relationship between specific actions and outcomes. That is, the ability of an organizational decision maker to specify an action or set of actions which will lead to a desired outcome under a given set of conditions. The effectiveness of a given action may be unrelated to this—it may succeed even if there is no knowledge. However, over time, if a given outcome is required frequently, as would be the case in the transformation process or in processes such as control or coordination, the effectiveness of a given action will be a function of how well this knowledge allows the decision maker to predict that a given action will yield a given result.

This knowledge must include more than the action-outcome relationship. It must also specify the conditions under which a given action will lead to a given outcome. If the conditions under which an action is undertaken change, a change in the action-outcome relationship may occur and effectiveness will be undermined.

This knowledge would include the technology of the organization. This would represent a general concept of technology, one that would include administrative as well as technical knowledge. The narrower concept of the technology of an organization, as the actual equipment used in the transformation process, would be included here as well, since this knowledge would specify some actions in terms of the use of a particular kind of machine and implicitly the knowledge that the machine itself represents.

This knowledge would also include knowledge of the external environment as well as the internal environment of the organization (cf. Duncan, 1972). This knowledge, to be useful, must specify how a given condition affects organizational action-outcome relationships. It is in itself an action-outcome relationship.

These kinds of knowledge would be relevant to both the specification of particular activities within the organization and for establishing long-range strategic actions. In both cases the problem facing the decision maker is the same—the choice of a given action which is intended to achieve a given outcome.

The organization will be limited in its effectiveness to the degree that a decision maker can make use of such knowledge to specify organizational actions which do indeed lead to intended outcomes, and to specify the outcomes needed to lead to more general objectives of the organization. It is our contention that such knowledge is the underlying basis for organizational action and by extension, organizational effectiveness.

Adaptation   The conditional nature of organizational action-outcome relationships implies that when the conditions under which actions are undertaken change, the action-outcome relationship may change. That is, a given action will not necessarily continue to yield a desired outcome. Adaptation is then defined as the deliberate change in organizational actions by decision makers in response to changed organization-environment conditions.

If we wish to keep adaptation in the range of deliberate purposeful organization actions, we must point out that such changes, at any level of the organization, require some level of knowledge. This knowledge must be sufficient to identify that conditions have changed and how these changes if at all affect the action-outcome relationships. Aquilar (1967) has suggested that organizational members scan the environments of the organization searching for such changes. This search, however, requires that there be some idea about what is being looked for. That is, that these members of the organization have some knowledge about the factors in the environment which are relevant for organizational actions. This then permits the identification of changes which, given the action-outcome knowledge, require changes in organizational actions.

The effectiveness of the organization over time will reflect the ability of organizational decision makers to identify when such changes in conditions require changes in organizational actions to maintain the achievement of desired outcomes, and when such changes do not require changes in action.

The change in organizational actions in response to changing conditions can be said to be effective to the extent that the intended organizational outcomes are obtained. In other words, to the extent that the organization remains effective or increases its effectiveness. Given our previous arguments, this will be dependent on the knowledge which is available to decision makers which makes possible the identification of *relevant* changes in the conditions affecting action-outcome relationships and the specification of change in organizational actions.

Consistency of action We suggested above that a major component of effectiveness was the consistency of organizational actions. The potential interdependencies which may exist among organizational activities suggest that any given organizational action may affect or be affected by a number of actions. The effectiveness of a given action will thus be in part dependent on these effects. This implies that organizational effectiveness will be a function of the degree to which decision-makers have knowledge about the nature of these interdependencies.

This remains a major issue regardless of where such decisions are made within the organization. If decision making is decentralized, decision makers require such knowledge to enable them to specify actions which do not undermine or conflict with actions undertaken in other parts of the organization. This requires more than an awareness of what actions other decision makers have undertaken. It must also involve the knowledge of the effects of these actions on each other.

Even if decision making is centralized, such knowledge is critical in establishing organizational actions. The ability to specify the effects of a given action on other actions is necessary if consistent, coordinated action is to be achieved and so to achieve the desired organizational outcomes.

Conclusion

In this section we have presented a conceptualization of the organization and its operation which is based on the idea of purposeful, coordinated action. We have argued that this action must be based on knowledge which identifies the actions to be undertaken if a given purpose is to be achieved. This conceptualization suggests that this knowledge is central in understanding the effectiveness of a given organization and in its ability to remain effective by adaptation with respect to changes in the organization's environment.

III. ORGANIZATIONAL LEARNING AND ORGANIZATIONAL KNOWLEDGE

Organizational Learning Defined

The concept of organizational learning can be defined in terms of the action-outcome concept of the organization. *Organizational learning is defined here as the process within the organization by which knowledge about action-outcome relationships and the effect of the environment on these relationships is developed.* The development of this knowledge need not imply any change in effectiveness or adaptation. Unlike individual learning, which is often defined as a process by which relatively permanent changes occur in a person's behavior as a result of some experience the person has had (Bass and Vaughn, 1966), in organizational learning we are concerned with the development of the knowledge which would make such a change possible or indeed unnecessary.

For example, organizational decision makers, over a period of time, may have learned that if they threaten to lay off large numbers of employees because of high U.S. labor costs relative to foreign labor costs, the government might be inclined to implement foreign import quotas to make U.S. firms more competitive. The organization itself as a result of developing this knowledge would not be changing; its members would have only a better understanding of how to manipulate the organization's environment.

Nor is this knowledge necessary for organizational survival. As Lindbloom (1965) has suggested, organizational members may "muddle through," making reasonable decisions which work at some satisfactory level. An organization may remain in operation without effective knowledge about the action-outcome relationship or environmental conditions which affect these. In the public sector, this is possible as a result of external conditions such as continued government funding. In the private sector, it is possible if competitors are in similar positions with respect to knowledge.

However, such knowledge is critical for organizations to be actively changed or adapted as a deliberate action on the part of decision makers. It is doubtful that a manager-dominant coalition may boldly and blindly enter a new venture or restructure an entire division without some outcome in mind and some expectation that this action will lead to this outcome given the costs involved.

The learning process then is that process in the organization which leads to the development of such knowledge. It is knowledge that is the outcome of organizational learning and not any particular action or change. Before the process itself can be addressed, this knowledge must be clarified and placed in a conceptual definition. Further, the effectiveness of such actions will depend on the ability to specify accurately the actions whch are required to obtain a given outcome.

The need for knowledge, then, in the organization is high. We suggest that organizational members are more or less aware of this need and when aware will actively seek such knowledge. To the degree that they do and are able to develop useful knowledge—that is knowledge that accurately specifies action-outcome relationships—the organization will be effective.

Organizational Knowledge

Up to this point we have not been specific with regard to the nature of the knowledge that is used in the organization as we conceptualize it. Before suggesting how this knowledge is developed, it is necessary to provide a specific idea of the kind of knowledge that is needed by an organization.

The kind of knowledge we will address here can be referred to as organizational knowledge. This is defined as that knowledge which is available to organizational decision makers and which is relevant to organizational activities. By the relevance of such knowledge we mean specifically that it can be used to determine organizational action (at any level from tasks to strategy) with respect to a specific outcome. Thus, such knowledge would include action-outcome relationships and the conditions which affect these relationships, as we have discussed.

Organizational knowledge need not be held by all decision makers in the organization. Indeed, the complexity of most organizations suggests that this is not likely and probably impossible. The role of staff personnel can be understood as the storage and explanation of specialized organizational knowledge. Furthermore, individuals in different parts of the organization will have specialized knowledge about the organization. It is the access to and use of knowledge and not the possession of it that is critical in this concept of organizational knowledge. We will now briefly discuss some of the attributes of organizational knowledge.

Organizational knowledge will therefore be distributed across the organization. Many different individuals will hold such knowledge as a result of specialization. However, since organizational activities will be interdependent, and organizational actions will be related to a common organizational purpose, individual decision makers will require access to a wide range of knowledge that they themselves do not necessarily hold. This need for access requires that knowledge, to be organizationally useful, be *communicable*—capable of being stated in terms that are in principle understandable to other members of the organization. This does not imply that such knowledge be stated in a common language, only that it not be private, that is, intuition or insight that cannot be explained or expressed in a meaningful way to other individuals.

To be organizational, knowledge must also be *consensual*. That is, there exists acceptance of this knowledge across members of the organization and agreement concerning the validity and utility of this knowledge. This is especially crucial, since we suggest that this knowledge is used to determine overall organization activities in such processes as planning, design, or strategy formation. Another way to state this is that when organizational members at any level of the organization specify a given action, they must be able to justify the actions taken (especially if these actions fail). Again, given the need for coordination within the organization, we suggest that actions undertaken in the organization will be based on knowledge that is agreed upon by a number of members of the organization as valid.

Finally, organizational knowledge must be *integrated*. That is, the body of knowledge in the organization we refer to as organizational knowledge is understood to be a set of interrelated statements of action-outcome relationships. The interrelationships of the statements are necessary for the use of this knowledge to generate coordinated, purposeful action. This would be the case regardless.of whether actions are chosen by the individuals engaged in the activity or if actions are to be programmed by individuals other than those performing the activity.

We do not mean to imply that all organizational action is based on organizational knowledge—that is communicable, consensual, integrated knowledge. Individual decision makers will use knowledge that is private, i.e., noncommunicable, as well as their own intuitive, informal understandings in decision making. However, the total absence of organizational knowledge would result in decisions being made on inconsistent models of the environment and on different assumptions. That there are potentially organizations in which this is the case we do not dispute. The degree to which the knowledge used in the organization has these characteristics will vary across organizations. Given the way in which we have conceptualized effectiveness, the effectiveness of the organization will be a function of the degree to which the knowledge used by its decision makers is organizational knowledge.

The Organizational Learning Process

The organizational learning process is specifically concerned with the growth and change of organizational knowledge. We have argued that organizational knowledge is critical for the effective operation and adaptation of the organization. While other kinds of knowledge are also important, it is, as we have suggested, necessary for there to be organizational knowledge to take advantage of the insights and intentions of individuals. Thus, the organizational learning process must be understood in terms of this particular kind of knowledge.

The growth and change of organizational knowledge can occur in three ways : (1) knowledge about new action-outcome relationships or new conditions which effect previously known action-outcome relationships may be obtained, (2) existing knowledge may itself be changed by replacing a given action-outcome relationship with a new one, by changing the probability associated with a given relationship or changing (either by increasing or decreasing) the uncertainty associated with that probability, and (3) organizational knowledge can increase in the sense that additional support or validity is associated with a given action-outcome relationship.

Organizational and individual learning   Before addressing the organizational learning process, it is necessary to distinguish organizational from individual learning. Individual and organizational learning differ in terms of the kind of knowledge produced. Individual learning may produce changes in the private (e.g., noncommunicable) knowledge held by an individual. Organizational learning is limited to public knowledge, but which is socially defined as valid, relevant, and available to other members of the organization. This distinction makes the current concepts of organizational learning inadequate. March and Olsen (1976) have limited their discussion of organizational learning to the learning by individuals within an organizational setting. While they attempt to tie this learning to organizational activities (especially decision making and problem solving) performed by individuals, they have not provided any basis for tying this to overall organizational activities or the role of knowledge in the organization.

Argyris and Schon (1977, 1978) have considered a more organizationally based process of learning in terms of the development of an individual's "theories of action." While their model includes the impact of organizational factors such as communication on the learning process, the theories which emerge are personal theories which underly an individual's own actions. While these theories can be understood as action-outcome knowledge, no constraint is placed on whether or not they represent communicable, consensual, integrated statements of action-outcome relations. Thus, this is *personal* rather than *organizational* knowledge.

The concept of organizational learning offered by both March and Olson (1976) and Argyris and Schon (1974, 1978) is limited to the individual's knowledge. This view of learning does not provide an understanding of how the kind of knowledge required for systematic organizational action can be developed. In other words, how an integrated, communicable consensual knowledge base can be developed. In fact, in our estimation, they have done little more than extract basic concepts of learning theory, problem solving, and theory construction at the individual level and placed these into an organizational context. The outcomes of this process can only be understood to affect organizational processes through the specific actions of individuals. If there is any organizational learning, it could only be understood as changes in some aspect of the organization reflecting the aggregation of changes in individual behavior. Organization-wide changes could only reflect the learning of individuals in the dominant coalition, which would imply that only these individuals can develop any knowledge of the entire system and must do so on their own, since this learning need not be communicable. Such a model cannot explain systematic organization action.

Organizational learning as an organizational process   Organizational learning must be understood as more than the simple aggregation of individual learning. If this were the case, the only knowledge in the organization would be fragmented—it would refer only to the particular activities of the individuals who produced it, given the nature of division of labor or specialization in organizations. This specialization limits the range of organizational/environmental phenomena (e.g., organizational activities on decisions, environmental events or responses) relevant to any particular organizational member. If learning takes place only in an experiential mode at the individual level, as March and Olsen's (1976) theory seems to indicate, the learner would be limited to producing knowledge about this limited range of activity. If this were the case, organizational knowledge would be fragmented and relevant only to the decision maker who produced it. There would be no way in which a general knowledge base could be produced. It would be difficult to establish any conceptual links between the knowledge produced by one person with that produced by another and to use it in integrated organizational activity.

Thus, organizational learning, if it is to produce the kind of knowledge that we have argued is important for organizational effectiveness, must involve an organizational process in which the learning done by a given individual can be shared, evaluated, and integrated with that done by others. Thus, while the individual is the only entity in the organization who can learn, this must be viewed as part of a system of learning with exchanges of what is learned among individuals.

This knowledge produced by individuals is organized only when it becomes exchanged and accepted by others. Thus, the exchange is necessary, although not sufficient, for organization learning. It is this exchange that makes it possible for individuals to integrate the fragments of specialized knowledge into an organizational knowledge base.

Thus, unless the knowledge offered by an individual is acceptable to others, it will not be integrated. This requires that the knowledge be subject to validation by some criteria and that others identify it as relevant to their own needs. This is necessary for organizational knowledge to be consensual and integrated.

This process need not be formal. Indeed we would argue that it is a natural process driven by the need for organizational knowledge. It reflects the communication of individuals with respect to their own needs for knowledge. The overall organizational knowledge base emerges out of this process of exchange, evaluation, and integration of knowledge. Like any organizational process, the only actors are individuals. But it is a social process, one that is extraindividual. It is comprised of the *interaction* of individuals and not their isolated behavior.

Argyris and Schon's (1978) model of organizational learning begins to ïocus on the organizational level as they discuss the modification and change in organizational theory in use as current strategies are challenged ïy organizational decision makers (1978, pp. 129–147). But, this model does not specify how the process of challenge and change in the theory used in the organization transcends the individual—how changes become accepted by others and integrated into other theories as a basis for organizational action.

Paradigmatic nature of organizational learning   As we have argued, organizational learning involves the creation of knowledge about factors defined by members of the organization to be relevant for organizational activities. Given the complexity of organizations, the environments they operate in, and the probable interaction of events and actions taken by organizational members, the range of possible questions asked in the organizational learning process is large. As March and Simon (1958) and many cognitive psychologists (cf. Carrel and Payne, 1976) have pointed out, people have limits on the complexity of the information they can process. Any given organizational member can handle only a fraction of the possible factors which might be considered in the organizational learning process.

It is possible to argue that individuals involved in organizational learning more or less randomly hit on factors which are relevant and that when they do, and manage to achieve a useful statement of cause and effect, organizational knowledge is increased. This is the implicit assumption in the March and Olsen (1976) model. Such a view provides a very limited notion of organizational learning. It cannot provide a basis for describing a systematic organizational learning process. But organizational learning should be understood as systematic. A view of organizational learning as systematic requires some mechanism to "simplify" the complexity of the world about which organizational members create new knowledge.

Given what we have argued about the social bases of organizational learning, we suggest that such a mechanism must also be socially defined. That is, it cannot simply be provided by the individual. Such a mechanism must establish or impose some order or structure on the world with which organizational members interact. We suggest that such a mechanism would be some form of framework comprised of concepts which group phenomena into classes or categories and make abstract thinking possible. There must exist within the organization some consensus about this framework in order to make communication among organizational members possible.

This concept is close to that of enactment as described by Weick (1977). Weick argues that organizational members share perceptions of what factors comprise the environment of the organization. This process of enacting the environment in a sense creates the reality of organizational environments. This then is similar to Berger and Luckman's (1967) concept of the social construction of reality. We would only add that the frameworks we are suggesting include the definition of the organization itself and of internal organizational processes.

From the perspective of organizational learning, these frameworks serve a function analogous to Kuhn's (1970a) concept of a paradigm. Masterman (1970) points out, and Kuhn (1970b) agrees, that one important use of the term paradigm is as "a set of beliefs, a way of seeing or organizing the principles governing perception."

Thus, these paradigmatic frameworks provide a basis for structuring the "world" about which organization members learn. We submit that such frameworks exist within organizations and are to a large extent particular to a specific organization. That is, that a given organization is characterized by a paradigm that is shared by organizational members. These are provided to organizational members in their socialization. Indeed, an organizational member must learn the system of concepts used within the organization if (s)he is to be able to communicate and understand the actions they are to take and the actions taken by others.

These paradigms are necessary for organizational learning. They provide a basis for abstracting general action-outcome relationships from specific events. They provide a way of determining the relevance or importance of questions within the organizational learning process. They provide a common language, which makes possible the sharing of experience and insights among organizational members.

Growth and change of organizational knowledge: The role of performance gaps So far we have discussed several concepts related to organizational learning. We now turn to a discussion of the process itself, the production of organizational knowledge as an ongoing organization process.

We will assume that there is an existing knowledge base within the organization. This is not an unrealistic assumption. Even if one argues that organizations could be created "from scratch," the creation of an organization implies that there exists a technology and some basis for organizing activities to implement that technology. This must be based on some form of knowledge. In a new organization, the founders can be assumed to have such knowledge. This knowledge base is used to formulate strategy and to make decisions. It provides some level of action-outcome understanding on the part of organizational decision makers or members about their activities, the overall processes within the organization, and the environment-organization interactions.

The question becomes: What causes the knowledge base of the organization to change? As long as the results of actions taken on the basis of organizational knowledge are consistent with the expectations of organizational decision makers, the existing knowledge base will be changed in that existing knowledge will be validated. However, organization knowledge can never be perfect or in fact anything more than an over-simplification of reality.

When a discrepancy exists between how the organization is performing and what its decision makers believe it ought to be performing, there is what Downs (1966, p. 191) defines as a performance gap. This performance gap increases organizational search for alternative courses of action that might reduce the performance gap. When organizational knowledge is incomplete, performance gaps exist. The search that is stimulated by performance gaps is guided toward generating knowledge that will help reduce the performance gap. For example, a manufacturing organization may find that it is not able to maintain product levels in amounts required to meet customer demand. Volume simply is not where it should be. The resulting performance gap provides a stimulus for the organization to seek out information on new technologies that can be used to increase production. The gap has stimulated the need for organizational learning to occur to generate new knowledge for the organization.

This search can thus be understood as the basic activity underlying the organizational learning process. The range of possible explanations for a given performance gap is wide. It may be the case that a decision was simply not implemented properly or that factors which were known about but are outside the control of the decision maker undermined the activity. On the other hand, the gap may reflect a failure of existing organizational knowledge. The action-outcome relationships which underlay the choice of action may have been inaccurate, or factors relevant to the activity were not identified because knowledge was incomplete.

In seeking to close the gap, the decision maker must take these factors into account. The search undertaken here is based on the paradigm and existing knowledge. If the gap can be attributed to external factors or implementation, it is not a matter of organizational learning. However, if it cannot be attributed to these factors, it must be considered a failure of organizational knowledge.

The attempt to solve these problems constitutes organizational learning. The decision makers attempt to account for failures of knowledge in terms of providing a refinement or change in existing knowledge by integrating the results obtained from specific organizational activities. In other words, the action-outcome relationships comprising organizational knowledge are modified or new ones are offered. This takes place in terms of the paradigm which suggests possible approaches to solving the problems. A new statement of an action-outcome relationship *may* become part of the organization's knowledge base if it provides a solution to problems which are relevant to the organization. Thus, organizational learning is conceived of here as an evolutionary process in which existing knowledge is changed and expanded. The existing knowledge base contains errors which become obvious only when the knowledge is used as a basis for organizational activities. When these errors appear, attempts are made to make corrections in the knowledge base. These corrections have implications for activities and for other statements in the knowledge base.

Sources of change in organizational knowledge   We have not addressed the ways in which these changes in knowledge in response to performance gaps can be obtained. These changes in the form of modified or entirely new statements of action-outcome relationships may be obtained in a number of ways. As we have suggested, they initially result from the search activities of individuals who may employ different methods to develop knowledge relevant to their own activities.

The most sophisticated method to generate such changes in knowledge would involve formal empirical methodology, such as that suggested by Staw (1977). In a variation on Campbell's (1971) experimenting society, Staw suggests that organizational members can facilitate their understanding of the organization and its environment by engaging in quasi-experiments based on organizational activities. While Staw does not suggest that this is actually being done, it is the case that the techniques of evaluation research are becoming more common in both the public and private sector (cf. Rieken and Boruch, 1974).

Changes in organizational knowledge may result from less formal, more experiential activities. Individuals engaging in organizational actions can formulate action-outcome knowledge if they have access to information about the outcomes of these actions. Individuals will be able to formulate such knowledge only to the extent that they have such information. Since the outcomes associated with an organizational action may reflect effects of actions in a number of different parts of the organization or its environment, this method is limited by the levels and patterns of communication within the organization. This experiential learning is close to the process suggested by March and Olsen (1976) of individual action leading to organizational action, which results in an environmental response and so impacts on individual beliefs. However, we suggest that rather than changes in beliefs, the results are changes in the statements concerning action-outcome relationships which potentially can be made public.

Another potential source of changes in this knowledge is what might be called "arm chair theorizing." Individuals within the organization may formulate new action-outcome relationships in a deductive process. These relationships may be derived from fundamental assumptions or premises held by individuals in the organization and then made public on this basis. Chandler (1962) implies that Alfred Sloan constructed his successful strategy for General Motors largely in this way and validated it in practice afterwards.

Knowledge may enter the organization from outside as well. Organizational members do not work in a vacuum. They have an awareness of the ideas and actions of individuals in other organizations. In addition, they have access to institutions that produce knowledge—the sciences, research and development laboratories, and other organizations which comprise what Havelock (1971) calls the knowledge production and utilization system. This system can supply organizational members with knowledge that is already developed. Another important outside source of knowledge are consultants in diverse fields who can be called upon to provide insights and theories of organizational action.

We will not speculate at this time on the relative effectiveness of these different methods, although this poses interesting theoretical and empirical issues. The important issue for us here is that there exist many potential methods for developing organizational knowledge. The method used is not in itself an issue here except to the extent that it is a factor in the transition of the knowledge derived by individuals from personal to organizational knowledge.

Regardless of the method by which an individual finds a new action-outcome relationship or modifies an existing relationship, this change in knowledge must be made public, communicated to and be accepted or legitimated by others before it can be considered a change in organizational knowledge. This does not mean that the individual has not yet learned, or that this new knowledge cannot be a basis for individual activities. It does mean that this knowledge cannot be used for organizational activities beyond the individual. In other words, at this point, no *organizational* learning has occurred.

Factors affecting the acceptance of knowledge The acceptance of new knowledge by organizational members will reflect their willingness to use this knowledge as a basis for organizational action. Organizational knowledge is instrumental knowledge. If we assume, following Thompson (1967), that organizations are operated under norms of rationality, these changes in knowledge will be accepted only if there is a belief by organizational members that this new knowledge will lead to increased organizational effectiveness at some level of the organization.

Ideally, this belief should reflect some rational criteria established within the organization. The method used and evidence supporting the change would provide such criteria. Individuals would accept these changes in knowledge only if, given such criteria, the knowledge can be demonstrated to be valid.

This view is limited and somewhat naive. Organizations are social systems comprised of individuals. March and Simon (1958) have argued that individuals are capable only of limited rationality given cognitive limits on information processing. Furthermore, it is widely accepted that organizations are characterized by political processes in which the differential power of individuals is an important factor (Crozier, 1963; Pettigrew, 1973). These considerations will affect the acceptance of new knowledge and so the organizational learning process.

New knowledge is not likely to be accepted if it conflicts greatly with the paradigm held by the organization's members. The cognitive limits suggested by March and Simon (1958), which affect decision making, will also affect this acceptance, which is essentially a decision itself. Even given evidence of the validity of new knowledge in terms of organizational criteria, individuals will be limited in their ability to make major changes in how they view the organization or its environment. If new knowledge raises issues beyond those included in the paradigm, involves new environmental or organizational factors, or specifies radically different kinds of action-outcome relationships, its acceptance will require such changes. Thus, new knowledge will be accepted only if it is consistent with the existing paradigm within the organization.

Since these paradigms are frameworks which simplify reality, they will be imperfect in that they will not include potentially relevant factors. Performance gaps may occur which cannot be resolved within the paradigm, because they in fact reflect such factors. Given this paradigmatic nature of organizational learning, such performance gaps will be analogous to what Kuhn (1970a) calls anomalies.

If the number of anomalies of this kind remains relatively small, this will not cause a major problem within the organization. If, however, the number of anomalies becomes great, their existence will begin to undermine the belief of organizational members in the validity of organizational knowledge. These anomalies will indicate that the knowledge on which action within the organization is based is inadequate. Under these conditions, nonparadigmatic knowledge will become increasingly acceptable. Indeed, there may occur a "paradigm revolution," to borrow Kuhn's (1970a) term.

The extent to which the members of an organization will tolerate anomalies will vary across organizations. However, at some point the existence of such anomalies will pose a major problem within the organization and there will be a discontinuity in the organizational learning process represented by a paradigm revolution.

The acceptance of organizational knowledge will also reflect the politics of the organization. New knowledge will not be likely to be accepted if it conflicts with the knowledge held by powerful individuals within the organization. Organizational power has been argued to reflect in part the abilities of certain individuals to control or reduce uncertainty for others (Crozier, 1963; Hickson et al., 1971). We suggest that this reflects the knowledge held by individuals concerning specific organizational activities and environments. Changes in organizational knowledge may threaten this power base and so will be resisted by these individuals. Given their power, they will be able to effectively accomplish this.

This raises a secondary issue, which can only be mentioned briefly here. New knowledge may be developed within the organization that results in changes in the distribution of power within the organization. This can lead to the acceptance of new knowledge which conflicts with the knowledge of previously powerful individuals. This change in power may result in conflicts concerning the validity of different aspects of the organizational knowledge base. The resolution of these conflicts may result in important changes in the knowledge used within the organization and therefore changes in organizational activities.

Organizational learning and communication   The organizational learning process will reflect the nature of communication within the organization. Given that no organizational learning occurs until knowledge is accepted by others in the organization, it must first be communicated. We have already argued that this requires that the changes in knowledge developed by individuals must be capable of being stated in a communicable form. However, it must also be transmitted to others. Organizational learning, therefore, will be possible only to the extent that there is communication.

The learning process will therefore be sensitive to the amount of communication within the organization. If communication is limited, the changes in knowledge developed by individuals will be less likely to become public. If other members of the organization are not aware of these changes, however useful and acceptable, they cannot be accepted and integrated into the knowledge base of the organization. The amount of organizational learning that occurs will therefore be low.

A high level of communication does not in itself imply high levels of organizational learning. Knowledge must still be accepted for this to occur. A high level of communication is, however, a necessary condition for this learning.

Conclusion
Organizational learning has been conceptualized here as an organizational process. It involves more than the simple aggregation of the learning of individual members of the organization. Rather it reflects changes in organizational knowledge through the communication of new statements of action-outcome relationships developed by individuals and the

acceptance of these new statements by others within the organization. Organizational learning is also a social process. It reflects a socially constructed and accepted paradigm, the political nature of the organization, and the nature of communication within the organization. The outcomes of the organizational learning process are changes in organizational knowledge—knowledge which is accepted by organizational members as a legitimate basis for organizational action.

As discussed here, these changes are occurring in a largely evolutionary process. The knowledge base is extended and refined in increments. These increments reflect the addition of new statements of action-outcome relationships which are added to or supersede existing statements. Occasionally, however, this evolutionary process is disrupted by paradigm revolutions which in a sense redefine aspects of the organization or its environment for the members of the organization.

As it has been conceptualized here, organizational learning is a general organizational process. The changes in organizational knowledge which it produces may be associated with any organizational activity. Like organizational control or coordination, organizational learning will occur at all levels and in all parts of the organization. While the organizational learning process can be understood in these general terms, it will be necessary to develop more "middle range" theories of organizational learning to understand its relationship to particular organizational activities such as organizational design. The next section of the paper will then develop this middle range theory of organizational learning as applied to the design process.

IV. ORGANIZATIONAL LEARNING: THE DESIGN FORMULATION PROCESS

In the first three sections of this paper, we have presented a conceptual definition of the organization and the conceptual framework for understanding organizational learning given this definition of the organization. In this section we present a more specific model of a particular type of organizational learning—organizational learning with respect to providing knowledge for organization design.

We will now briefly describe the learning process as it applies here to the design formulation process. The components of the learning process involved in organizational design are presented in Figure 1. Again, organizational learning is defined here as the process within the organization by which knowledge about action-outcome relationships and the effect of the environment on these relationships is developed.

In developing this model describing action-outcome relationships be-

Figure 1.   Organizational learning model for the design process.

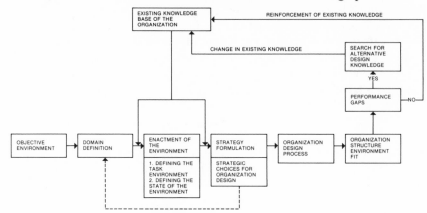

tween the organization and its environment, we make the assumption that there is an existing knowledge base in the organization. This existing knowledge base is developed from past learning of the organization.

Domain Definition

Domain definition is the first step in the learning process. The organization's environment is infinitely complex and it is in most cases impossible for organizational decision makers to be aware of and respond to the complete environment. Therefore, organizational decision makers have to define that subpart of the environment it is going to deal with. The organization's domain as Levine and White (1961) have defined it consists of claims which an organization stakes out for itself relative to its environment. The domain defines what kinds of outputs the organization will produce and for whom in the environment.

Obviously, domain definition and strategy formulation are very much interdependent. In the present model we are stating that the initial step is for organizational decision makers to define their domain and then, as the model indicates, strategy formulation comes later. However, as the dotted line indicates in Figure 1, strategy does affect the domain definition process as the organization evolves over time. In fact, one quite frequent organizational strategy is, because of changes in the environment, to change the organization's domain. For example, with the development of the polio vaccine, the March of Dimes was faced with going out of business—unlike most organizations it had attained its goal of reducing polio as a threatening disease (Sills, 1957). However, the organization was not disbanded; rather it redefined its domain to include dealing with birth defects. Thus, the organization redefined its domain in order to stay in existence.

The Enactment Process

Enactment is the second step in the learning process. Before organizational decision makers can respond to the environment, they must identify and define the organization's environment. The process of enactment is one of structuring the situation so as to define and create some meaning and sense to it so that organizational decision makers can react to it. Enactment is a creative process in that it is the way meaning is attached to the environment (Weick, 1977). There is no decision environment until it is created by organizational decision makers. Enactment is the "definition of the situation" for the organization.

Organizational decision makers need different kinds of information during enactment. This required knowledge focuses on information that organizational decision makers need before strategies for organizational design can be developed to deal with the environment. The kind of information required about the environment is defined for organizational members from the existing knowledge base in the organization. This existing knowledge base indicates how the environment is likely to affect action-outcome relationships in the organization and how events in the environment have an impact on the organization. In the enactment phase of the learning process, then, two types of knowledge are needed. First, it is important that decision makers become aware of the make-up of their environment. This is the definition of the task environment. Second, the state of the environment, its salient characteristics, must be defined before the organization can design a response to it. Here the concern is with the complexity and dynamics of the environment. We will now briefly discuss the nature of these two kinds of knowledge.

Before action-outcome learning can occur, decision makers must define their task environment. The task environment comprises those parts of the environment which decision makers define to be relevant or potentially relevant to goal setting and goal attainment (Dill, 1958).

Table 1 identifies a fairly comprehensive list of environmental components as developed by Duncan (1972). Clearly, no decision unit in the organization would consider all of these factors at one point in time. Therefore, the key issue is defining that part of the environment—the task environment—that is relevant for problem solving and decision making. Organizational decision makers cannot begin to learn to deal with their environment until in fact they have identified just what is the relevant environment. For example, has the organization defined-identified the right customers to develop and market their products to? Recent comparisons of General Electric's success and Westinghouse's lack of success have focused on this task environment problem (*Business Week,* January 31, 1977). Westinghouse engaged in a huge gamble in uranium contracts that have turned bad, entered into low-cost multifamily housing on gov-

Table 1.   Components of Organizational Environments

Internal environment

(1) Organizational personnel component
  (A) Educational and technological background and skills
  (B) Previous technological and managerial skill
  (C) Individual member's involvement and commitment to attaining system's goals
  (D) Interpersonal behavior styles
  (E) Availability of manpower for utilization within the system

(2) Organizational functional and staff units component
  (A) Technological characteristics of organizational units
  (B) Interdependence of organizational units in carrying out their objectives
  (C) Intra-unit conflict among organizational functional and staff units
  (D) Inter-unit conflict among organizational functional and staff units

(3) Organizational level component
  (A) Organizational objectives and goals
  (B) Integrative process integrating individuals and groups into contributing maximally to attaining organizational goals
  (C) Nature of the organization's product service

External environment

(4) Customer component
  (A) Distributors of product or service
  (B) Actual users of product or service

(5) Suppliers component
  (A) New materials suppliers
  (B) Equipment suppliers
  (C) Product parts suppliers
  (D) Labor supply

(6) Competitor component
  (A) Competitors for suppliers
  (B) Competitors for customers

(7) Socio-political component
  (A) Government regulatory control over the industry
  (B) Public political attitude towards industry and its particular product
  (C) Relationship with trade unions with jurisdiction in the organization

(8) Technological component
  (A) Meeting new technological requirements of own industry and related industries in production of product or service
  (B) Improving and developing new products by implementing new technological advances in the industry

(Duncan 1972)

ernment contracts, and entered into the mail order record club business, all of which have led to substantial losses. General Electric, on the contrary, used systematic long-range planning and careful analysis before entering markets, with the result that they have been extremely successful.

After key decision makers have identified their task environment, the next step is to be aware of the state of the environment. What are its salient characteristics? Various theorists (Emery and Trist, 1965; Duncan, 1972; and Jurkovitch, 1974) have discussed the different characteristics the environment can take on. Here we will focus on two such dimensions, simple-complex and static-dynamic (see Table 2).

The simple-complex dimension of the environment focuses on the degree to which factors in the environment considered for decision making are few in number and are similar to one another, versus large in number and very different from one another. An example of a simple unit would be a lower-level production unit whose decisions are affected only by the parts department and materials department, upon which it is dependent for supplies, and the marketing department, upon which it is dependent for determining its output volume. An example of a complex environment would be a programming and planning department. This group, when making a decision, must consider a wide variety of environmental factors. They may focus on the marketing and materials department, customers, suppliers, etc. Thus, this organizational unit has a much more heterogeneous group of environmental factors to deal with in decision making—their environment is more complex than the production unit.

The static-dynamic dimension of the environment is concerned with the degree to which the factors of the environment remain the same over time or are changing. A static environment would be characterized by a production unit which has to deal with a marketing department whose output requests remain the same and a materials department which is able to supply a steady rate of inputs to the production unit. However, if the marketing department were continually changing its requests for different production outputs and the materials department were variable in its ability to supply parts, this would characterize a more dynamic environment.

Once key decision makers have identified the make-up and state of their environment, the design process can begin. The relevant decision environment has to be identified before organizational decision makers can start developing action-outcome knowledge regarding that environment.

The Relationship Between Organizational Learning, Strategy, and Organization Design

It is important at this point to discuss the interrrelationship between strategy, design, and organizational learning before discussing the learning model further.

Table 2. Environmental State Dimensions

	Simple	Complex
	Cell 1: low perceived uncertainty	Cell 2: moderately low perceived uncertainty
Static	(1) Small number of factors and components in the environment (2) Factors and components are somewhat similar to one another (3) Factors and components remain basically the same and are not changing	(1) Large number of factors and components in the environment (2) Factors and components are not similar to one another (3) Factors and components remain basically the same
	Cell 3: moderately high perceived uncertainty	Cell 4: high perceived uncertainty
Dynamic	(1) Small number of factors and components in the environment (2) Factors and components are somewhat similar to one another (3) Factors and components of the environment are in continual process of change	(1) Large number of factors and components in the environment (2) Factors and components are not similar to one another (3) Factors and components of environment are in a continual process of change

(Duncan 1972)

Strategy formulation   Once the organization's environment has been defined, the strategy formulation process occurs. The key decision makers are making a number of decisions regarding organizational objectives. This is the development of the organization's long-term goals and objectives and the adoption of courses of action and the allocation of resources necessary for carrying out these goals. Chandler (1962, p. 13), Andrews (1971), McNichols (1977), Mintzberg (1977) have indicated that the strategy formulation process involves the interplay between three forces: (1) the environment that may change irregularly, (2) the operating system of the organization that seeks to stabilize organization action despite the changing environment, and (3) leadership in the organization whose task is to mediate between the changing environment and the organization's desire to maintain stability. Organizational strategy, then, is concerned with how to deal with the environment. The kind of organization structure that is designed is the mechanism by which the strategy is implemented.

For example, an organization's strategic decisions over time might include the following: (1) picking a product market it sees as being attractive, (2) selecting the technology and financial resources needed to enter this market, (3) moving out of other market areas that did not meet the organization's criteria for successful return on investment. Again, this action does not imply organizational learning. Organizational decision makers may not be aware why these particular actions have the impact they do. However, organizational learning, because of its action-outcome knowledge, should lead to more effective strategy formulation. For example, Westinghouse in the early 1970s made a strategic decision to increase its sales volume in generating equipment by selling this equipment at a fixed price contract. The unanticipated result is that Westinghouse now has cut itself off from profits in the steam turbine industry until 1979 (*Business Week*, January 31, 1977, p. 61). Had key organizational decision makers at Westinghouse had action-outcome knowledge of the pricing structure effects on long-range profits, they perhaps would have developed a different, more effective strategy.

Organizational design   Organizational design is the allocation of resources and people and the structuring of the organization to carry out its objectives. The design of the organization ideally is constructed to fit the nature of the organization's environment. To make design decisions, then, organizational decision makers need information about the environment. If the environment is relatively simple and stable, the traditional functional organization is most appropriate. When the environment becomes more complex and dynamic and the organization can organize around particular markets or products, the decentralized structure is more appropriate.

Effective organizational design does not necessarily imply organizational learning. An organization can construct its design without having action-outcome knowledge about what impact that design may in fact have on the environment. For example, currently the most prevalent organizational design is the decentralized structure. Most organizations with complex environments that can be segmented into more homogeneous product or geographical market areas implement this type of design (Rumelt, 1974). However, this does not mean that these organizational decision makers really understand the impact of their structure on the environment. One of the important problems many decentralized organizations face is the lack of integration. There tends to be growing interdependence between the different parts of the organization and no way to coordinate these units. Decision makers have not "learned" that their decision environment has increased the interdependence among units and now the organization has to modify its structure to satisy the needs for coordination.

At this point we must distinguish between learning and a more evolutionary process of adaptation. Organizational decision makers may develop an effective organizational design to fit the environment through an evolutionary process. Here the appropriate structure is "selected out" by effective adaptation to the environment. The design "seems to work," so the organization continues to implement it. However, organizational decision makers here are not exhibiting any understanding of causation of their behavior by linking their actions and resulting outcomes and their impact on the environment. Thus, as we have defined it, learning has not occurred. The organization is muddling through on a trial-and-error basis rather than demonstrating that it has knowledge about the relationships between its actions and outcomes and their impact on the environment and the environment's impact on the organization.

The relationship between learning, strategy formulation, and organizational design is thus complex. Learning is not required for strategy formulation or design. However, for a more effective strategy and design, the ideal cycle between these factors would be shown in Figure 2. Ideally organizational learning develops the knowledge for organizational decision makers so that they can develop an organizational strategy that is responsive to the particular task environment the organization faces. Learning also provides decision makers with accurate information regarding the appropriate design given the nature of the environment. Also, the strategy that the organization develops affects the design it implements. For example, if an organization decides to produce a variety of consumer products in the food and recreational areas, it would probably implement a multidivisional organizational structure organized around these different product markets.

Strategy Formulation

The key issue in strategy formulation that we will focus on in the learning process revolves around the issue of strategic choice (Child, 1972). These "choices" focus on how organizational decision makers decide to deal with their environment given their objectives. Here we are going to focus on the choices organizational decision makers have to design their organization in order to deal with the environment. Also, we are going to develop a decision heuristic that explicitly identifies the choice process in design and provides a framework for understanding this strategic choice process.

At any point in time these options for organization design are limited. Currently, the dominant forms of organization are the functional and the decentralized organization or some variation of the two. While these options are by no means fixed, they are the set from which existing designs

Figure 2.   Relationship between learning,
strategy, and design.

are selected. The objective of our discussion of the design process is to more clearly identify the decision process involved in selecting a particular design option given the organization's environment.

We are thus concerned with strategic choice as it effects organization structure. Strategy as we view it here then focuses on how organizations carry out their activities through the design of structure.

Organizational Design Process

The organizational design question in organizational learning, then, is one of implementing the organizational structure most appropriate to the environment with which the organization must deal. The objective of the kind of organization structure that is implemented is twofold: (1) to generate information for decision making that reduces uncertainty, and (2) to generate information that will help coordinate the diverse parts of the organization.

In our discussion here, we will be developing a theory of organizational design in a deductive way from current theory and research in organizational theory. Our intention in this discussion is to more clearly identify the design process as it relates to organizational learning. The structures that we discuss are what currently exist in effective organizational practice. Our intention is to begin to describe the processes involved in order to better understand them. However, this discussion will have a normative flavor at times in that we are attempting to describe what appears to be the most effective organizational structure given certain environmental characteristics. However, our objective is to develop a middle range theory based on existing theory and practice of organizational design from a learning perspective.

A most useful way to conceptualize organization structure is from an information processing view. The key characteristic of the structure of the organization is that it links the various elements of the organization through the transformation of information. The structure of the organization provides the channels of communication through which information flows in the organization. Research (Duncan, 1973) has indicated that when organizational structure is more formalized and centralized information flows are restricted, the system is not able to gather and process a lot of information when faced with uncertainty. For example, when the organization's structure is very centralized, decisions are made at the top of the organization and information tends to get filtered out as it moves up the chain of command. When the decision involves a lot of uncertainty, it is unlikely that the few individuals at the top of the organization will have all the necessary information. Having more of the subordinates participate in the decision-making process (decentralization) may generate more information to help reduce the uncertainty associated with the decision.

The question now becomes, what are the different structures available to organizational decision makers to choose from in dealing with their environment? Contingency theories of organization (Luthans and Steward, 1977; Child, 1974) have elaborated that there is no one best structure. However, they have been less clear in elaborating the decision process organizational members can work through in deciding which structure to implement. In this section, we will first briefly identify the different structures available and then discuss the design decision process as it relates to organizational learning.

Functional organization   When the organization's environment is relatively simply (not a lot of factors to consider in decision making) and stable (environment is not changing, nor are the demands made by the environment), the functional organization is most appropriate for the information and coordination requirements of the organization. In this kind of environment, uncertainty is low and coordination needs are low. Specialization by functional area is efficient, because coordination efforts can be handled at the top of the organization. The top-level managers resolve disputes and solve problems the subordinate managers are unable to solve. Table 3 presents a summary of the characteristics and strengths and weaknesses of the functional organization. The primary weakness of the functional organization is that when the organization's environment begins to become more dynamic and uncertainty tends to increase, decisions pile up at the top of the organization. Individual units do not have all the information required for decision making, with the result that mana-

Table 3.   Characteristics of the Functional Organization

Organizational functions	Accomplished in functional organization
Goals	Functional sub-goal emphasis (Projects lag)
Influence	Functional heads
Promotion	By special function
Budgeting	By function or department
Rewards	For special capability

Strengths	Weaknesses
1. Best in *stable* environment	1. Slow response time
2. Colleagueship ("Home") for technical specialists	2. Bottlenecks caused by sequential tasks
3. Supports in-depth skill development	3. Decisions pile at top
4. Specialists freed from administrative/coordinating work	4. If multi-product, product priority conflict
5. Simple decision/communication network excellent in small, limited-output organizations	5. Poor interunit coordination
	6. Stability paid for in less innovation
	7. Restricted view of whole

gers push decisions upward. Top-level managers then become overloaded with decisions and thus are slow in responding to the changing environment. Thus, the functional organization is limited as the organization's environment becomes more dynamic and the information requirements become greater for decision makers.

Organizational decision makers are now faced with a dilemma. They have two design strategies (Galbraith, 1977) that can be implemented. First, designs can be instituted that reduce the amount of information required for decision making. Decentralization is the principal strategy here. Second, organizations can institute lateral relations along with computer-based vertical information systems to increase the amount of information available for decision making.

Decentralized organization   The decentralized organization is the result of the creation of self-contained tasks. In decentralization, the organization is designed around product areas, project groups, or market areas. Each group has all the resources needed to perform its particular task. For example, each product area has its own organization (Figure 3) and resulting resources to carry out its particular mission. The information required for decision making is reduced, because by organizing around a product or market area, the output diversity of the organization is reduced. Decision makers only have to worry about their particular product or area; they have their own resources to carry out these activities and thus they don't have to compete for shared resources or schedule shared resources. Decision makers have a more homogeneous environment and can concentrate on their particular market. The decentralized structure is particularly effective when the organization's environment is very complex and can be segmented or broken down into product or market areas around which the organization can structure itself. For example, a health products organization producing drugs and medical equipment might start out by being in a functional organization, but this structure would be ineffective in trying to manage a diverse range of products. It would be difficult for one manufac-

Figure 3.   Decentralized organization.

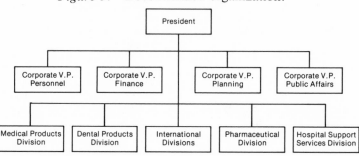

turing unit to have the diversity of information to coordinate and produce the wide range of products. It would also be difficult for one marketing unit to promote this diverse group of products. Different kinds of information and skills would be required to market the different products. Segmenting this complex environment in product areas or market areas can simplify the information needs and allow each division to concentrate on its own speciality.

Table 4 summarizes the characteristics, strengths, and weaknesses of the decentralized organization. The major shortcoming of the decentralized organization is coordination across divisions. If there is little interdependence between the units (pooled), the decentralized organization is very effective. However, if the environment changes—becomes more uncertain, which increases the interdependence (sequential or reciprocal) between divisions—the decentralized organization is problematic in that it does not have the coordinating mechanisms required to integrate across divisions. For example, in the hospital products company shown in Figure 3, it might happen that several of the decisions might have to work with the International Group to help develop some product lines for overseas markets. In the existing structure, there is no mechanism to facilitate getting information to coordinate these activities.

Thus, the most serious weakness of the decentralized organization is its deficiency in facilitating coordination among organizational units when

Table 4. Characteristics of the Decentralized Organization

Organizational functions	Accomplished in decentralized organization
Goals	Special product emphasis (Technologies lag)
Influence	Product, project heads
Promotion	By product management
Budgeting	By product, project, program
Rewards	For integrative capability

Strengths	Weaknesses
1. Suited to fast-change	1. Innovation/growth restricted to existing project areas
2. High product, project or program visibility	2. Tough to allocate pooled resources (i.e., computer, lab)
3. Full time task orientation (i.e., $, schedules, profits)	3. Shared functions hard to coordinate (i.e., purchasing)
4. Task responsibility, contact points clear to customers or clients	4. Deterioration of in-depth competence— hard to attract technical specialists
5. Processes multiple tasks in parallel, easy-to-cross functional lines	5. Possible internal task conflicts, priority conflicts
	6. May neglect high level of integration required in organization

they become interdependent. The organization can now implement the second general design strategy of increasing the organization's information-processing capabilities. Here the organization implements computer-based information systems and lateral relations. We will discuss lateral relations here as the major strategy for increasing information for reducing uncertainty and facilitating coordination.

Lateral relations   Lateral relations is really a process that is overlaid on an existing functional or decentralized organization structure. It is an additional design that is added to the existing organization structure. Lateral relations as a process moves decision making down to where the problem exists in the organization. It differs from decentralization as there is no creation of self-contained tasks.

There are various types of lateral relations. *Direct contact* can be used by managers of diverse groups as a mechanism to coordinate their different activities. Here managers can meet informally and discuss common problems that they encounter. *Liaison roles* are a formal link between two units to handle the communication between them (Galbraith, 1977). The best example of a liaison role is the engineering liaison in a manufacturing department. In this instance this engineering liaison may be located in the production organization as a way of coordinating engineering and production activities.

When coordination between units becomes more complex, an *integrator role* may be established. The integrator role is particularly used when organizational units that must be coordinated are differentiated from one another in terms of their structure, subgoals, time, orientation, etc. (Lawrence and Lorsch, 1967). In this type of situation, there is the possibility that conflict might exist between the various units. For example, production, marketing, and R & D units in an organization may be very differentiated from one another. Marketing is concerned with having products to sell that are responsive to customer needs. R & D may be concerned with developing products that are very innovative and that shape customer needs. Production, on the other hand, may want products to remain unchanged so that production set-ups don't have to be modified. There thus are some differences among the three units in terms of their subgoals. The integrator role is instituted to coordinate and moderate these diverse orientations. The integrator could be a materials manager or group executive whose additional function would be to coordinate and integrate the diverse units so that the organization's common objectives can be met.

To be effective as an integrator, a manager needs to have certain characteristics. First, he needs wide contacts in the organization so that he has relevant information for the different units he is attempting to integrate. Second, the integrator needs to understand and share some of the goals and orientations of the different groups. He cannot be seen as being a partisan to one particular group's perspective. Third, the integrator has to be rather broadly trained technically so that he can talk the language of the different groups. By being able to demonstrate that he has some expertise in each area, he will be viewed as more credible by each group and will also be better able to facilitate information exchange between the units. The integrator can in effect become an interpreter of each group's position to the others. Fourth, the groups that the integrator is working with must be able to trust him. Again, the integrator is trying to facilitate the information flow and cooperation between the groups and thus the groups must be able to trust the integrator in that he is working for a mutually acceptable solution for all the groups. Fifth, the integrator needs to exert influence on the basis of his expertise rather than through formal power. The integrator can provide information and identify alternative courses of action for the different units as they attempt to coordinate their activities. The more he can get them to reach agreement on solutions and course of action rather than having to use his formal power, the more committed they will be to implementing the solution. Finally, the conflict-resolution skills of the integrator are important. Because differentiation between the units exists, conflict and disagreement are inevitable. Therefore, it is important that confrontation is used as the conflict-resolution style. By confrontation we mean that parties to the conflict identify the causes of conflict and are committed to problem solving to find a mutually acceptable solution to the conflict and then work to implement that solution.

When coordination requires the coordination of six or seven different units, task forces or teams can be established. Here a group of managers are put together in a group to jointly work out coordination problems their diverse groups may be having. Members of the task force team represent their different groups and meet to work out problems of coordination and integration. For example, in a manufacturing organization, the marketing, production, R & D, finance, and engineering managers may be scheduled to meet twice a week (or more when required) to discuss joint problems of coordination that they may be having which require their joint assistance to solve. This mechanism is a problem solving group whose task is to facilitate coordination.

Lateral relations, as we alluded to them in discussing the integrator role, require certain design considerations and skills if this process for increasing information for reducing uncertainty and increasing coordination is going to be effective. From a design perspective, four factors are required (Galbraith, 1977). First, the organization's reward structure must support and reward cooperative problem solving that leads to coordination and integration. Will managers' performance appraisal reflect their successful or unsuccessful participation in coordination and integration efforts? If the organization's reward system does not recognize joint problem-solving efforts, lateral relations will not be effective. Second, in assigning managers to participate in some form of lateral relations, it is important that they have implementation responsibility. Line managers should be involved in that they understand the problems perhaps more intimately than staff personnel, and, more importantly, they are concerned about implementation. Staff members can be used, but line managers should have the dominant involvement, as this will lead to more commitment on their part to implementing solutions that come out of lateral relations problem-solving efforts. Third, participants must have the authority to commit their units to action. Managers who are participating in an effort to resolve problems of coordination must be able to indicate what particular action their unit might take in trying to improve coordination. For example, in the manufacturing company task force example mentioned earlier, the marketing manager should be able to commit his group to increasing the lead time for getting information to production for deadlines for delivering new products to customers. Four, lateral processes must be integrated into the vertical information flow. In the concern for increasing information exchange *across* the units in the organization, concern must also exist for vertical information exchange so that the top levels in the organization are aware of coordination efforts.

Certain skills are also required on the part of participants for lateral relations to work. First, individuals have to be able to deal with conflict effectively. They must be able to confront conflict in the sense of identifying the sources of conflict and then engaging in problem solving to reach a mutually acceptable solution to the conflict situation. Second, participants need good interpersonal skills. They have to be able to communicate effectively with one another and avoid making other participants defensive. The more they can learn to communicate with others in a descriptive, nonevaluative manner, the more open will be the communication progress (Argyris, 1976). Third, participants in lateral relations will have to understand that influence and power should be based on expertise rather than formal power. Because of the problem-solving nature of lateral relations, an individual's power and influence will change based on the particular problem at hand and the individual's ability to alter key information to solve the problem. At various points in time different members will have more influence because of their particular expertise.

Lateral relations, then, is a process that is overlaid onto the existing functional or decentralized organization structure. Lateral relations require various skills, so it is imperative that an organization never adopt this without some training for those individuals who are going to be involved. Before implementing lateral relations, team building might be utilized to develop the interpersonal skills of the managers involved (Dyer, 1977).

Organizational design decision tree heuristic   Having discussed the different kinds of organization structure, it is now important to summarize the decision-making process the organizational designer can utilize in selecting the appropriate organization structure to "fit" the environment. Figure 4 presents a decision tree analysis for selecting the appropriate structure. In utilizing the decision tree there are a number of questions the designer needs to ask.

The first question focuses on whether the environment is defined by the dominant coalition as *simple* (i.e., there are few factors to consider in the environment) or *complex* (i.e., there are a number of different environmental factors that need to be considered). If the environment is defined as *simple*, then the next question focuses on whether the factors and components of the environment are static (remain the same over time) or are dynamic (change over time). If the environment is defined as *static*, there is likely to be little uncertainty associated with decision making with resulting low information requirements for decision making. In this *simple-static* environment, the *functional organization* is most efficient. It can most quickly gather and process the information required to deal with this type of environment.

At this point the question might be raised: Are there any organizational environments that are in fact both simple and static or is this a misperception on the part of the dominant coalition that oversimplifies the environment? The case can be made that there are in fact environments like this, but the key is that these environments may change—they may become

Figure 4.   Organizational design decision tree heuristic.

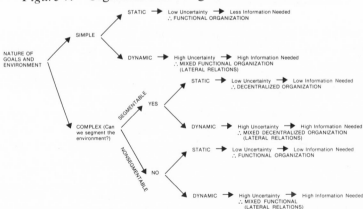

more dynamic as the market place changes, as resources become scarce, or the organization's domain is challenged. For example, the motor home–recreational vehicle industry was very successful in the early 1970s. Its market was relatively homogeneous (simple) and there was a constantly high demand (static) for its products. Then the oil embargo of 1973 hit and the environment suddenly became dynamic and the industry had a very difficult time changing because it had done no contingency planning regarding "what would happen if" demand shifted, resources became scarce, etc. The important point here is that an organization's environment may in fact be simple and static, but organizational decision makers should continually scan the environment and be sensitive to the fact that things can in fact change and thus some contingency planning may be useful. The dominant coalition should not become seduced that just because the environment is static now it might not become more dynamic in the immediate future.

If this simple environment is defined as *dynamic,* with some components in the environment changing, some uncertainty may be experienced by decision makers. Thus, information needs will be a bit greater than when the environment was static. Therefore, in this *simple-dynamic* environment the *mixed functional organization with lateral relations* is likely to be the most effective in gathering and processing the information required for decision making. Because the organization's environment is simple, the creation of self-contained units would not be efficient. It is more economical to have central functional areas responsible for all products and markets as these products and markets are relatively similar to one another. However, when uncertainty arises and there is the need for more information, some form of lateral relations are added—overlaid onto the existing functional organization.

Figure 5 shows the functional organization of a manufacturing organization. The organization may suddenly be faced with a problem associated with its product. Competitors may have developed a more attractive replacement for the company's product. As a result of this unique problem, the president of the firm may set up a task force chaired by the vice president of sales to develop new products. The task force consists of members from manufacturing, sales, research, and engineering services. Their task will be to develop suggestions for new products.

If the organization's environment is defined by the dominant coalition as *complex* (i.e., there are a large number of factors and components that need to be considered in decision making), the next question to ask is: Can the organization segment its environment? Can the environment be segmented into geographic, market, or product areas? If the environment is defined as *segmentable,* the next question focuses on whether the environment is static or dynamic. If the environment is defined as *static,*

Figure 5.   Functional organization with task force.

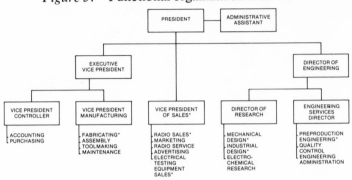

*Members of task force

there is going to be low uncertainty and thus information needs for decision making are not going to be high. Thus, in the *complex-segmentable-static* environment the *decentralized organization* is most appropriate. The organization can break the environment apart in the sense that it can organize around products or markets, etc., and thus information, resources, etc., are only required to produce and market these more homogeneous outputs of the organization.

In the *complex-segmentable-dynamic environment* there is a change in the components of the environment in terms of the demands they are making on the organization, or in fact the organization has to now consider different factors in the environment that it had not previously considered in decision making. Uncertainty and coordination needs now may be higher. The result is that decision makers need more information to reduce uncertainty and provide information to facilitate coordination. The *mixed decentralized organization with lateral relations* is the appropriate structure here.

Figure 6 presents the design of a multidivision decentralized health products organization. Some form of lateral relations may be added to this structure to help generate more information. For example, the International Division may be attempting to develop some new products, but may be encountering some problems, with the result that the entire organization, stimulated by the president's concern, may be experiencing a lot of uncertainty as to how to proceed in this division. As a result, a task force might be set up of the manager of the International Group and the Dental Group and the Pharmaceutical Group to work together to come up with some ideas for new products in the International Division. The lateral relations mechanism of the task force facilitates information exchange *across* the organization to reduce uncertainty and increase coordination of efforts between the divisions. By working together in the task force, the

Figure 6.   Decentralized organization.

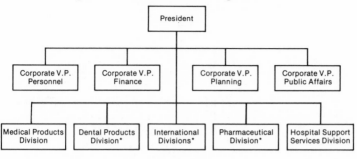

*Members of task force

division managers will be exchanging information and will hopefully be gaining a better understanding of their common problems and how they need to work and cooperate with one another in order to solve these problems.

If the organization's complex environment is defined by the dominant coalitions as *nonsegmentable,* the function organization will be appropriate because it is not possible to break the environment up into geographic or product-service areas. There simply may be too much interdependence among environmental components or the technology of the organization is so interlocked that again it is not possible to create self-contained units organized around components of the environment. A hospital is a good example of this type of organization. The environment is clearly complex. There are numerous and diverse environmental components that have to be considered in decision making (i.e., patients, regulatory groups, medical societies, third party payers, suppliers, etc.). In the *complex-nonsegmentable-static* environment, environmental components are rather constant in their demands. Thus, here the *functional organization* is most appropriate. The organization can specialize in the different functional areas (Figure 7).

In the *complex-nonsegmental-dynamic* environment, the environment is changing, uncertainty is higher, and information needs are greater. The existing functional organization, through its very specific rules, procedures, and channels of communication, will likely be too slow in generating the required information. Therefore, some form of lateral relations may be added to the functional organization (i.e., mixed-functional organization). For example, community groups may suddenly become very concerned about hospital costs and start putting pressure on the hospital to reduce its costs. The community groups may start advertising campaigns to get the public at large aroused to put pressure on the environ-

Figure 7. Functional organization with lateral relations.

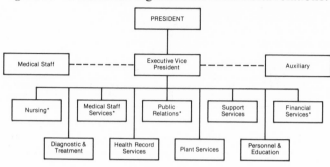

*Members of task force

ment, the hospital might set up some task forces made up of representatives from nursing, medical staff, and financial services under the chair of public relations to respond to changes and to try to influence these groups in the environment.

Organizational Design and Effectiveness

Throughout the previous discussion of organization design we have continually alluded to "effective fit," that one structure is more effective than another under certain instances. Therefore, we now must define effectiveness so that organization decision makers know what kind of information they need to get to determine if in fact they have implemented the appropriate design for gathering information to reduce uncertainty and solve coordination problems. Effectiveness as defined here will consist of *three* dimensions, two at the organizational level and one at the individual level. *First,* the organization has to successfully attain its goals and objectives in terms of productivity, profit, return on investment, etc. (Steers, 1977). *Second,* the organization needs to adapt to its environment. We do not mean adaptation with respect to organization design. This is discussed in the next section. Here we mean adaptation in terms of changes in activities within the organization in response to changing conditions that influence action-outcome relationships. As we discuss adaptation here we mean the organization needs to be able to: (1) identify problems in the environment, (2) generate information about these problems and transfer this information to that part of the organization that can do something about the problems, (3) take corrective action, (4) get feedback on the corrective action in order to determine if the problem was solved. *Third,* at the individual/organizational interface, individual roles should be defined so that they experience low role conflict and low role ambiguity. If role conflict and ambiguity are high, the individual is likely to experience high role stress, which may lead to more coping behavior on the part of the individuals so that they are less able to perform their job effectively.

In attempting to determine if the organization had implemented the appropriate structure, there are certain kinds of information-feedback that should indicate the structure may not be appropriate. Implementing the appropriate structure may have some direct impact on goal attainment, but it is likely to have its biggest impact on adaptability and role behavior.

Certain kinds of symptoms regarding ineffective *adaptability* such as the following may begin to occur:

1. Organizational decision makers may not be able to anticipate problems before they occur. There is a tendency in the organization to wait until problems occur and then react to them. The organization simply does not have enough information to attempt to contingency plan.

2. Decision makers may err in trying to predict trends in their decision environment. For example, Heublein (*Business Week,* July 4, 1977) headquarters staff lost some control over the different divisions in their multidivisional firm. The result was that its Kentucky Fried Chicken Corp. and its Smirnoff, Popov, and Relska vodka brands were not able to keep up with their markets. There was not the proper coordination across the divisions, with the result that the organization in essence lost control over what was happening in relation to its environment.

3. The organization may not be able to get key information for decision making to the right place for effective decision making. For example, divisional managers from different product groups may have information that quality and liability standards on their respective products are unrealistically high. However, because of decentralization and no effective coordination through some form of lateral relations, this information may not get to the staff groups in the organization that are responsible for setting corporate policy in this area.

4. The organization, having identified a problem vis-à-vis its environment, may simply not be able to take corrective action quickly enough. In the Heublein example cited above, the organization was attempting to deal with a marketing problem in the fast food industry which is so volatile that it is difficult to do any long-range planning (*Business Week,* July 4, 1977, p. 65). The corporate staff was simply not linked into the day-to-day problems of the Kentucky Fried Chicken Corp. and had not realized that more and more people prefer to dine out rather than take out food.

There are also symptoms of poor structure fit at the level of the individual-organizational interface in terms of some *increase* in either role conflict or role ambiguity. It is important that the organization monitor the level of role conflict and role ambiguity and the resulting stress that participants experience so that the system has a base line for comparison. If there is a significant increase in conflict and/or ambiguity and resulting stress from this base line, then the organization may consider that this increase is a symptom of a design problem. For example:

1. Individuals may be experiencing higher role conflict. This may occur when the organization is implementing a functional organization when the environment is more dynamic. The environment may be changing and the individual may be required to make quick response to this changing environment. Having to wait for new policy changes to come down the hierarchy may delay the organization from responding to this changed environment. Decision makers at the top of the organization will also suffer from role conflict when the environment is rapidly changing. In the functional organization, when new situations occur, they are pushed up the hierarchy for action. The result is that top-level decision makers become overloaded and the organization's response to the environment slows down. In a dynamic environment the functional organization constrains the decision making adaptation process.

2. Individuals in the organization may also be experiencing higher role ambiguity; they may be unclear as to what is expected of them in their roles. For example, role ambiguity is likely to occur when the decentralized organization is implemented without some effective use of lateral relations. Individuals may feel they don't have the needed information for decision making. As in the Heublein example, the divisional managers may not know what corporate's staff's policy is on various issues and corporate staff may have lost touch with the divisions.

Organizational Learning and Redesign

The above examples indicate the kind of information decision makers should be aware of as these symptoms indicate that in fact the structure implemented is not appropriate. If the structure of the organization is not working in the sense that the above symptoms exist, a performance gap is likely to occur somewhere in the organization. When the symptoms of an inappropriate structure fit occur, organizational decision makers are likely to perceive a discrepancy between the organization's specific expectations regarding how the organization should be operating and how it actually is operating. The existence of the above symptoms thus identifies performance gaps in terms of the organization's structure. The performance gaps then initiate a search process by organizational decision makers for the underlying causes of these gaps. The critical factor is that if corrective action is to be taken, it is essential that the dominant coalition is aware of the performance gaps and the causes of the gaps. This is true because the dominant coalition is the major influence in the utilization of the existing knowledge base of the organization in designing the structure of the organization. This search process then focuses on trying to develop knowledge which will be used to identify alternative organizational designs.

This search process then is the initial activity in organizational learning as described in Part III of this paper. The knowledge produced by a particular decision maker with respect to performance gaps will result in changes in organizational knowledge which can be used in organizational design. However, this occurs only if the dominant coalition is aware of this new knowledge and accepts it as valid. Assuming this happens, there is then a permanent change in the knowledge base of the organization.

The search process is a stimulus to seeking new organizational knowledge that can be used in redesigning the organization. If organizational decision makers understand the organization design decision tree presented above, they have a heuristic for deciding what their structure should be. If they get feedback which indicates some of the symptoms of poor adaptation (can't identify problems in a changing environment— don't have enough information, etc.), they can check the decision tree heuristic to see if in fact they have implemented the right structure. If their environment has all of a sudden become very *dynamic* while still remaining relatively *simple* and the organization is using a functional organization, the organization may try using some form of lateral relations to increase the information gathering and processing capabilities for the organization. A task force might be utilized to bring manufacturing, marketing, and R & D managers together to identify what new products the organization might develop to compete in a changing environment. Organizational designers should be aware that any change in structure is a major organizational change that will be accompanied by resistance. Therefore, designers should take measures to reduce resistance by getting those affected by the change involved in the change process (Zaltman and Duncan, 1977).

The organization decision tree heuristic provides organizational decision makers with an understanding of how to match their organization structure to the demands of their environment. By understanding the design decision tree and using it as a guide to making design decisions in terms of dealing with their environment, decision makers will be more aware of the design process in terms of the choices they have in using organization structure and to the feedback cues they get as to how responsive their structure is to the environment. They now have a model to use and if it does not fit the environment, they have some alternative choices to make to see how they work. Thus, they will be able to more explicitly choose their structure and they will have a guide as to how it is working. Thus, as they go through this process, they will be developing action-outcome knowledge about what impact their structure is having; organizational learning will be taking place.

In a very important sense, organizational learning also occurs when there are no performance gaps. This is an indicator of the adequacy and validity of existing organizational knowledge. This reinforces the existing knowledge base of the organization. This constitutes learning in that organizational members, and particularly the dominant coalition, will be more likely to assume that this knowledge is correct and so be less likely to change it.

It is important to note that performance gaps are, for organizational decision makers, only defined in terms of organizational criteria. Inappropriate organizational designs may not result in performance gaps if the organizational criteria used are inadequate. In other words, decision makers may have no basis for identifying that the design is inappropriate. This will result in the reinforcement of existing organizational knowledge, even if the knowledge is invalid. Ultimately this can result in major problems within the organization and possibly a total failure of the organization.

The existence of an organizational knowledge base makes possible what may be called proactive organizational changes. Given an understanding of the nature of the environment, its characteristics and components, and how these affect organizational activities, the members of an organization can monitor or scan the environment (Aguilar, 1967). By identifying changes in the environment, the members of the dominant coalition can undertake to redesign the organization before these changes affect the organization. This implies that performance gaps are being forecast and based on these forecasts changes are made in the design to prevent the performance gap from occurring.

V. CONCLUSION

The primary objective of this paper has been to develop a conceptual framework for understanding the organizational learning process. We have presented a concept of organizational learning which specifies that it is an organizational process concerned with the development of organizational knowledge. This knowledge development occurs within a social context and involves more than the aggregation of individual learning.

Having developed a framework for organizational learning, we have then presented a "middle range theory" of the organizational learning process as it relates to organizational design.

There is no question that the concept of organizational learning is complex and difficult to specify with any great detail. However, organizational learning remains central to our understanding of how organizations and their members behave over time in the strategic design of organizations.

Organizational learning has been implicit in many macro theories— models of the organization. Thompson (1967) has described organizations as operating under norms of rationality. The dominant coalition in the organization strives to behave in a deliberately rational way. Terreberry (1967) has briefly introduced the notion of learning, but has never identified the process.

The objective of this paper has been to identify the components of the learning process. Clearly, the next step now is for organizational theorists to go into organizations and begin to examine the process by which organizational learning occurs.

REFERENCES

1. Aguilar, F. (1967) *Scanning the Business Environment,* New York: Macmillan, Inc.
2. Andrews, K. (1971) *The Concept of Corporate Strategy,* Homewood, Ill.: Dow Jones-Irwin, Inc.
3. Argyris, C. (1976) *Increasing Leadership Effectiveness,* New York: Wiley/Interscience.
4. Argyris, C., and D. Schön (1974) *Theory in Practice,* San Francisco: Jossey-Bass Inc., Publishers.
5. ———. (1978) *Organizational Learning,* Reading, Mass.: Addison-Wesley.
6. Bass, B., and J. Vaughn (1966) *Training in Industry: The Management of Learning,* Belmont, Calif: Wadsworth Publishing Co., Inc.
7. "Behind the profit plunge at Heublein," *Business Weeek,* July 4, 1977.
8. Berger, P., and T. Luckmann, (1967) *The Social Construction of Reality,* London: Allen Lane.
9. Campbell, D. T. (1976) Methods for the Experimenting Society, Evanston, Ill.: Northwestern University (Evaluation Research Program Series, No. 5).
10. Carrol, J. S., and J. W. Payne (1976) *Cognition and Social Behavior,* Hillsdale, N.J.: Lawrence Erlbaum Associates.
11. Chandler, A. (1962) *Strategy and Structure,* Boston: MIT Press.
12. Child, J. (1972) "Organizational structure, environment and performance: The role of strategic choice," *Sociology 16,* 1–22.
13. ———. (1974) "What determines organization performance: The universals versus the it all depends," *Organizational Dynamics,* Summer 30–40.
14. Crozier, M. (1963) *The Bureaucratic Phenomenon,* Chicago: University of Chicago Press.
15. Cyert, R. and March J. (1963) *Behavioral theory of the firm.* New York: Prentice-Hall.
16. Dill, W. (1958) "Environment as an influence on managerial autonomy," *Administrative Science Quarterly 2,* 409–443.
17. Downs, A. (1966) Inside Bureaucracy, Boston: Little, Brown and Company.
18. Duncan, R. (1972) "Characteristics of organizational environments and perceived environmental uncertainty," *Administrative Science Quarterly 17,* 313–327.
19. ———. (1973) "Multiple decision making structures in adapting to environmental uncertainty: The impact on organizational effectiveness," *Human Relations 26,* 273–292.

20. Dyer, W. (1977) *Team Building: Issues and Alternatives*, Reading, Mass.: Addison-Wesley.
21. Emery, F. and E. Trist (1965) "The causal texture of organizational environments," *Human Relations 18*, 21–32.
22. Galbraith, J. (1977) *Organization design*, Reading, Mass.: Addison-Wesley.
23. Havelock, R. (1971) *Planning for innovation*, Ann Arbor, Mich.: The University of Michigan (Center for Research on the Utilization of Scientific Knowledge).
24. Hickson, D. J., C. R. Hinings, C. A. Lee, Schneck, and J. M. Pennings (June, 1971) "A strategic contingencies theory of intraorganizational power," *Administrative Science Quarterly 16*, 216–229.
25. Jurkovitch, R. (1974) "A core typology of organization environments," *Administrative Quarterly 19*, 380–394.
26. Kuhn, T. S. (1970a) *Structure of Scientific Revolutions*, Chicago: University of Chicago Press.
27. ———. (1970b) "Reflections on my critics," in I. Lakatos and A. Musgrave (eds.), *Criticism and the Growth of Knowledge*, London: Cambridge University Press.
28. Lawrence, P., and J. Lorsch (1967) *Organization and environment*. Boston: Harvard University, Graduate School of Business, Division of Research.
29. Levine, S., and P. White (1961) "Exchange as a conceptual framework for the study of interorganizational relationships," *Administrative Science Quarterly 5*, 583–601.
30. Lindbloom, C. E. (1965) *The Intelligence of Democracy*, New York: The Free Press.
31. Luthans, F. and T. Steward (1977) "A general contingency theory of management," *Academy of Management Review 2*, 181–195.
32. March, J. G., and J. P. Olsen (1976) *Ambiguity and Choice in Organizations*, Bergen, Norway: Universitetsforlaget.
33. March, J., and H. Simon (1958) *Organizations*, New York: John Wiley & Sons, Inc.
34. Masterman, M. (1970) "On the nature of a paradigm," in I. Lakatos and A. Musgrave (eds.), *Criticism and the Growth of Knowledge*, London: Cambridge University Press.
35. McNichols, T. (1977) *Executive Policy and Strategic Planning*, New York: McGraw-Hill.
36. Mintzberg, H. (1977) "Policy as a field of management theory," *Academy of Management Review 2*, 88–103.
37. Pettigrew, A. (1973) *The Politics of Organizational Decision Making*, London: Tavistock.
38. Reicken, H., and R. Boruch (1974) *Social Experimentation*, New York: Academic Press, Inc.
39. Rumelt, R. (1974) *Strategy, Structure and Economic Performance*, Boston: Harvard Business School, Division of Research.
40. Sills, D. (1957) *The Volunteers*, New York: The Free Press.
41. Silverman, D. (1971) *The Theory of Organizations*, New York: Basic Books, Inc. Publishers.
42. Staw, B. (1977) "The experimenting organization: Problems and prospects," in B. Staw (ed.), *Psychological Foundations of Organization Behavior*, Pacific Palisades, Calif.: Goodyear Publishing Co. Inc.
43. Steers, R. (1977) *Organizational Effectiveness: A Behavioral View*, Santa Monica, Calif.: Goodyear Publishing Co. Inc.
44. Terreberry, S. (1967) "The evolution of organizational environments," *Administrative Science Quarterly 12*, 590–613.
45. "The opposites: GE grows while Westinghouse shrinks," *Business Week*, January 31, 1977.

46. Thompson, J. (1967) *Organizations in Action*, New York: McGraw-Hill.
47. Weick, K. (1977) "Enactment processes in organizations," in B. Staw and G. Salancik (eds.), *New Directions in Organizational Behavior*, Chicago: St. Clair Press.
48. Zaltman, G. and R. Duncan (1977) *Strategies for planned change*. New York: Wiley Interscience.

ORGANIZATION DESIGN AND ADULT LEARNING[1]

Douglas T. Hall, NORTHWESTERN UNIVERSITY

Cynthia V. Fukami, NORTHWESTERN UNIVERSITY[2]

ABSTRACT

This paper is an analysis of the ways in which organization design and adult learning affect each other. Several factors are identified which are making employee learning or development an increasingly critical area of concern for effective modern organizations. Principles of learning are discussed, and factors which facilitate learning at work are discussed. The greatest gaps in our knowledge about adult learning are shown to be in the "macro" area of organization design. Conversely, it is also argued that a critical gap in the organization design literature is in the "micro" area of how individuals learn to make and effectively implement (or fail to implement) design changes.

It is shown individual learning can be a potent limiting or facilitative factor in

Research in Organizational Behavior, Vol. 1, pp. 125–167.
Copyright © 1979 by JAI Press, Inc.
All rights of reproduction in any form reserved.
ISBN 0-89232-045-1

organization design activities, and that organization design can play the same formative role vis-a-vis individual learning processes. A series of propositions relating learning to design is presented. It is concluded that the critical link between the individual
and the organization is the *organizational role*. Specifically, organization design
changes create new roles for individuals; unless adequate provision is made for individuals to learn to function effectively in these new roles, the design changes will fail.
On the other hand, attempts to stimulate individual learning, without changing organization design factors (tasks, structure, rewards, etc.) which define these new roles,
will also fail.

Organization design literature doesn't tell us much about how to design an
organization. Adult development literature tells us even less about how to
keep people learning throughout their careers. However, the way an organization is designed has great potential for facilitating adult learning.
And in the process of building better human learning opportunities, the
design of the organization can be strengthened. Briefly put, the purpose of
this paper will be to explore the "macro" area of organization design and
the "micro" topic of adult learning and to show how each can be
strengthened with inputs from the other. By confronting the practical
problem of how we can improve organization and indiviudal effectiveness, we can further our theoretical understanding of organization design
and adult learning. In these areas, Kurt Lewin's dictum can be inverted:
"Nothing stimulates new theory like an important practical problem."

THE PROBLEM

What exactly is the problem we are addressing? It is nothing less than the
knotty business of how we can design organizations and develop individuals to maximize the effectiveness of each. This is the classic issue that
has been with us from the "beginning of time" in the field of organizational behavior (i.e., the 1950s): How can we integrate the goals of a
productive organization with the needs of the healthy adult employee?
This issue has been variously addressed as the fusion process by Bakke
(1950), integration by Argyris (1957, 1964), reactions to bureaucracy by
Merton (1957), and the psychological contract by Barnard (1938), Simon
(1957), Levinson et al. (1962), and Schein (1971).

The difficulty with this integration idea is that although its importance
seems clear enough, it has appeared either *(a)* too idealistic to achieve, or
(b) too difficult to measure (e.g., in Simon's terms, what exactly does an
"inducement" or a "contribution" look like?) It is much easier to measure individual properties alone (such as job attitudes and perceptions of
work environment) or to measure properties of organizations alone (e.g.,

centralization, formalization, environmental states) than it is to work across levels and relate organization properties to individual properties. In fact, the field of organization behavior, like any social entity, became differentiated as it grew over time until researchers developed specialized interests either at the individual level of analysis or the organization level. At present, the field is literally split into micro and macro camps. One result is that we have lost sight of the fact that each organization is made up of people with great individual differences. The effectiveness of the organization and the effectiveness of its members are highly interrelated and organization specific. Explaining and exploiting the nature of this interrelationship will be the central purpose of this paper.

Why is the issue of organization design and adult learning relevant today? Why resurrect a 25-year-old issue? Several issues seem to be converging now to give this problem area new urgency:

1. Annual surveys in large corporations are showing that employees' identification with their organization has been steadily decreasing (Smith, Scott, and Hulin, 1977).

2. Perhaps related to the decline in organizational identification, employees show increased concern for their own career development, self-fulfillment, and the quality of their work lives. There is an increasing awareness that responsibility for career development lies with the employee, not the company, resulting in more self-directed careers (Yankelovich, 1974; Renwick and Lawler, 1978).

3. As a result of this increase in employee self-direction, more people are rejecting corporate personnel decisions which affect their careers, such as transfers, promotions, and job assignments. In some cases, this means turning down a promotion or accepting a lower-level job to be in a desirable geographical area such as San Diego, San Francisco, or Minneapolis. The increasing incidence of dual career couples also adds to this career independence, since a member of a working couple may have to turn down a promising geographical move if it would be incompatible with the partner's career. Or one member may have to quit a job in order to relocate in a new area with the partner.

4. As the population bulge caused by the post-World War II "baby boom" moves into middle age, problems of fighting obsolescence and aiding mid-career learning will become increasingly important. People in this age group will also be moving into middle-management positions, raising the spectre of obsolescing people, 30 years from retirement, blocking the advancement of younger employees.

5. The strong likelihood of later, rather than earlier, retirement, and of the elimination of mandatory retirement at a particular age will add further to the age range of employees in responsible organizational positions. The

easy solution of early retirement or delaying action until retirement may simply not be available much longer in dealing with obsolete employees. The only alternative may be continuing learning and development.

6. Organizations continue to face increasingly turbulent environments caused variously by capital problems, difficult union relations, foreign competition, energy, government regulation, and other stresses. This calls for increased adaptability and flexibility, which in turn require increased creativity and flexibility from individual employees. A dynamic organization cannot afford stagnant employees.

Now that we have identified the problem and its importance, let us examine what the literature to date can tell us about the processes of effective organization design and human learning.

WHAT IS EFFECTIVENESS?

The first step in understanding effective organization design and adult learning is to understand what we mean by effectiveness, both for the individual and the organization. We are aided greatly in our examination of organizational effectiveness by a review on the topic by Steers (1975). Steers points out that a sizable body of research has employed univariate models of organizational performance, an assumption that there is an "ultimate criterion" of effectiveness. Campbell's (1973) review identified nineteen variables which have been most often used in single-criterion studies, of which the most widely used were: (1) overall performance, (2) productivity (i.e., objective output data), (3) employee satisfaction, (4) profit or rate of return, and (5) employee withdrawal (turnover and absenteeism). Taking issue with the univariate approach, Steers argues that since organizations are complex systems, with multiple objectives and functions, effectiveness is necessarily a multifaceted (or multivariate) construct.

In a representative sample of seventeen studies which employed multivariate models of effectiveness, Steers found "surprisingly little overlap across the various approaches" (1975, p. 549). To illustrate this conclusion he summarized the frequency of the dimensions used in the seventeen studies as follows:

Dimension	Times Mentioned
Adaptability-flexibility	10
Productivity	6
Satisfaction	5
Profitability	3
Resource acquisition	3

Dimension	Times Mentioned
Absence of strain	2
Control over environment	2
Development	2
Efficiency	2
Employee retention	2
Growth	2
Integration	2
Open communications	2
Survival	2
All other criteria	1

The most commonly used (found in over half of the studies) dimension of organizational effectiveness was adaptability-flexibility, the capacity to respond quickly to changing needs and conditions. (This seems analogous to the individual's need to be adaptive and to continue one's development as a means of avoiding obsolescence.) Productivity and satisfaction were used in about one-third of the studies. The other dimensions were used in only two or three studies.

Steers reviewed eight problems which complicate the study of organization effectiveness (e.g., construct validity, criterion stability, time perspective, generalizability). Two issues he covers which seem especially promising are theoretical relevance and level of analysis. On the first point, he argues that a theoretical approach to effectiveness may be more useful than a strictly empirical or normative approach. He argues for a "model [which] looks at *relationships* between important variables and does so within a system's framework capable of increasing our understanding of organizational dynamics" (1975, p. 554).

Considering the level of analysis studied, he argues that examining the interaction between individual and organization behavior will aid our power to understand both levels of behavior:

> . . . many models of effectiveness deal exclusively on a macro level, discussing organization-wide phenomena as they relate to effectiveness, but ignoring the critical relation between individual behavior and the larger issue of organizational effectiveness. Thus, there is little integration between macro and micro models of performance and effectiveness. If we are to increase our understanding of organizational processes—and, indeed, if we are to make meaningful recommendations to managers about effectiveness—models of organizational effectiveness must be developed which attempt to specify or at least account for the relationships between individual processes and organizational behavior [Steers and Porter, 1974]. So far, little has been done on this critical problem (1975, p. 554–555).

An attempt to bridge the individual and organization levels in defining effectiveness has been offered by Hall (1976). In a review of the research on individuals' career effectiveness, he found that studies have generally defined the construct in terms of four dimensions: (1) *performance* over time (measured variously as rank or salary in relation to length of service or age, average salary increases, supervisory ratings, etc.), (2) *career attitudes* (the way the career is perceived and evaluated by the individual; e.g., career satisfaction, career involvement), (3) *adaptability* (capacity to acquire new skills and knowledge—the opposite of obsolescence), and (4) *identity* (awareness of one's values, interests, abilities, training, and career goals and the sense of "wholeness" these create over time). By far the most often used career effectiveness criterion was performance over time. This is consistent with Steers' (1975) finding that productivity (at the organization level) was one of the most common measures of organizion effectiveness. Career-related attitudes were the second most frequent criterion of individual career effectiveness. Identity and adaptability, that variable most frequently used at the organization level, were used less often for individual career success.

As Steers (1975) has argued, there should be parallels between individual and organization effectiveness, since by definition organization effectiveness is an aggregate of individual performances. In fact, Hall reported a strong similarity between the four most common dimensions of career effectiveness and four theoretically derived dimensions of organization effectiveness as proposed by Parsons (1960). Based on systems theory and an analysis of the functions which an organization must perform if it is to survive over a period of time (the theoretical approach recommended by Steers, 1975), Parsons identified four critical functions: goal attainment, adaptation, integration of individuals into work roles, and pattern maintenance (preservation of cultural patterns, norms, and values over time). The first three functions were also posited by Argyris (1964).

How can these dimensions of organization effectiveness be related to individual effectiveness, as recommended by Steers (1975)? First, goal attainment at the organization level is analogous to performance or task accomplishment over time at the individual level. The adaptation of the organization is clearly related to the adaptability of its employees; both measure the capacity to respond to changing demands, opportunities, and constraints in the environment.

Maintenance of the organization's cultural patterns is analogous to the development and maintenance of the individual's identity over the years. The integration of people into work roles is related to career attitudes in that positive attitudes toward one's work and career role are in fact the "glue" which binds the person to the organization. The more positive the person's career attitudes (e.g., job involvement, career commitment), the less likely she is to leave the organization (Porter, Steers, Mowday, and

Figure 1. Relationship between individual career effectiveness and organization effectiveness (from Hall, 1976, p. 96).

Individual Career Effectiveness	Organization Effectiveness
1. Work performance ⬅————————➡	1. Goal attainment
2. Adaptability ⬅————————➡	2. Adaptation
3. Positive attitudes toward career ⬅————————➡	3. Integration of people into work roles
4. Sense of identity ⬅————————➡	4. Cultural pattern maintenance

Boulian, 1974). The relationships between organizational and career effectiveness is shown in Figure 1.

Since the four Parsons dimensions are conceptually derived from open systems theory, they are more generic than those reported by Steers. (In Lewin's terms, the Parsons dimensions are *genotypic* effectiveness categories, while Steers' are *phenotypic* dimensions.) Thus, if we recode the organization effectiveness criteria reported by Steers in terms of Parsons' more general categories, a different picture emerges:

Effectiveness Dimensions		Number of Times Mentioned
Adaptation:		16
Adaptability-flexibility	10	
Development	2	
Growth	2	
Control over environment	2	
Goal attainment:		14
Productivity	6	
Profitability	3	
Resource acquisition	3	
Efficiency	2	
Integration:		13
Satisfaction	5	
Absence of strain	2	
Employee retention	2	
Open communications	2	
Integration	2	
Pattern maintenance:		1
Sense of identity	1	

One dimension reported by Steers, survival, could not be included. The reason for this is that by definition the four Parsons functions are those which are required for long-term organizational survival. Therefore, survival is implicit in all of the above dimensions. Put another way, survival is *the* ultimate indicator of organizational effectiveness, at least from the systems perspective.

The above results are in agreement with Steers that adaptation is the most frequently used dimension of organization effectiveness, but not by such a wide margin as he found. The Parsons dimensions show a more even distribution of criteria across the adaptation, goal attainment, and integration dimensions. Pattern maintenance was almost never used. However, now that many organizations are coming out of the rapid growth of the 1960s and early 1970s (which perhaps was why adaptation has been so important in the literature) into a period of slow growth and consolidation, pattern maintenance (maintaining the core, identity-defining components of the enterprise) may become more important. As organizations divest themselves of unwise acquisitions and attempt to identify their traditional core markets and mission, they are performing a critical pattern-maintenance function.

These four dimensions of individual/organization effectiveness can also be conceptualized in terms of two dimensions: (1) the time span they cover (long-term or short-term) and (2) the focus of the activity (task or person). When the four functions are examined on these two dimensions, a 2 × 2 matrix emerges, as shown in Figure 2.

In conclusion, what do we mean by "development" and "effectiveness"? For continuing effectiveness, both a long-term and a short-term focus are necessary, as are concerns for both tasks and people. The *effective organization* is one which performs the short-term functions of goal attainment and integration and the long-term functions of adaptability and pattern maintenance. The *effective person* is one who can do the short-term activities of performing well and developing positive attitudes and the longer-term functions of maintaining adaptability and a strong sense of identity. *Development* or *learning* will be defined as improving

Figure 2. Effectiveness dimensions as a function of time span and focus of activity.

Focus of Activity

	Task-Oriented	Person-Oriented
Short-Term	Goal Attainment/ Performance	Integration/ Attitudes
Long-Term	Adaptation/ Adaptability	Pattern Maintenance/ Identity

Time Span

the functioning of the person or the organization on these four dimensions. Furthermore, organizations which operate effectively on all four dimensions will tend to develop individual members who are effective on the four analogous individual dimensions. And members who are individually effective will tend to contribute to effective organization.

Thus, effectiveness from both the organization's and individual's point of view requires a consideration of the dynamics and processes of both levels operating in the same system. Our paper will therefore discuss organization design and how it may relate to the individual, individual development in relation to the organization, and will then develop several propositions and implications for future research combining these areas.

ORGANIZATION DESIGN

It is the purpose of this section to review the existing limitations of present research in organization design, and to suggest directions which will expand the concept of "organization design" to include the individual.

What Is Organization Design?

Organization design entails the manner in which the organization brings about a congruence between the goals of the organization and the achievement of those goals (Galbraith, 1977). The existing literature on organization design, however, reflects a divergence in terms of the general thrust of how organizations achieve this congruence; in short, how design occurs and what variables design embraces.

The Design Process: Static or Dynamic?

In one of the most complete discussions of organizations, Galbraith (1977) summarizes his definition of design as a *decision process* within which a coherence is achieved between ". . . the goals or purpose for which the organization exists, the patterns of division of labor and interunit coordination and the people who will do the work." He sees design as achieving a good fit among five strategic variables: task structure, information and decision processes, reward systems, and people, as shown in Figure 3. However, while Galbraith defines organization design in terms of process, he then continues his discussion by presenting the manager with "alternative designs" from which to choose, on the basis of costs and benefits, a design best suited to specific organizations. Thus, when Galbraith refers to design as a process, he is referring to the process of deciding among alternatives, which seems to be a rather static matching activity. This reflects a major gap in the organization design literature: Is organization design a process or is it static?

Figure 3. Galbraith's concept of organization design (1977, p. 31).

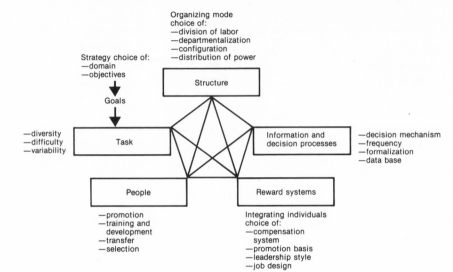

Although recent work (Leavitt, 1976; Nystrom, Hedberg, and Starbuck, 1976; Weick, 1977) has conceptualized organization design as an intuitive, subjective, continuous process, a second group in the literature focuses on genotypic design alternatives (Kilmann, Pondy, and Slevin, 1976) based on large-scale, macro perspectives. While an examination of the genotypic alternatives is useful in investigation of organization level processes, we propose that in order to include the perspective of the individual and his development, we must focus on design as an interacting process (Fox, Pate, and Pondy, 1976).

Argyris (Argyris and Schön, 1974, Argyris, 1977) provides such an analysis in describing how organizations can become more like self-correcting learning systems through the process of double-loop learning. His work on professional education describes how practitioners can learn to behave in a way more consistent with their stated values and provides a promising lead in linking individual learning to organization design.

How, specifically, does the learning of the central power figure affect the design of the organization? Argyris and Schön (1974) describe two models by which executives create organizational environments: Model I and Model II. Model I is governed by personal goals of achievement, emphasis on winning, avoidance of feelings, and rationality. Attempts to implement these goals lead to the following structural characteristics: unilateral management of the environment, task control, self-protective behavior, and unilateral protection of others. These structural characteristics tend to lead to defensive behavior, hinder upward communication, and result in self-sealing theories and single-loop learning. (Single-loop learning involves the detection and correction of error that does not question the basic objectives being sought.) Ironically, then, by avoiding feelings and questing for pure rationality, less valid information is generated and communicated, which *reduces* rationality.

Under Model II learning, underlying norms stress valid information, free choice, and internal commitment to action. These norms lead to quite different structural characteristics: environments in which participants can be self-directed and can experience psychological success, mutual task control, concerns for personal growth, and bilateral protection of others. This structure produces more open, learning-oriented norms and double-loop learning (learning which questions the underlying control system in the organization.)

Argyris' (1957) early work describes graphically how the self-sealing learning methods of senior executives lead them to perceive employees' attempts to grow and maintain self-esteem as deviant behavior. In an attempt to increase the predictability of employees' behavior, higher management increases formal structural controls. These controls then lead employees to find ways to thwart the controls. This leads management to add more controls, which leads to more unanticipated behaviors, etc. Thus, a self-reinforcing cycle of organizational decay is produced.

With double-loop learning methods, in which top executives can question their basic method for learning, they become open to information about how their own methods of control actually affect the behavior of subordinates. By learning about their own behavior in this way, they can learn to design environments with more *mutual influence,* which actually increases the trustworthiness and reliability of subordinates' behavior. By sharing control they gain more control. This cycle of executive learning and organization design takes place as follows:

Top executives' model of learning	→	Structure of behavior	→	Employee response	→	Reinforcement of learning model

In the Galbraith view of design, the focus on specific decisions among alternatives leads the manager to conceptualize design as a static rather than a continuous process. Galbraith seems to assume goals are fixed. As a result, managers may tend to view design as a series of discrete decisions rather than viewing the series of decisions as a set which feed back on each other and in which the basic goals and control structure may be changed (a la Argyris). In this way, design becomes merely a *reconstruction* rather than an *evolution* of the organization. It is proposed here that we need to view design as an ongoing process. We need to know more about how organizations learn to make the design choices described by Galbraith. Our view is that the missing link here is *individual learning*.

Rationality vs. Individual Learning in Organization Design

Another issue which Galbraith's work raises is the question of the rationality of decisions between design alternatives. Although Galbraith and others accept organizational goals as given, in fact the organization may not have conscious goals before the design decision is made (Pfeffer and Salancik, 1977; Pfeffer, 1978; F. Hall, 1975). This assumption of a priori goals and rationality was examined by F. Hall (1975).

Based upon a case survey study[3] of how goals are determined in public schools, she rejects the notion that organizational goals are fixed or pre-determined. *In fact, organizational goals themselves result from individual learning.* Hall's data indicate that goals tend to *emerge* over time as senior administrators interact with the environment and make resource commitments. Through their actions, then, these administrators by inference clarify what the goals for the organization are.

Hall (1975) concludes with several implications for how administrators can better learn to clarify goals. Her recommendations are consistent with the self-confrontation assumptions of Argyris' Model II:

> The first implication is that education administrators and policy makers might improve efforts to establish and implement goals by exploring discrepancies between "stated goals" and policy commitments. Where such discrepancies exist, adjustments should be made . . .
>
> . . . a second implication is that administrators and policy makers might improve goal setting by treating it as a confrontation of values, demands, and expectations held by people and groups who interact with the ystem. . . .
>
> Finally, a third implication is that existing goals may best be clarified through the inferential analysis of past decisions, budgets, and standard policies. That is, rather than asking ad hoc committees to write goal statements, board members and administrators might better ask themselves: "What can we infer from our actions?" If it is through such actions that goals are established, then an analysis of these acts may well prove to be a better method for identifying what the real goals in the system are (1975, p. 71).

Another counterview of organizational goals is provided by Georgiou (1973), who follows Barnard (1938) in arguing that the purpose of the organization, rather than being the formal end toward which the organization is oriented, is in fact the means to personal satisfaction of its members.

In Barnard's analysis, purpose is subordinate to the contributors' demands; to be ignored, discarded or altered if this is what the satisfaction of the contributors' incentives required. Organizational success in this view cannot be judged by the degree to which any organizational goal is achieved, but only by the "absolute test" of an incentive system—its capacity to survive through being able to elicit sufficient contributions from its participants by providing them with sufficient rewards (1973, p. 300).

Thus, Georgiou argues that people and incentives are the key sources of energy in organizations. The organization design literature's disregard for the individual and the way people learn to develop new organization designs represents its critical gap.

As Galbraith notes, information processing is an important facet of organization design. However, he fails to note that information is not always available to decision makers. If information is available, it may not be in its original form, free of distortion. Information enters organizations through individual people, often key opinion leaders or gatekeepers (Allen and Cohen, 1969).

Design Is Atheoretical

Perhaps some of these more general issues of process vs. static and rationality can be traced to a lack of a serious theoretical framework on which design is based. Galbraith's (1977) book describes design alternatives available to organizations in various cases, but he presents little informaiton about how the responsible executives learned which design alternative was appropriate. The process is portrayed as an unrealistically rational matching of alternatives to needs.

In his review of Galbraith's book, Argyris (1978) points out this gap (which exists in all of the design literature, not just in Galbraith's work). He discussed a reorganization at the Boeing Company:

If the Boeing design decisions were made without the benefit of the Galbraith theory—and since the results were consonant with his theory—then what theory did the Boeing designers have in mind that led to these actions? Could it have been a simpler and yet equally useful theory? . . . Maybe the Boeing people have a theory of design that permits less a priori designing than does the Galbraith theory but encourages more on-line iterative learning and hence winds up making the decisions Galbraith would also recommend. If so, then the Boeing theory would also have the advantage of creating internal commitment to the design while it is being created (a quality that Galbraith values). And this, in turn, would increase the integration of the individuals with the organization, which is a key component of the Galbraith theory. . . . Galbraith is aware that his perspective is weak on how to get from here to there. Perhaps, if he had attempted to construct the theory the Boeing designers had in mind that moved them from here to there, he would have gained insights into this important gap in his theory (Argyris, 1978, p. 164).

The Domain of Design

An additional problem of the organization design literature is that it fails to set up boundaries for itself in terms of separation from other processes as well as recognizing overlaps with other bodies of literature on organizations. For example, what is the difference between organization design and organization change and development? From the decision process framework of organization design, design is discussed in terms of costs and benefits of alternative designs which involve change and development of the present organization structure, the creation of the "integrating role" to facilitate lateral relations. However, recent work at the micro level in organization behavior has focused on the development of work roles to achieve the same ends. For example, work in the area of job design and enrichment has centered on developing individual jobs so that there are more links between individuals with each other as well as clients outside the organization (Hackman, 1977).

In a similar vein, what is the difference between organization design and organization structure? Although structural configuration is certainly a part of organization design (Gerwin, 1975), the structure of the organization tends to evolve from other variables which have different origins, such as size, which is not a factor to be "designed."

Thus, while the concept of organization design needs further definition, it also needs "boundary" work, which sets the scope and limits as to what other organization concepts and processes are part of organization design and which are not. This problem is compounded when one follows the analytical, static focus of organization design rather than focusing on process issues.

Integrating the Individual into Organizational Design

The literature pays little attention to the individual in the organization. While Nystrom, Hedberg, and Starbuck (1976) discuss the importance of interacting processes, they fail to include the process of individual behavior.

The behavior of individuals within the organization represents a degree of uncertainty with which the organization must cope. Lorsch and Morse (1975) refer to people as the "internal environment" of the organization. This human uncertainty, when viewed from the bottom up, is usually met with the organization's increased exercise of control. Rather than reducing this uncertainty, or attempting to control it (which may be impossible anyway), it is posited here that organizations should instead attempt to *learn* from the variance of individual behavior and use that uncertainty as feedback or a "signal" with which to monitor the organization's design just as it monitors the external environment.

Likewise, when the uncertainty in the internal environment is viewed from the top down, it is seen as a lack of goal clarity. This, then, creates additional uncertainty at lower organizational levels. Thus, the notion of the internal environment provides a way of filling a gap between the macro and micro perspectives on organizational behavior. By including the learning individual as an active part of the design process, organization design can then free itself from the classical school of formal organization, which treats the variances of individual behavior as "error" variances, and instead investigate the interaction of the individual and organization together as a system.

If we are to develop more understanding of organizational behavior, we must bridge the individual and organizational level. In terms of the macro side of the topic at hand, organization design should be expanded to include the individual. However, the "micro" side of organization behavior is just as guilty for excluding organization level processes from its models of the individual, particularly in the study of individual development. We will continue our investigation by focusing now on the other half of our topic: individual development.

INDIVIDUAL LEARNING

Principles of Learning

Now let us turn to the literature on learning in work settings. The extent to which organization-individual integration occurs depends upon learning as both parties develop the ability to influence and adapt to each other. Most of the writing in this area suggests that learning environments created by the job itself are crucial if learning is to be transferred and utilized. These principles of learning are based on the experimental research in this area and are summarized by Campbell, Dunnette, Lawler, and Weick (1970).

1. *Distributed learning activities seem more effective than massed learning.* In general, learning activities which are concentrated in a short time period (say, 30 hours in 3 days) seem less effective than the same number of total hours spread out over a longer time period (e.g., 30 hours over 5 weeks). There is more consistent evidence for this conclusion regarding motor skills whereas for verbal learning and more complex skills the findings are less clear-cut (Bass and Vaughan, 1966). However,

Bass and Vaughan conclude that "the advantages of spaced practice generally outweigh its disadvantages" (1966, p. 49).

2. *Learning which is rewarded (positively reinforced) will tend to be repeated, remembered, and utilized in other situations.* On the other hand, punishment for poor performance does not aid learning so well as rewards for good performance. The more important danger, though, with most development activities in industry is that they may lead to *no* reinforcement of any kind, positive or negative, which tends to extinguish the new learning. Probably the most potent sources of positive reinforcement for new learning would be the immediate superior and the work group.

3. *Knowledge of results (feedback) enhances learning.* Feedback aids learning in several ways: (1) it clarifies expectations and learning goals, (2) it helps direct and shape effort by correcting mistakes, (3) it helps define and evaluate performance (since in many work situations the exact definition of "good performance" is ambiguous), and (4) it serves as an intrinsic reward for good performance. In general, the sooner the feedback comes after the learner's response, the more impact it has.

4. *The more the learner is ready and motivated to learn, the more learning will occur.* One popular way of increasing motivation is to provide for the *participation and active involvement* of the learner in the learning process. Other factors which add to motivation are being dissatisfied with present performance and/or feeling a desire for improvement (i.e., being "unfrozen"), seeing the learning activity as intrinsically interesting, and seeing the learning as leading to some valued outcome (e.g., greater mastery of the job, a possible promotion).

5. *Learning must be transferable to the job and future job situations to be of practical value.* The person may learn a great deal from a two-hour classroom activity, and may be very excited about it, but if this learning cannot be utilized on the job, its practical value to the organization may be minimal.

6. *For learning to be most effective, the new behaviors must be practiced.* In fact, new behaviors must be *overlearned* through practice so that smooth performance can be maintained under stress and other difficult conditions.

So much for what the theory and research on learning have to say. Let us consider next how learning looks in practice in work settings.

The Technology of Learning at Work
In their classic review of training theory and practice in industry, Bass and Vaughan (1966) described a variety of ways learning is managed. *On-the-job techniques* include job-instruction training, apprentice training, internships and assistantships, job rotation, and coaching. The *off-*

the-job techniques identified by Bass and Vaughan include lectures, university courses, films, television, conference or discussion, case study, role playing, simulation, programmed instruction, laboratory training, and group exercises.

How does learning theory compare with learning practice? How well do the learning methods used in industry apply the principles of learning we reviewed earlier? Bass and Vaughan (1966) conclude that the off-the-job techniques, which are often the most frequently used in industry, do have limitations as learning tools. While off-the-job activities can create motivation and active participation of the learner (especially simulations, cases, and role plays) and can often be controlled and organized more readily than on-the-job activities, the critical deficiency of off-the-job learning concerns *transfer of learning.* Because on-the-job learning builds learning into the job, transfer from the learning activity to later job activities is also designed in. There is no doubt about whether the learning situation is realistic because it is *real.*

James O'Toole, in *Work, Learning, and the American Future* (1977), also sees on-the-job becoming more important in the future. He argues that integrating learning and work has significant advantages over formal schooling:

> . . . one hopes, the changing demands placed on managers by the young workers they must motivate and retain will lead to a shift in employer attitudes in the long run, making human resources development as natural a task for managers as planning, coordination, and control are today.
>
> Most clearly this shift implies a greater role for on-the-job training than for vocational schooling. Training employees at the work place has several fairly obvious advantages: It is a long-range employer investment rather than a temporary measure or public expense; trainees are paid while they learn; there is less emphasis on educational credentials; problems of forecasting do not arise because employers know their needs and can meet them quickly; the skills developed are really put to use by employees; and workers are positively motivated to learn (1977, p. 1970).

The irony here is that even though on-the-job activities seem to have greater potential for long-term learning transfer, the most frequently used learning in industry seems to be off the job. Training specialists love to create fancy training rooms with all kinds of technology: rear-screen projectors, videotape cassette learning packages, two-way mirrors, libraries, etc. Unfortunately, these materials may be meeting the trainers' needs more than the trainees'!

In contrast, learning activities carried out on the job, especially on a permanent, full-time job (such as job-instruction training and coaching) seem to have the best potential for development. The more that learning is integrated with actual work activities, the more effective it is. Let us consider in more detail one important type of on-the-job learning, career development.

Learning and Career Stages

A type of learning which is currently receiving a great deal of attention in work settings is career development. By "career" we refer to the sequence of work-related experiences which occur over the course of the person's work life. By "development" we refer to an increase in any of the four effectiveness dimensions described earlier: performance, adaptability, attitudes, and identity. These increases in effectiveness often result from new knowledge and skills which can be applied to work and career tasks.

A review paper by Super and Hall (1978) provides a useful overview of research on organizational factors which facilitate career development. Our purpose here will be to summarize briefly what research has been done and to identify gaps that exist in our understanding of career development. The review was organized in terms of successive career stages—exploration, organizational entry, establishment and advancement, middle career, and late career.

Stage 1. Exploration

Perhaps the most critical part of a person's work career is the first stage, the transition from school to work. Unemployment among teenagers tends to be three times that of the national average, and the rate of unemployment among minority youth is five times that of the national average. The needs of the young worker at this stage include the development of skills as well as feelings of support and autonomy. What factors in the school and work environment can help make this a smoother transition?

1. *Career counseling* in schools can be of some value (Pincus, Radding, and Lawrence, 1974), but counseling centers often use practicum students as counselors, employ few innovative practices, and conduct little evaluation of their activities (Graff, Rague, and Danish, 1974). Counseling in the work organization, dealing with actual job problems, may be of greater value. Unfortunately, little research has been done in this area.

2. *Internships* can be a way of providing experiential learning about one's chosen occupation (McGovern and Tinsley, 1976).

3. *Self-directed career planning* can be useful ways of increasing decision-making, insightfulness, interpersonal growth (Robinault and Weisinger, 1973), and vocational maturity (Feldman and Marinelli, 1975).

4. *Training programs for the hard-to-employ* are effective to the extent that:

   a) the *job itself* is rewarding (O'Leary, 1972).

   b) Cross-cultural communication skills are developed (Triandis, Feldman, and Weldon, 1974).

   c) Supervisors' behaviors reflect high consideration and low structure (Beatty, 1974).

   d) The training program focuses on developing *job skills* rather than attitudes (Salipante and Goodman, 1976).

   e) *Counseling* is used in long programs (Salipante and Goodman, 1976).

   f) *Organizational rewards* (early pay raise, promotions, high job status, and counseling) are given. These have more impact on retention than program characteristics such as the content of training.

Stage 2. Organization Entry

Entry refers to the process of joining an organization and becoming involved in it. Individual needs at this stage involve developing a specialty as well as finding support within the organization. As with the school-to-work transition, retention is an important issue in this career stage. *Realistic job previews* have been the most frequently studied career facilitating factor here (e.g., Wanous, 1973). Such previews tend to result in the following: more moderate expectations, fewer thoughts of leaving, no reduction in recruiting success, and reduced turnover.

Stage 3. Establishment and Advancement

This is the period of most rapid growth during the person's career. Needs at this particular stage include the development of skills in training and coaching others, and updating skills. The following factors have been found to facilitate career development in this stage:

1. Career development courses (Sellman, Miller, Bass, and Mihal, 1975).

2. Behavior modeling training for supervisors (Burnaska, 1976).

3. Challenging initial job assignments (Berlew and Hall, 1966; Bray, Campbell, and Grant, 1974).

4. Supervisors who provide support with autonomy (Hall and Schneider, 1973) and who set high expectations for new employees (Livingston, 1969).

(Factors 3 and 4 above are also important in the entry stage.)

Stage 4. Middle and Late Career

This stage involves the last part of an individual's career, where the shift occurs from a power role to one of guidance and consultation and where there is a gradual detachment from the organization. Less training and development activity is directed at older workers than at younger employees, and research in these career stages is therefore sparse. Certain factors, however, have been found to facilitate continued growth in middle and late career:

1. Continued career exploration (Morrison, 1977).
2. Effective personnel policies (e.g., selection for long-range needs, good assessment and training, and career counseling).
3. Challenging initial jobs (i.e., early job challenge has a long-term impact on career effectiveness).
4. Periodic job changes.
5. Good climates for communication.
6. Rewards for performance.
7. Participative leadership.
8. Matrix structures (2–8 are all from Kaufman, 1974).
9. Lateral and downward job moves (Thompson and Dalton, 1976).

Now that we have examined the individual development literature, let us turn to the gaps it contains.

Gaps in the Individual Learning Literature

There seem to be five general factors which affect the learning of individuals in organizations: the entry process and early socialization, the job itself, the boss and work group, organization structure, and organization policies (especially personnel policies). The status of applied research on each of these career-affecting factors is summarized in Figure 4. Most of the research to date has focused on entry and socialization, the job itself, and the boss/work group.

Figure 4.   Summary of the extent of applied research on various influences on career development.

Career influence	Level of analysis	Amount of applied research
1. Entry and socialization (including exploration)	Individual	Considerable
2. Job characteristics	Individual	Considerable
3. The immediate superior and work group	Leader–Member	Moderate
4. Organization structure	Organization	Little
5. Organizational personnel policies	Organization	Little

The real gap in the individual development literature comes in two "macro" areas, organization structure and organizational personnel policies. *We know very little about how a particular structure, such as a matrix, product, or functional structure, affects individual development.* Also, macro issues of organizational structure, policies, and rewards largely determine the extent to which the two most direct influences on the employee's learning—the immediate supervision and the job itself—can be utilized. Many companies have found that it is fruitless to invest resources in redesigning jobs or in training supervisors to facilitate employee development if supervisors are not rewarded and supported for these activities or if subsequent jobs in the employee's career path are not also challenging. The employee may get "turned on" to his career in his present job and through his present boss, but it is the future career path, the possible future job sequence, which will provide long-term career development (or lack of same). And what controls the sequence of future jobs? The *structure* of the organization and the organization's *policies* about how people should move through these positions.

It would seem then that macro variables, such as structure and policies, would have a potent effect on employee development, but we have very little data or theory available at this time. (This fact was brought home very strongly to one of the authors, who was asked to indicate what effect a guaranteed lifetime employment policy would have on employees' career development and productivity. After employing three computer literature search systems, covering 2080 studies, we found that none addressed this important policy issue.)

Therefore, we would conclude that macro variables are the crucial missing element in research on individual development in organizations, much as we had previously concluded that micro variables were the missing element in the organization design literature. It follows, then, that bridging of the micro and macro literature would greatly enhance our understanding of both individual learning and organization design.

How might these two areas be better linked? The remainder of this paper will present a model and hypotheses dealing with the ways individual learning and organization design facilitate each other.

HOW ORGANIZATION DESIGN AFFECTS ADULT LEARNING

Now that we have identified some of the gaps in the organization design literature and in the adult learning literature, let us consider how those gaps might be filled. Our basic argument is that each area (organization design and adult learning) contains the key to covering the gaps in the other. We will present this argument as a series of propositions about development in organizations.

Propositions about Design and Learning

Individual behavior   Let us start with the individual and consider the basic process of learning and development. We make certain assumptions about the nature of the person.

First, people are motivated to seek rewards from work. These needs include both extrinsic rewards (such as pay and promotion) and intrinsic rewards (Porter and Lawler, 1968). Important intrinsic rewards which relate to development are psychological success (feelings of achievement and growth), increased competence, and increased self-esteem (Lewin, 1936; Argyris, 1957; White, 1959).

Proposition 1.   Psychological success, feelings of competence, and increased self-esteem result from the attainment of a goal which satisfies the following criteria:

1. The person participated in setting the goal.
2. The person had autonomy to work toward the goal in his or her own way.
3. The goal was relevant to the person's self-identity.
4. The goal was perceived as relevant to the career role (Hall, 1971).

Proposition 2.   The basic cycle of adult development at work is as follows: goals → effort → performance → rewards → psychological success → self-esteem → involvement → future goals (Hall and Hall, 1975; Hall and Foster, 1977; Goodale, Hall, and Rabinowitz, 1977).

These two propositions represent a very "micro" view of development; i.e., they describe what goes on inside a person as he or she responds to work goals and experiences, varying degrees of success or failure. In the process of working toward new goals, the person acquires new knowledge, new attitudes, and new skills, which represent respectively the cognitive, affective, and behavioral elements of learning. How does this success cycle affect the four dimensions of effectiveness described earlier? Achieving new goals leads to increased performance (goal attainment), changed attitudes (psychological success), and identity changes (increased self-esteem), which are three of our four dimensions of individual effectiveness.

However, this psychological success cycle does not necessarily lead to the fourth dimension of effectiveness, increased adaptability, unless the goal represents a different *area* of performance for the person. If the goal reflects simply a new *level* of performance, little adaptability would result. In fact, one way people become overspecialized and nonadaptive is by focusing on increasingly higher goals in one particular area of endeavor. The more competent and successful they get in that area, the harder it becomes (both psychologically for the person and economically for the organization) to move the person into a new area of work. Thus, this cycle may facilitate three dimensions of development, but over time it may retard development in the fourth area, adaptability. Let us now consider how adaptability can be increased.

Proposition 3.   Adaptability is maintained by moving periodically to new areas of work and new types of goals (Kaufman, 1974; Dalton and Thompson, 1976; Pelz and Andrews, 1976).

This movement generally entails transfers into different job assignments, new projects, or a different functional area. A shift approximately every 5 years can be helpful to keep the person fresh and adaptive (Pelz and Andrews, 1976).

Movement into new jobs is also necessary to keep increasing the challenge and difficulty of one's work goals. Within a single job assignment, the person will eventually reach the limit to the challenges which are possible there. At that point, a new job will be necessary. A *job history* containing an ongoing sequence of new challenges and requiring new skills is necessary for long-term development (Morgan, 1977).

Proposition 3 reflects the fact that employee development is a long-term process which takes place over a sequence of job assignments. Important development can occur within a particular job, but it will eventually stop if new assignments requiring continuing development are not provided.

Leader-member behavior   If the factors which most directly affect individual development are the job and the work goals it provides, we must consider next: What influences the job? Who is responsible for the way a job is defined and for the assignments the person is given? Generally, the person most directly responsible for job assignments is the immediate superior. The superior is responsible for communicating work goals and objectives to the employee. It is the boss who either sets high expectations for the employee or doesn't (Livingston, 1969, Graen et al., 1977). The boss is also in a position to provide coaching and assistance to help the person attain his or her work goals. The feedback the superior provides is important in helping the person assess performance. Furthermore, feedback from the boss is an important reward in itself (recognition and praise).

Proposition 4. Supervisory behavior influences career development through its effects on the psychological success process. Particularly important are the autonomy, feedback, high expectations, and support which the supervisor provides. Career counseling and nominations (or sponsorship) for future positions are also critical developmental tasks which may be performed by a superior.

Although the superior can have a strong impact on the person's career, he or she does not always engage in the activities included in Proposition 4. Usually there are two reasons why managers do not spend much time on employee development: (1) Managers are not rewarded for developing employees (which also means they have no time for developing employees, since they spend all their time on activities which *are* rewarded), and (2) managers lack the skills necessary for employee development (goal setting, coaching, counseling, planning, and feedback). Thus, we are led to proposition 5:

Proposition 5. Managers will engage in employee development activities to the extent that they are *(a)* rewarded and *(b)* trained for these activities.

Organization behavior If managers will engage in development activities to the extent to which they are trained and rewarded for doing these tasks, let us now look to the factors which influence the manager's behavior: the structure and operating policies of the organization. The structure of the organization also reflects the sequences of possible job moves which are available to the person as per Proposition 3. The personnel policies of the organization, furthermore, will determine which of the possible job sequences will in fact be used to develop the person (e.g., is there a policy favoring interdepartmental transfers? development within one functional specialty? moves only with promotions?).

Proposition 6. The structure of the organization will determine what job moves and career paths are available for employee development. Structure is the limiting organizational characteristic affecting individual development. Therefore, changing the structure of the organization (part of the design process) will influence employee learning.

To illustrate this proposition further, an organization with a functional or departmental structure will generally have single-function career paths. This means that people will tend to stay in one department and develop into functional specialists, such as marketing, finance, production, etc. In a product-oriented organization, people will become specialized in their own product area, but will be also exposed to various functional expertise that relates to their product. Geographically organized companies develop people who identify with their region. And so forth. Whatever the basis of the organization's structure, people will tend to identify with and learn the unit with which they have the most sustained contact.

Furthermore, the shape of the organization also affects development. A tall organization provides more job levels, more promotion opportunities, and therefore more expectations for upward mobility. A flat organization, on the other hand, offers fewer opportunities for advancement and thus yields lower expectations regarding promotion. In this way, organizations with objectively faster promotions may actually contain people who are *less satisfied* with promotion prospects than an organization with slower promotions, through relative deprivation (Stouffer et al., 1949).

The flat, decentralized organization, if it is to be effective, also requires greater autonomy for the individual manager, because higher-level managers simply cannot supervise closely the large number of subordinates who report to them. However, in order to make decentralized decisions which are at once (1) autonomous, and (2) in the best interests of the organization, it is necessary to develop local managers who are highly aware of and identified with the organization's goals (Kaufman, 1964). Thus, the local manager in an effective organization would be likely to develop simultaneously greater self-sufficiency and organizational identification than his or her counterpart in a tall organization.

The matrix organization, on the other hand, tends to develop people who are sensitive to the goals and problems of other departments in the organization, as well as their own. The reason is that in the matrix structure, people work on projects, task forces, or programs that are staffed with people from several different departments working toward a common goal. Common goals tend to reduce intergroup conflict and increase cooperative behavior (Sherif et al., 1961). In the process of working together with people from other functions, the person learns about these other areas, preventing excessive specialization and facilitating adaptability.

Furthermore, working in a matrix structure exposes an employee to managers in other departments. Assuming these other managers are satisfied with the person's performance, this would make them more willing to accept that person into their department at some future data if an interdepartmental transfer were being considered. One of the great barriers to cross-functional moves in functionally oriented organizations is the manager's lack of knowledge about the person who might be moved into his department and the resulting suspicion that he might come out on the short end of a "turkey trade."

Another way the structure of the organization can support employee development is through the creation of new organizational units with responsibility for development, such as employee development departments. This could also take the form of adding responsibility for employee development to existing departments, such as personnel, training, or organizational development. Providing a budget and financial resources to support these activities would be essential to the success of these new functions.

The importance of macro influences has been stressed recently by Uri Bronfenbrenner (1977), a developmental psychologist. He points out that not only does social structure influence development strongly, but that changes in the social environment produce profound effects in their interaction with the developmental stages and role transitions of the person. Two of his major hypotheses follow:

> A fruitful framework for developmental research is provided by the *ecological transitions* that periodically occur in a person's life. These transitions include changes in role and setting as a function of the person's maturation or of events in the life cycle of others responsible for his or her care and development. Such shifts are to be conceived and analyzed as changes in ecological systems rather than solely within individuals. . . .
>
> Research on the ecology of human development requires investigations that go beyond the immediate setting containing the person to examine the layer contexts, both formal and informal, that affect events within the immediate setting (Bronfenbrenner, 1977, pp. 526–527).

As was said before, *changing structures provide new potential career paths and learning opportunities* in an organization. The *personnel policies* set forth by top management *determine how the structure will be utilized to develop people.* Unfortunately, many organizations do not even have formal policies relating to people, even though clear personnel policies, implemented, supported, and controlled by top management, are the most important part of improving the development of human resource (Foulkes and Morgan, 1977).

Proposition 7.   Personnel policies, particularly policies related to the movement of people, have a direct impact on the individual's job history and therefore upon his or her career development. Therefore, altering organizational policies (another part of the design process) can strongly facilitate learning.

Let us consider briefly what kinds of policies can affect development.

Policies regarding skill progression   Since each job has such a strong impact upon a person's development of skills, knowledge, and attitudes, a rationally planned sequence of job moves can be a good way of managing continuing development in these areas. For example, in a company with a job-evaluation system (such as the Hay system), it would be possible to plot career paths that would systematically move people through job changes that would provide some optimal level of stretching, such as 15 percent increases in total job points (i.e., overall job challenge). Without the job evaluation system, it would be more a matter of intuition to select job sequences with properly graded increments. A description of how planned skill progression is being used in one company is found in Wellbank, Hall, Morgan, and Hamner (1978).

Policies regarding broadening and cross-functional movement Many companies are finding that their middle and top executives have become too narrow by "growing up" in only one department. Then when it is necessary to move them into general management positions, they lack sufficient understanding of the other departments to operate effectively as general managers. For this reason, it may be desirable to have a policy of moving people between departments throughout their careers. In addition to providing a broad range of skills, this movement also increases the person's identification with the organization, reducing narrow departmental identification (Kaufman, 1964).

Policies regarding downward movement The clear norm in most organizations has been that the only good move is an upward move. Lateral moves are suspect, and downward moves spell failure. However, with slower growth and more senior employees in organizations (if mandatory retirement laws change), promotions will become rarer. Alternative directions of employee movement will have to be found. Geographical movement is becoming less popular with younger employees as dual careers and concerns for quality of life increase; this reduces promotion opportunities even more. One option that remains is more lateral and downward movement within a particular geographical location.

The more controversial of these two directions of movement is obviously downward movement. Relaxing constraints against demotions would increase the flexibility of organizations drastically as each demotion would give many lower-level people an opportunity to move up. Downward movement does seem feasible under certain conditions:

1. If the person realized that he or she were being moved down to have a better chance for moving up in another area or department (i.e., if the move is temporary). This sort of temporary move is a useful way of extricating a person who has reached a high-level "dead end" in a career path.

2. If the person were given the option of either moving away from a particularly attractive geographical area (such as Denver, San Francisco, Atlanta, San Diego, etc.) or staying, but moving down. For many people,

especially those in later career stages, quality of life is more important than job level. In fact, for an older person with fewer financial pressures, quality of life may be further enhanced by being put in a lower-level, lower-pressure job. However, few people would ever request such a move; the company would have to initiate it.

3. If the person's present position were clearly agreed to be a temporary one, after which he or she would move back down to a lower-level job. Organizations do not give tenure, but in effect they really do: When a person is promoted, the implicit assumption is that the person will not be moved back down again. One way to avoid this "ratchet effect" is to have *fixed-term promotions*. For example, a person could be made a district manager for 5 years, after which he would go back to his previous job. Universities and schools use these rotational assignments for department chairmen. Bell Laboratories finds that the only way a person will accept a management job in some areas is if it is a temporary assignment (e.g., for 3 years). Dalton and Thompson (1977) describe another R & D organization which also employs rotational promotions. This policy could be a useful way to combat the "Peter Principle" in other types of organizations as well.

4. If position descriptions and organizational levels were sufficiently ambiguous so that some downward moves appeared to be lateral moves. In fact, such ambiguous transfers occur all the time, and a favorite indoor sport is figuring out just which moves are real promotions and which are demotions. There are a variety of ways a person can in effect be moved down and helped to save face: *(a)* changing the job title so it appears upgraded, *(b)* presenting the new assignment as a critical "trouble-shooting" assignment for this experienced person, *(c)* claiming that even though it's a lower level job, the area or division is larger or more important, and *(d)* creating a new job with less important responsibilities.

Policies regarding hiring, development, and promotion practices Organizations with a policy of promotion-from-within grow different types of people than do organizations which hire from the outside for middle and top management positions. People in promotion-from-within organizations are more likely to remain in that organization and to identify more strongly with the organization over time (Hall and Schneider, 1972). On the other hand, a risk is that they may develop "blinders" from spending too much time in the same environment, and the organization may experience low innovation caused by too few new external inputs.

Promotion-from-within organizations, such as the Bell system, Sears Roebuck, and the Roman Catholic Church, also tend to hire younger people, with less formal education, preferring to provide their own in-house development. For example, Sears and AT&T do not hire MBAs, preferring to grow their own management talent. Pursuing the same line of reasoning, then, one would hypothesize that the commitment to employee development (in terms of financial resources invested in development) would be higher in promotion-from-within organizations than in organizations which have more entry at middle and top levels.

Policies regarding permanent employment   A growing number of organizations are showing interest in the idea of a guaranteed employment policy (Drucker, 1977). Under such a policy, employees who have been with a company for a certain number of years (usually around 2) become "permanent employees," who are protected from layoffs in the event of economic downturns. (The employee can still be discharged for disciplinary or performance problems.) A guaranteed employment policy is a way of providing increased security and increasing the employee's commitment to the organization.

Perhaps more important, however, are the side-effects of a security program. With guaranteed employment, employees are more likely to accept downward moves (since they know the move down is not the first step toward being discharged). Furthermore, if employees are being considered for permanent employment, managers are forced to do a better job of selecting, developing, and evaluating employees. Activities such as performance appraisal and feedback, which are often neglected (Hall and Lawler, 1969), *must* be performed if the person is being considered for permanent employment. Once the employee has employment security, this places more of a burden on the supervisor to provide motivation through effective leadership, stimulating job assignments and good communications.

Policies regarding family involvement in career decisions   Many contemporary problems in the human resource area entail the family: dual-career families, negative attitudes toward transfers, quality-of-life concerns, mid-career changes, resistance to retirement, etc. Although employers are careful about imposing on the employee's privacy, they are already doing so with many personnel decisions, such as transfers and retirement. When an unpleasant change is being contemplated, some organizations will invite the spouse to participate in a discussion of the move and why the organization thinks it is necessary.

Such family meetings could also take the form of counseling and orientation sessions if the decision has already been made. One major oil company has learned through experience to involve the spouse and let the employee and spouse make the final decision regarding a transfer to Saudi

Arabia. This "realistic job preview" about the problems of living in the Middle East is important in easing the family's adjustment and decreasing turnover in these foreign assignments.

Examples of Organization Design Factors Limiting Individual Development

Career planning problems   Although job characteristics and individual goals and plans may have strong effects on the person's development, it may not be possible to implement one's goals or to move on to future challenging jobs if the structure of the organization and its policies are not supportive. For this reason, the authors question the value (indeed, the ethics) of the current wave of career planning programs in industry. Most are aimed at the individual employee and are carried out by the employee alone. He or she is encouraged to conduct a self-assessment (through workbooks, tape cassettes, or small workshop groups), after which career goals are identified and a career plan is developed. Little information is provided about career paths and opportunities available, either inside or outside the organization. (In the better programs, realistic information about career opportunities is provided, and the immediate supervisor is involved as a resource to help the person implement the plan.) The implicit assumption seems to be that once the person identifies a career plan, he or she will be able to implement it. Related to this is the assumption that the organization will provide the person with some freedom to work toward his or her own goals.

The authors know of several organizations which encourage employees to do career planning, but which provide neither the freedom nor the job opportunities to let the majority of employees implement those plans and goals. Either educational requirements, tough competition, or actual discrimination bar many desirable career paths, or company policies prevent the employee from exercising choice over job assignments and training experiences needed to achieve one's goals. In such a situation, the employee runs a real risk of being "all planned up with no place to go." Furthermore, the structure of the organization and the industry in which it operates influence the extent to which performance or ascriptive characteristics (socioeconomic origins) affect advancement; this places a very real limit on the amount of control a person would have over his own advancement (Pfeffer, 1978). A case of corporate policies limiting the impact of development activities occurred in a large energy company. This company has a policy of not giving employees choice of assignments. In fact, the policy states that turning down a transfer is grounds for dismissal. Personnel officials were interested in developing career awareness and planning programs for employees. However, the problem is that if the career planning programs are effective, the employees will be doomed to frustration since company policy prohibits them from exercising choice in their future job assignments (i.e., their future career path.)

How structure limits development An example of the limitations provided by organization structure occurred in a manufacturing company which started a policy of enriching entry-level jobs as a means of reducing turnover among new hires. This change succeeded in reducing turnover in the first year. Then in the second year, the employees were promoted to second-level jobs (which had not been enriched). The new employees experienced great dissatisfaction in these second jobs, and turnover went up. So the second-level jobs were changed. Then it was found that the managers of these jobs had to be trained to supervise enriched jobs more effectively. Then third-level jobs had to be redesigned. And eventually the design of the entire organization was changed because of the chain reaction resulting from the initial change aimed at facilitating the learning of first-year employees.

Another example is found in major accounting firms, in which the turnover of experienced managers is a serious problem. (Managers are people who have been in the firm for 5 to 10 years and are just below the level of partner). Accountants move up through the ranks very quickly and have a steep learning curve in the first 3 or 4 years: from junior staff accountant to senior to manager. Then they hit a 5- or 6-year plateau at the manager level, and they feel their learning stops. In this case, then, the pace of learning is limited by the structure of positions in the firm.

Therefore, macro organizational design characteristics (structure and policies) seem to be critical to the effective implementation of any employee development activities. If these conditions at the top are not favorable, no amount of effort directed at individuals will have lasting impact (Argyris, 1977). As Goodman and Salipante (1976) point out regarding the development of the hard-to-employ individual, there is a dilemma here. Organizational factors have the most impact on development, but they are harder to implement; changing the effects may be short-lived if the organizational environment does not support these changes.

How Individual Learning Affects Organization Design

Now let us reverse the causal arrow. How does individual learning affect organization design? On first blush, this may have overtones of tails wagging dogs, but be patient.

We started with the individual in the previous section and then worked our way to the top of the organization (structure and policy). Now let's stay at the top and work our way back down.

Proposition 8. Individual (micro) behavior at the top of an organization is organization (macro) behavior.

Because people in the highest ranks in an organization represent the organization (in their institutional, external activities), and since they are in a position to determine the structure and policies of the organization, their behavior, individually and collectively, *is* the behavior of the organization. Thus, when we study the learning and development of people at the top of an organization (Argyris, 1977; Duncan and Weiss, 1978, this volume), we are studying the development and design of the organization.

When Pfeffer and Leblebici (1973) examined executive recruitment and mobility of executives between firms, they argued that "the movement of executives between organizations is one form of interorganizational structures of behavior" (1973, p. 449). Therefore, not only does a process (job movement) involved in individual development affect organizational structure. but so also does *inter*organizational structure (a *very* macro design variable). Baty, Evan, and Rothermel (1971) made the same argument.

Further support for Proposition 8 is found in transfer behavior among managers in multinational companies. Galbraith and Edstrom (1977) argue that transfers result in the learning of new managerial behavior (organizational identification and the acquisition of new organizational information) as transferees are exposed to new parts of the organization:

> . . . transfer affects the transmission of verbal information in an organization and may be an approach to the design of the verbal information system. Verbal information processing has been shown to be critical in organizations performing uncertain tasks and to consume large amounts of managerial time (1977, p. 307).

Galbraith and Edstrom (1977) model this process as follows:

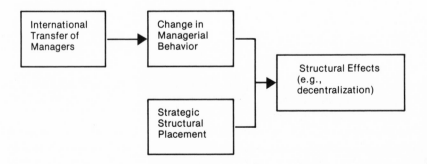

They explain the process in this way:

> Briefly, we believe that transfer changes managerial behavior, and that, collectively, changes in behavior change organization structure. More specifically, we hypothesize that transfer influences verbal contact with colleagues in other units and therefore amounts to designing the organization's verbal information system. The result is believed to be greater local control in the presence of interdependence (1977, p. 289).

The authors also show how this lateral movement results in high levels of identification with the organization; as the person takes various geographical and functional moves, the one constant factor in his life is the organization, and much of his social life centers around the organization, resulting in strong feelings of belonging.

Now let us consider how another "micro" variable, group behavior, affects the design process:

Proposition 9. Group behavior (micro) at the top of an organization is organization (macro) behavior.

Influential members at the top of the organization, acting collectively, can affect the direction of the organization's actions and commit its resources accordingly. Thompson (1967) described how a critical aspect of the design of an organization, its goals, were determined by members of the "dominant coalition." He describes the process of goal determination as follows:

> The view of organizational goals presented here overcomes both of the traditional problems with goals; we have not reified the organization, nor have we simply added the preferences of all members. In this view, organizational goals are established by individuals—but interdependent individuals who collectively have sufficient control of organizational resources to commit them in certain directions and to withhold them from others (1967, p. 128).

In his chapter on "The control of complex organizations," Thompson lists many propositions about the behavior and characteristics of the dominant coalition in controlling the organization. A few examples follow:

- Potential for conflict within the dominant coalition increases with interdependence of the members (and the areas they represent or control) (p. 138).
- When power is widely distributed, an *inner circle* emerges to conduct coalition business (p. 140).
- In the organization with dispersed power, the central power figure is the individual who can manage the coalition (p. 142).

Thus we have the following proposition:

Proposition 10.   The process by which organization design decisions are made is influenced by the learning of the central power figure and/or the dominant coalition.

Examples of Individual Learning Limiting Organization Design Effectiveness
How does individual learning affect organization design in practice? Again, let us use the accounting firm as an example. One firm is currently considering adding a new level in the organization chart between manager and partner, either a "senior manager" or a "junior partner." The reason this new design is being considered is to provide for additional learning and career growth between the person's fifth and tenth year with the firm. To the authors' knowledge, this concern for individual learning (and the reduced turnover it might cause) is the *sole reason* for the possible change.

A case in which individual learning affected the *physical design* of a company as well as the structural design was reported in the press several years ago. A company had built a new headquarters designed in the shape of a staircase, like so (side view):

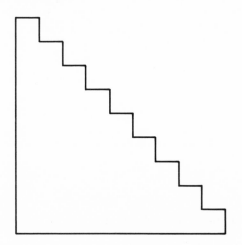

The explanation for this design was as follows:

At Allgemeine Rechtsschutz AG, a West German insurance firm, employees will have unmistakable evidence of their standing on the corporate ladder when a new $2.8 million headquarters opens in Düsseldorf. Each story will be occupied by a progressively higher echelon of workers, from 360 typists and clerks on the ground floor to

president Heinz G. Kramberg alone at the top of the twelfth floor. Kramberg says he
ordered the staircase design to "encourage ambition and provide a visual image of our
organizational structure (*Newsweek,* October 24, 1966, p. 99).

Another example of individual learning limiting system design is found
in the job enrichment and innovation areas. Companies involved in organization development, such as TRW (Davis, 1967), Corning Glass
Works (Beer and Ruh, 1976), and Lawerence Livermore Labs (Brewer,
1977) found that organizational change could only go so far, and then
individual learning, training, and development had to be provided to help
people operate effectively in the new system. Furthermore, major job
redesign efforts may stall because of the career concerns of innovative
system members; they may fear that managers in higher (and more traditional) levels of the organization may reject them for promotion, and this
anxiety diverts energy from the innovative activity. This appeared to be
one of the problems with internal resistance to change in General Foods at
their Topeka dog food plant. The failure of managers in more traditional
parts of the organization to learn how to operate effectively in the innovative climate is another limiting factor in the diffusion process (Ketchum,
1975).

In a similar vein, employees in restructured jobs find it takes time to
learn to work effectively with the high status staff people (e.g., engineers)
and managers with whom they now interact. Dowling (1973, p. 54)
quotes a staff psychologist on this issue: "This . . . phase has had its
problems. Typically, it's taken about six months for the groups to settle
down—adjust to the increased pressures and responsibilities."

Walton (1975) addresses the issue of why organizational redesign experiments which eventually failed because they could not be "diffused," or
integrated with the rest of the organization and its ongoing management
processes. (In terms of Lewin's model of the change process—unfreezing,
changing, and refreezing—we are seeing that many redesign efforts are
successful in unfreezing and changing some subsystem of the organization. They fail at the stage of refreezing or diffusing the change.) Examples of successful work redesign change followed by unsuccessful diffusion occurred at Shell U.K., General Foods' Topeka plant, Corning's
Medfield plant, Northern Electric, Hunsfos, Norsk Hydro, and alcan.
Walton argues that diffusion failed because the redesign effort failed to
carry out the following necessary steps:

Step 1.   Initiation of the pilot experiment.
Step 2.   Pilot experiment declared early success.
Step 3.   Recognition and resources provided for further restructuring.
          (Example: Volvo's Kalmar plant was provided a 10% higher

capital investment budget to accommodate the desired work structure.)

Step 4. More general interest in work restructuring aroused. (E.g., dissemination communications, news media recognition, site visits by outsiders.)

Step 5. Change agents' interventions extend throughout the corporate system.

Step 6. Facilitative networks develop.

Step 7. Personnel movement occurs.

From Walton's account, it appears that Step 6 (facilitative networks) and Step 7 (personnel movement) were the major stumbling blocks in these redesign efforts. And these are precisely the steps which require new learning by significant others in the organization. Note the individual learning involved in Step 6:

> [Step 6] is a step taken in only a few of the change programs studied. An interunit network of personnel involved in work restructuring is created to exchange ideas, to provide a supportive reference group for its members, and to build a constituency for change in corporate policies and procedures more favorable to work restructuring. In GF, networks of plant managers and their personnel managers are evolving to a point where many of their members can serve others generally as outside consultants (1973, p. 10).

And regarding personnel movement:

> The transfer of experienced personnel from an innovative unit is a way of exporting the knowledge, values, and skills at the heart of work restructuring. The innovative unit then can educate the new managers who transfer into the unit (1973, p. 10).

What all this amounts to, then, is that *organization redesign creates new work structures, which in turn create new roles for people.* And people must learn to perform effectively in those new roles. If this learning of new roles does not occur, the change fails to spread and therefore fails to survive. Examples of new roles are the boundary-spanning roles discussed above by Walton (managers of innovative units dealing with managers of traditional units), various integrator roles described by Lawrence and Lorsch, matrix member roles (lateral relations, rotation of members, interunit communication), etc.

In many of the redesign studies in the literature the organization reduces organizational uncertainty and conflict by the creation of these various types of integrating and linking roles. However, how does it perform this change? Have the conflict and uncertainty disappeared? No. *The conflict and uncertainty are now being absorbed by individuals and groups rather than by the organization.* And people have to learn how to cope effectively with uncertainty and conflict. If the organization does not take this need for individual learning into account, these new structures will fail.

One of the authors worked with a construction company which has just created a new role to serve as a link between the estimating department and the operating (i.e., construction) department. This link is critical if the company is to produce realistic cost estimates (in winning bidding competitions) and to stick close to this cost budget during construction. The role looks terrific on the organization chart. However, the people in the new role feel they have *(a)* no clear understanding of their new duties, and *(b)* no formal authority to carry them out. The change is quite rational, but unless training, role clarification, management support, and rewards (all aspects of learning) are provided, the new structure will fail.

In summary, it is possible to identify a chain of events whereby organizational redesign efforts can be slowed if continued individual learning is not facilitated (adapted from Hall, 1976, p. 175):

Step 1: Organization design interventions aimed at enhancing employee performance often *enrich people* and develop human resources in the process of enriching the work environment.

Step 2: The person may remain enriched (i.e., possess new skills and aspirations), even if the organizational interventions do not continue.

Step 3: The person may come to expect continuing rewards as a result of his or her development, in the form of opportunities for future growth, advancement, better pay, more leisure, or other extrinsic benefits.

Step 4: If these expectations for the future are not met, the result may be either (1) attempts to influence the organization to provide these rewards, or (2) greater alienation than existed initially, before the intervention. Unless further growth opportunities are provided, the long-term positive outcomes of an intervention may yield longer-term negative outcomes.

Step 5: This increased frustration and alienation may make the individual resistant to new interventions aimed at intrinsic motivation—"once burned, twice cautious."

Thus, we are led to our final proposition:

Proposition 11. The new roles which emerge in the process of organization redesign are the crucial link between the individual and the new design. The learning of the individual to perform effectively in that new role represents the integration of individual learning and organization learning.

The person's role, then, represents the "missing link" between micro and macro research on organizations. To be more precise, role is a critical link which has been available in the organizational literature as a key integrating concept for some time (cf. Kahn, Wolfe, Quinn, Snoek, and Rosenthal, 1964), but it is one which has been overlooked to date in much of our research on design and development. The organizational role, with its interpersonal "fishnet" of role senders, its set of formal and informal behavioral expectations, and its mechanisms for cognitively processing organizational, interpersonal, and personal inputs to create the self-perceived role provides a natural analytic scheme for integrating individual and organizational learning. We would strongly advocate that future research on organization design and individual learning pursue the issue of how behavior in new roles is learned.

CONCLUSION

The major argument of this analysis is that individual learning and organization design each represent the critical gap in the study of the other. Also, in practical applications, each represents the major stumbling block in the success of the other. As long as "organization design" is defined in terms of static organizational attributes (e.g., structure), rather than the process by which key members of the organization *learn* to perform new organizational roles, we will never understand why and how organizations can be diffused. On the other hand, as long as programs to develop individuals consider only the person and factors which directly affect the person (e.g., career planning workbooks, boss, job, work group), we will continue to see development efforts stalled by lack of structural support or by incongruent organizational policies. By applying what we know about how individuals learn, we can greatly extend our mastery of the organization design process. And by intervening at the level of organization structure and policy, our leverage on individual learning is greatly enhanced.

We would revise Galbraith's definition to say that the *ultimate objective* of the design process is to increase the fit among tasks, structure, information, decision processes, reward systems, and people (Galbraith, 1977). However, we would add that organization design is the process by which people learn over time to increase this fit. Therefore, by this definition, we conclude that the following are important (and previously neglected) features of the organization design process:

1. Any design strategy will entail *new roles* and therefore new learning for people at some point in the implementation of the design (to make

people congruent with the other four elements of the organization). More research on the emergence of and learning in new roles is strongly needed.

2. Any attempts to change people, without changing tasks, structure, information/design processes, and reward systems, will fail. (This is the "micro gap.")

3. Any attempts to change organizations by changing structure and other macro design elements alone, without also developing the people who must operate in new roles in that new structure, will also fail. (This is the "macro gap.")

4. Effective organizational and individual adaptation knows no micro/macro boundaries. Adaptation will be most successful when *(a)* the process of redesign is monitored and various "spin-off" needs and unintended consequences are identified and dealt with, and *(b)* allowance is made for new forms of individual learning and organizational structures and policies, regardless of the level at which the original intervention was made.

FOOTNOTES

1. The comments of Chris Argyris, Barry Staw, and Richard Steers on an earlier draft of this paper is greatly appreciated. Work on this paper was supported by the Earl Dean Howard Fund, Graduate School of Management, Northwestern University.

2. Now at the State University of New York at Buffalo.

3. The case-survey method involves content analysis of numerous case studies. Content analysis generates quantitative data, and with a large sample of cases, statistical analyses can be conducted. This new method is described more fully in F. Hall (1975) and Yin and Heald (1975).

REFERENCES

1. Allen, T. J., and S. I. Cohen (1969) "Information flow in research and development laboratories," *Administrative Science Quarterly 14*, 12–19.
2. Argyris, C. (1957) *Personality and Organization,* New York: Harper & Row, Publishers.
3. ———. (1964) Integrating the Individual and the Organization, New York: John Wiley & Sons, Inc.
4. ———. (1976) "Leadership, learning and changing the status quo," *Organizational Dynamics,* Winter, 29–43.
5. ———. (1978) "Review of Jay Galbraith, *Organization Design,*" *Administrative Quarterly 23* 163–165.
6. ———, and D. Schön (1975) *Theory in Practice: Increasing Professional Effectiveness,* San Francisco: Jossey-Bass, Inc., Publishers.
7. Bakke, E. W. (1950) *Bonds of Organization,* New York: Harper.

8. Barnard, C. (1938) *The Functions of the Executive*, Cambridge, Mass.: Harvard University Press.
9. Bass, B. M., and J. A. Vaughan (1966) *Training in Industry: The Management of Learning*, Belmont, California: Wadsworth Publishing Co. Inc.
10. Baty, G. B., W. M. Evan, and T. W. Rothermel (1971) "Personnel flows as interorganizational relations," *Administrative Science Quarterly 16*, 430–433.
11. Beatty, R. W. (1974) "Supervisory behavior related to job success of hard core unemployed over a two year period," *Journal of Applied Psychology 59*, 38–42.
12. Beer, M., and R. Ruth. (1976) "Employee growth through performance management," *Harvard Business Review 54*, 59–66.
13. Berlew, D. E., and D. T. Hall (1966) "The socialization of managers: Effects of expectations on performance," *Administrative Science Quarterly 11*, 207–223.
14. Bray, D. W., R. J. Campbell, and D. L. Grant (1974) *Formative Years in Business: A Long-Term AT&T Study of Managerial Lives*, New York: John Wiley & Sons, Inc.
15. Brewer, J. (May 1977) "Career development at Lawrence Livermore Labs," address delivered at Metro State University, Conference on Careers, Minneapolis.
16. Bronfenbrenner, U. (1977) "Toward an experimental ecology of human development," *American Psychologist 32*, 513–531.
17. Burnaska, R. F. (1976) "The effects of behavior modeling training upon managers' behaviors and employees' perceptions," *Personnel Psychology 29*, 329–335.
18. Campbell, J. P. (1973) "Research into the nature of organizational effectiveness: An endangered species?" working paper, University of Minnesota.
19. ———, M. D. Dunnette, E. E. Lawler, III, and K. E. Weick, Jr. (1970) *Managerial Behavior, Performance, and Effectiveness*, New York: McGraw-Hill.
20. Davis, S. (1967) "Organic problem-solving method of organizational change," *Journal of Applied Behavioral Science 3*, 3–21.
21. Dowling, W. F. (1973) "Job redesign on the assembly line: Farewell to blue-collar blues," *Organizational Dynamics*, Autumn, 51–67.
22. Drucker, P. F. (1977) "Thinking about retirement policy," *Wall Street Journal*, 20.
23. Duncan, R. B., and A. Weiss (1979) "Organizational learning," in B. Staw (Ed.), *Research in Organizational Behavior* (Vol. 1). J.A.I. Press, pp. 75–123.
24. Feldman, H. S., R. P. Marinelli "Career planning for prison inmates," *Vocational Guidance Quarterly, 23*, 358–362.
25. Foulkes, F. K., and H. M. Morgan (1977) "Organizing and staffing the personnel function," *Harvard Business Review 55*, 142–154.
26. Fox, F. V., L. E. Pate, L. R. Pondy (1976) "Designing organizations to be responsive to their clients," in R. Kilman, L. Pondy, and D. Slevin, (eds.), *The Management of Organization Design*, New York: North-Holland.
27. Galbraith, J. (1977) *Organization Design*, Reading, Mass.: Addison-Wesley.
28. ———, and A. Edstrom (1977) "Creating decentralization through informal networks: The role of transfer," in L. Pondy, R. Kilmann, and D. Slevin (eds.), *Managing Organization Design*, New York: American Elsevier Publishing Co., Inc., pp. 289–310.
29. Georgiou, P. (1973) "The goal paradigm and notes toward a counterparadigm," *Administrative Science Quarterly, 18*, 291–310.
30. Gerwin, D. (1976) "A systems framework for organization structural design," in R. Kilmann, L. Pondy, and D. Slevin (eds.), *The Management of Organization Design*, New York: North-Holland.
31. Goodale, J., D. T. Hall, and S. Rabinowitz (1977) "A longitudinal test of a model of the development of job involvement," unpublished manuscript, York University.
32. Goodman, P., and P. Salipante (1976) "Organizational rewards and retention of the hard-core unemployed," *Journal of Applied Psychology, 61*, 12–21.

33. Graen, G., J. F. Cashman, S. Ginsburg, and W. Schiemann (1977) "Effects of linking-pin quality upon the quality of working life of lower participants," *Administrative Science Quarterly 22*, 491–504.

34. Graff, R. W., D. Raque, and S. Danish (1974) "Vocational-educational counseling practices: A survey of university counseling centers," *Journal of Counseling Psychology 21*, 579–580.

35. Hackman, J. R. (1977) "Designing work for indiviudals and for groups," in J. R. Hackman, E. E. Lawler III, and L. W. Porter (eds.), *Perspectives on behavior in organizations*, New York: McGraw-Hill, pp. 242–256.

36. Hall, D. T. (1971) "A theoretical model of career identity development in organizational settings," *Organizational Behavior and Human Performance 6*, 50–76.

37. ———. (1976) *Careers in Organizations*, Santa Monica, Calif.: Goodyear Publishing Co. Inc.

38. ———, and L. W. Foster (1977) "A psychological success cycle and goal setting: Goals, performance, and attitudes," *Academy of Management Journal, 20*, 282–290.

39. ———, and F. S. Hall (1976) "The relationship between goals, performance, success, self-image, and involvement under different organization climates," *Journal of Vocational Behavior 9*, 267–278.

40. ———, and E. R. Lawler (1969) "Unused potential in research and development organizations," *Research Management 12*, 339–354.

41. ———, and B. Schneider (1972) "Correlates of organizational identification as a function of career pattern and organization type," *Administrative Science Quarterly 17*, 340–350.

42. ———, and ———. (1973) *Organizational Climates and Careers: The Work Lives of Priests*, New York: Academic Press, Inc.

43. Hall, F. S. (1975) "Goal setting in public schools: A critical analysis," *Journal of Educational Administration 13*, 62–72.

44. Kahn, R. L., D. M. Wolfe, R. Quinn, J. D. Snoek, and R. A. Rosenthal (1964) *Organizational Stress: Studies in Role Conflict and Ambiguity*, New York: John Wiley & Sons, Inc.

45. Kaufman, H. (1960) *The Forest Ranger*, Baltimore: Johns Hopkins Press.

46. Kaufman, H. G. (1974) *Obsolescence and Professional Career Development*, New York: AMACOM.

47. Ketcham, L. D. (1975) "A case study of innovation," in L. E. Davis and A. B. Chernes, (eds.), *The Quality of Working Life, Volume Two: Cases and Commentary*, New York: The Free Press.

48. Kilmann, R., L. Pondy, and D. Slevin (1976) "Patterns and emerging themes in organization design," in R. Kilmann, L. Pondy, and D. Slevin (eds.), *The Management of Organization Design*, New York: North-Holland.

49. Lawrence, P. R., and J. W. Lorsch (1969) *Organization and Environment*, Homewood, Ill.: Richard D. Irwin, Inc.

50. Leavitt, H. J. (1976) "On the design part of organization design," in R. Kilmann, L. Pondy, and D. Slevin (eds.), *The Management of Organization Design*, New York: North-Holland.

51. Levinson, H., C. Price, K. Munden, H. Mandel, and C. Solley (1962) *Men, Management, and Mental Health*, Cambridge, Mass.: Harvard University Press.

52. Lewin, K. (1936) "The psychology of success and failure," *Occupations 14*, 926–930.

53. Livingston, J. S. (1969) "Pygmalion in management," *Harvard Business Review 47*, 81–89.

54. Lorsch, J., and J. Morse (1974) *Organizations and Their Members: A Contingency Approach*, New York: Harper & Row, Publishers.

55. McGovern, T. V., and H. E. Tinsley (1976) "A longitudinal investigation of the graduate assistant work-training experience," *Journal of College Student Personnel 17*, 130–133.
56. Merton, R. K. (1957) *Social Theory and Social Structure*, Second Edition, Glencoe, Ill.: The Free Press.
57. Miller, J. A., B. M. Bass, W. L. Mihal (1973) "An experiment to test methods of increasing self-development attitudes among research and development personnel," publication TR-43, University of Rochester, Management Research Center, Rochester, New York.
58. Morgan, M. (1977) *The Effect of Job History on Managerial Career Success*, Ph.D. Dissertation, Northwestern University.
59. Morrison, R. (1977) "Career adaptivity: The effective adaptation of managers to changing role demands," *Journal of Applied Psychology 62*, 549–558.
60. Nystrom, P. C., L. T. Hedberg, and W. H. Starbuck (1976) "Interacting processes as organization designs," in R. Kilmann, L. Pondy, and D. Slevin (eds.), *The Management of Organization Design*, New York: North-Holland.
61. O'Leary, V. E. (1972) "The Hawthorne effect in reverse: Trainee orientation for the hard-core unemployed woman," *Journal of Applied Psychology, 56*, 491–494.
62. O'Toole, J. (1977) *Work, Learning, and the American Future*, San Francisco: Jossey-Bass, Inc., Publishers.
63. Parsons, T. (1960) *Structure and Process in Modern Societies*, New York: The Free Press.
64. Pelz, D. C., and F. Andrews (1976) *Scientists in organizations* (2nd ed.). Ann Arbor, Mich.: Institute for Social Research.
65. Pfeffer, J. (Dec. 1977) "Towards an examination of stratification within organizations," *Administrative Science Quarterly*.
66. —— (1978) *Organizational Design*, Arlington Heights, Ill.: AHM Publishing.
67. ——, and H. Leblebici (1977) "Executive recruitment and the development of inter-firm organizations," *Administrative Science Quarterly 18*, 449–461.
68. ——, and G. Salancik (1977) "Organization design: The case for a coalitional model of organizations," *Organizational Dynamics 6*, 15–29.
69. Pincus, C., N. Radding, and R. Lawrence (1974) "A professional counseling service for women," *Social Work 19*, 187–194.
70. Pondy, L., R. Kilmann, and D. Slevin (eds.), (1977) *Managing Organization Design*. New York: American Elsevier Publishing Co., Inc.
71. Porter, L., R. Steers, R. Mowday, and P. Boulian (1974) "Organizational commitment, job satisfaction, and turnover among psychiatric technicians," *Journal of Applied Psychology 59*, 603–609.
72. Porter, L. W., and E. E. Lawler III. (1968) *Managerial Attitudes and Performance*, Homewood, Ill.: Irwin-Dorsey.
73. Renwick, P., and Lawler, E. E. III (1978) "What You Really Want From Your Job," *Psychology Today 11*, 53–65.
74. Robinault, I. P., and M. Wisinger (1973) "Leaderless groups: A tape cassette technique for vocational education," *Rehabilitation Literature 34*, 80–84.
75. Salipante, P., and P. Goodman (1976) "Training, counseling, and retention of the hard-core unemployed," *Journal of Applied Psychology 61*, 1–11.
76. Schein, E. (1971) *Organizational Psychology*, Englewood Cliffs, N.J.: Prentice-Hall, Inc.
77. Sellman, W. S. (1970) *Effectiveness of Experimental Training Materials for Low Ability Airmen*, U.S. Air Force, Human Relations Laboratory Technical Report No. 70-16.
78. Sherif, M., J. Harvey, B. J. White, W. R. Hood, and C. Sherif (1961) *Intergroup*

Conflict and Cooperation: The Robbers Cave Experiment, Norman, Okla.: University Book Exchange.

79. Simon, H. (1957) *Administrative Behavior,* New York: Macmillan, Inc.

80. Smith, F. J., K. D. Scott, and C. L. Hulin (1977) "Trends in job-related attitudes of managerial and professional employees," *Academy of Management Journal 20,* 454–560.

81. "Stairway to success," *Newsweek* October 24, 1966, p.999

82. Steers, R. M. (1975) "Problems in the measurement of organizational effectiveness," *Administrative Science Quarterly 20,* 546–558.

83. ———, and Porter, L. W. (1974) "The role of task-goal attributes in employee performance," *Psychological Bulletin 81,* 434–452.

84. Stouffer, S. A., E. A. Suchman, L. C. DeVinney, S. A. Star, and R. M. Williams (1949) *The American Soldier: Adjustment During Army Life,* New York: John Wiley & Sons, Inc.

85. Super, D. E., and D. T. Hall (1978) "Career development: Exploration and planning," *Annual Review of Psychology 29,* 333–372.

86. Thompson, J. D. (1967) *Organizations in Action.* New York: McGraw-Hill.

87. Thompson, P. H., and G. W. Dalton (1976) "Are R & D organizations obsolete?" *Harvard Business Review 54,* 105–116.

88. Triandis, H. C., J. M. Feldman, and D. E. Welden (1974) "Designing preemployment training for the hard to employ: A cross-cultural psychological approach," *Journal of Applied Psychology 59,* 687–693.

89. Van Maanen, J., E. Schein, and L. Bailyn (1977) "The shape of things to come: A new look at organizational careers," in J. R. Hackman, E. E. Lawler, and L. Porter (eds.), *Perspective on Behavior in Organizations,* New York: McGraw-Hill, pp. 153–162.

90. Walton, R. E. (1975) "The diffusion of new work structures: Explaining why success didn't take." *Organizational Dynamics,* Winter, *3,* 3–22.

91. Wanous, J. (1973) "Effects of realistic job preview on job acceptance, job attitudes, and job survival," *Journal of Applied Psychology 58,* 327–332.

92. Weick, K. (1977) "Organization design: Organizations as self-designing systems," *Organizational Dynamics, 6,* 30–46.

93. Wellbank, H. L., D. T. Hall, M. A. Morgan, and W. C. Hamner (1978) "Planning job progression for effective career development and human resources management," *Personnel,* March–April, 54–64.

94. White, R. (1959) "Motivation reconsidered: The concept of competence," *Psychological Review 66,* 297–323.

95. Yankelovich, D. (1974) *The New Morality,* New York: McGraw-Hill.

96. Yin, R. A., and K. A. Heald (1975) "Using the case survey method to analyze policy studies," *Administrative Science Quarterly 20,* 371–381.

ORGANIZATIONAL STRUCTURE, ATTITUDES, AND BEHAVIORS[1]

Chris J. Berger, PURDUE UNIVERSITY

L. L. Cummings, UNIVERSITY OF WISCONSIN-MADISON

ABSTRACT

This paper reviews and evaluates the research published since 1965 that examines the relation between organizational structural characteristics and participant attitudes and behaviors. The effects of organizational level have received the widest attention, with particular emphasis being given to the relation of level to participant satisfaction. Other structural characteristics examined include line-staff function, size, span of control, complexity, and shape. An explicit attempt is made to evaluate progress since the last major review of this literature (Porter and Lawler, 1965). Progress continues to be constrained by the lack of clearly formulated, sophisticated conceptual networks,

Research in Organizational Behavior, Vol. 1, pp. 169–208.
Copyright © 1979 by JAI Press, Inc.
All rights of reproduction in any form reserved.
ISBN 0-89232-045-1

by limited methodological designs, and by inappropriate analyses. Suggestions for improvements are offered along with cautious conclusions concerning present knowledge.

This paper represents what might be called a pretheoretical analysis of the relation of organizational structure to attitudes and behavior. As such it aims toward a "cleaning-up" and interpretation of the relevant literature. It distinguishes between the relevant and irrelevant studies, given a priori standards of relevancy, and critiques and interprets the relevant ones.

A decade ago, Porter and Lawler (1965) published a review of the literature concerning the relationship of organizational structural characteristics to job attitudes and job behaviors. Drawing the distinction between variables which were appropriate for intra- and total organizational analysis, Porter and Lawler focused on the following seven structural characteristics.

Organizational levels
Line and staff hierarchies
Span of control
Subunit size
Total organization size
Tall or flat shape
Centralized or decentralized shape

The dependent variables examined were job attitudes (broadly defined as an opinion concerning some job aspect) and job behavior (performance, turnover, absenteeism, and employee-grievance rates). After reviewing research published through 1964, they concluded that "at least 5 of these 7 structural variables (with possible exceptions being span of control and centralized/decentralized shape) were found to be significantly related to one or more attitude or behavioral variables" (p. 23).

Since the Porter and Lawler review, more than forty empirical studies investigating the relationship between some aspect(s) of organization structure and members' attitudes and/or behavior have been reported. While James and Jones (1976) have recently reviewed the *conceptual* literature that theorizes about the relation between structural characteristics and individual attitudes and behaviors, the *empirical* literature of the past ten years has not been reviewed, integrated, and critiqued. The purposes of the present paper are to: (1) review this literature of the past decade, (2) indicate empirical generalizations which may be drawn from the literature and compare these with the generalizations derived by Porter and Lawler, and (3) discuss a number of conceptual and methodological problems in this literature.

The scope of this review has been affected by several considerations. First, the time period covered is 1964 to the present. Second, unpublished studies have been omitted. Third, only empirical studies which measured more than one level of a given structural variable have been included, thus focusing on comparative research. Fourth, in contrast to the Porter and Lawler review, studies which sampled individuals in organizations other than business and industry have been included. Fifth, studies which focused on dependent variables at the organizational level of analysis have been excluded. Thus, for example, a study which related organization size to performance was included if individual members' performance was measured, but was excluded if performance was measured across an entire department, plant, or organization. This was done for two reasons: Our primary interest centers on the behavior of persons within organizations, and we wish our review to be cumulative with the earlier review by Porter and Lawler (1965).

A final category of studies excluded from this review is that focusing on organizational climate. Only those studies researching climate in which the authors specifically addressed the relationship of organizational structure (as distinct from climate) to members' attitudes and/or behaviors have been included. Three reviews of the organizational climate literature have appeared recently (Hellriegel and Slocum, 1974; James and Jones, 1974; Schneider, 1975).

The paper is broadly organized into two sections. The first of these examines the empirical evidence. Here, an attempt is made to create a review which is cumulative with the Porter and Lawler (1965) review and thus utilizes a similar categorization scheme wherever possible. Within the seven structural categories, studies are separated by attitudinal and behavioral dependent variables. Important studies are described and examined in detail, and analytical comments appropriate only to a particular study are included. Brief summary statements appear at the end of each of these sections.

The second major section of the paper attempts to provide a broader critique and perspective on the methodology of the studies reviewed, with the explicit purpose of detailing major methodological flaws which continue to plague us and severely limit our ability to generate useful cumulative evidence in this area.

REVIEW OF EMPIRICAL LITERATURE

SUBORGANIZATIONAL: ORGANIZATIONAL LEVELS

Organizational level refers to an individual's position in the vertical hierarchy of authority and ranges from nonsupervisory workers at the lower end of the scale to the chief executive at the upper extreme.

While the definition of organizational level would appear straightforward, most researchers have not explicitly defined this variable at a conceptual level and none have discussed its theoretical relationship with other organizational and/or individual variables. Operationally, most studies have simple asked respondents for a self-report of their position within their organization's hierarchy of authority. A few studies have categorized individuals according to their position in a hierarchy relative to the total number of levels in their particular organization, thus facilitating comparison between equivalent positions in organizations with differing numbers of levels.

Relation to Attitudes

Need and job satisfaction   After reviewing more than a dozen studies relating two or more orgnaizational levels to individuals' job satisfactions, Porter and Lawler (1965) concluded that the literature as of 1964 showed increasing job satisfaction across increasing hierarchical levels. While many of the early studies they reviewed sampled blue-collar workers, Porter (1962, 1964) reported the results of a survey of nearly 2,000 managerial personnel at various organizational levels in different business and industrial organizations. These two reports combined with Porter and Lawler's review apparently set the stage for subsequent researchers' attempts to broaden our knowledge with respect to the organizational levels/job satisfaction relationship in different types of organizations. One alternative sample employed by researchers since 1965 has been the military. Porter and Mitchell (1967), for example, surveyed 1,279 commissioned and noncommissioned U.S. Air Force officers. The Porter Need Satisfaction Questionnaire (PNSQ: Porter, 1962) was used to measure individual's perceived need fulfillment and need satisfaction. The thirteen scales comprising the PNSQ were designed to measure the need fulfillment and need satisfaction of five basic needs (security, social, esteem, autonomy, and self-actualization), following Maslow (1954). With respect to need fulfillment, each higher level of commissioner officers reported significantly higher need fulfillment. However, noncommissioned officers did *not* show a consistent increase in need fulfillment with increasing rank.

Concerning need satisfaction, in the noncommissioned officer group there was a positive and consistent relationship between organizational level and satisfaction, while the commissioned officers did not show this pattern. Thus, over all organization levels, the hypothesis of consistently increasing need fulfillment and/or satisfaction was not supported.

Johnson and Marcum (1968) related organizational level to need satisfaction in a sample of career U.S. Army officers. Officers were divided into three groups according to their miltiary rank. Need satisfaction was measured by a nine-item questionnaire similar in format to the PNSQ. The data was analyzed with Kruskal-Wallis one-way analysis of variance by ranks. In all, thirteen pair-wise comparisons were made, less than half of which showed significant differences across all three military ranks. Of those which were significantly different across ranks, need satisfaction did not consistently increase with increasing organizational level.

A third study exploring the need satisfaction/organization level relationship within the military was reported by Mitchell (1970), who surveyed 675 commissioned U.S. Air Force officers. The variables included in this study were need fulfillment and need satisfaction (PNSQ), organizational level (three military ranks), and line or staff assignment. Within the six line and staff assignment categories, data were analyzed by signed-rank tests. With differences between line and staff assignments controlled, there were fairly consistent increases in need fulfillment with increases in military rank, but inconsistent changes in need satisfaction with increases in military rank.

A second group of studies relating to the need satisfaction/ organizational level hypothesis focuses upon samples drawn from public sector organizations. Rhinehart, Barrell, DeWolfe, Griffin, and Spaner (1969) for example, reported survey results of 2,026 employees of the federal Veterans Administration. Respondents were medical, professional, and nonprofessional supervisory personnel, who were classified into four managerial levels. Need satisfaction was measured by the PNSQ. Data were analyzed with signed-rank tests and indicated that need satisfaction increased consistently with organizational level.

A second study using a sample of government employees to examine the organizational level/job satisfaction relationship has been reported by Lichtman (1970). In a sample of ninety-four middle managers, first-level supervisors and "working-level technical employees" from two offices of the Internal Revenue Service, organizational level was related to job satisfaction, need achievement, organizational knowledge, job-related tension, internal control, and productivity (the latter five of which will be discussed subsequently). Global job satisfaction was measured with an instrument developed by Harris (1949). Overall differences in mean job satisfaction across the three organizational levels were statistically significant ($p < .01$) with job satisfaction increasing with organizational level.

Miller (1966a) studied the organizational level/job satisfaction relationship in a randomly selected sample of national-level officials of large craft and industrial unions. Organizational level was dichotomized into "higher"- and "lower"- level positions based on the individual's job title.

Need satisfaction was measured by the PNSQ. In craft unions, higher-level officials were more satisfied than lower-level officials. Within industrial unions, no significant differences in satisfaction were found.

Slocum (1971) utilized the PNSQ to study the need satisfaction of 210 managers of a Pennsylvania steel plant. He found higher satisfaction among middle and top managers (versus first-line supervisors) across esteem, autonomy, and self-actualization needs. Similar trends, though not statistically significant, were noted for security and social needs.

The seven studies reviewed above researching the organizational level/ job satisfaction hypothesis in predominantly nonbusiness samples show mixed results. Studies sampling military personnel failed to show strong support for a positive relationship between level and satisfaction. While two studies (Porter and Mitchell, 1967; Johnson and Marcum, 1968) showed mixed support, a third (Mitchell, 1970) reported no significant increases in need satisfaction across organizational levels. Clearly, none of these three military studies provides support for the statement that need satisfaction increases consistently with organizational level.

On the other hand, two studies which sampled federal government employees (Rhinehart et al., 1969; Lichtman, 1970) did support the hypothesis of consistent and significant increases in need and job satisfaction with increases in organizational level. Finally, a study sampling national union officials (Miller, 1966a) and one study of business managers (Slocum, 1971) were not inconsistent with the hypothesis.

More complex studies   The following studies, while examining the impact of organizational level, differ from the preceding studies in terms of their design and/or analytical procedures. They include one or more of the following characteristics: *(a)* additional predictors of job or need satisfaction in terms of additional organizational and/or individual variables, *(b)* multivariate statistical analysis, and *(c)* multiple instruments used to measure job satisfaction.

None of the studies reviewed above related individual differences to differences in job or need satisfaction. Moreover, there were no estimates of the net relationship of organizational level and the individual level variables to satisfaction. In contrast, several studies have attempted to answer these questions within appropriate multivariate frameworks.

Lawler and Porter (1966) for example, examined the relationship between several organizational and individual variables and managerial pay satisfaction. Reporting data from Porter's (1962) survey of 1,916 managers, organizational level, line/staff position, organization size, tenure, age, education, and actual salary were used to predict managers' satisfaction with their pay. Multiple correlation indicated that organizational level showed a small but significant negative net relationship to pay satisfaction. Except for tenure, the demographic variables were not significantly related to pay satisfaction.

A more recent study considering the relationship of organizational level and pay satisfaction in a multivariate framework has been reported by Schwab and Wallace (1974). The sample consisted of 273 randomly selected employees (stratified by pay system) from one large firm. Individuals' age, sex, tenure with firm, type of pay system, pay level, and organizational level were used to predict satisfaction with pay using zero-order and partial correlational analyses.[2] Pay satisfaction was measured by appropriate items from the Minnesota Satisfaction Questionnaire (MSQ; Weiss, Dawis, England, and Lofquist, 1967) and the Cornell Job Descriptive Index (JDI; Smith, Kendall, and Hulin, 1969). When multiple correlation was used to control for the linear effects of other variables, all predictors except age and tenure in the case of the MSQ, and age in the case of the JDI were significantly related to satisfaction with pay. Organizational level showed small but significant negative net relationships with pay satisfaction as measured by both the JDI and the MSQ.

In a national random sample of 911 solid-waste management employees, Locke and Whiting (1974) found white-collar employees to be more satisfied with their jobs than blue-collar employees. In addition, the white-collar employees were more likely to associate variations in satisfaction with intrinsic rewards than with extrinsic sources and exhibited a greater tendency toward this distinction than the blue-collar employees.

MacEachron (1977) has recently reported findings indicating an interaction between field independence [as measured by Oltman's (1968) portable rod-and-frame test] and organizational level in relation to the JDI on a sample of seventy female nurses. Organizational level was indexed at three levels: nursing supervisors and RNs, licensed practical nurses, and nursing aides. While there was not a significant correlation between level and four of five JDI dimensions for the total sample, field independence did moderate the relation. Field-independent nurses exhibited a positive correlation between level and satisfaction on each of the five JDI dimensions while field dependents exhibited only one significant, negative correlation between level and satisfaction (co-workers). In addition, it is interesting to note that field independence was positively correlated with organizational level ($r = .38$).

A second group of studies using multivariate techniques has focused on job or need satisfaction in relation to organizational level and one or more additional structural variables. Two associated articles (El Salmi and Cummings, 1968; Cummings and El Salmi, 1970), for example, related several structural properties to various need measures. The sample consisted of 425 managers from a number of organizations. Need fulfillment, need importance, and need satisfaction were measured with the PNSQ.

Subjects also responded to an item concerning their "perceived chance of attaining the level of need fulfillment—for each item—they thought should exist in their present jobs" (Cummings and El Salmi, 1970, p. 2). Structural variables were organizational level, line or staff position, total organization size, and subunit size. All of the organizational variables were trichotomized, with organizational level and tall/flat shape being standardized across organizations.

Cummings and El Salmi (1970) found organizational level to be significantly related to need fulfillment and perceived possibility of need fulfillment. While need fulfillment increased consistently across all three organizational levels, the possibility of need fulfillment appeared to decrease with increasing level. Organizational level was not significantly related to either need importance or need satisfaction. Further, none of the six organizational variables measured were reported to be significantly related to need importance or need satisfaction in the univariate analysis.

El Salmi and Cummings (1968) reported the effects of three interactions (organizational level by line/staff position; organizational level by total organization size; and organization level by tall/flat shape) on several of the four need variables. The organizational level by total size interaction was significantly related both to need fulfillment and perceived possibility of need fulfillment. In the case of the former relationship, the highest-level managers reported more need fulfillment in small than in large organizations. However, middle and lower-middle managers in larger organizations reported more need fulfillment than did their counterparts in small companies. The organizational level by tall/flat shape interaction was significantly related to need satisfaction. A final interaction term (organizational level by line/staff position) was not significantly related to need fulfillment. Overall, then, these two reports (El Salmi and Cummings, 1968; Cummings and El Salmi, 1970) show no support for the univariate effect of organizational level on need satisfaction, although need fulfillment did increase with organizational level.

A third category of multivariate studies has used multiple measures of job satisfaction. One such study has been reported by Waters and Roach (1973). Their sample consisted of 101 managerial personnel in a national insurance company. Organizational levels ranged across all managerial ranks. Need satisfaction was measured by the PNSQ. In addition to the thirteen items contained in the PNSQ, three items concerning satisfaction with pay, being informed, and "pressure" experienced in the management position were measured using the PNSQ format. Also included was a measure of overall job satisfaction which was rated on a 12-point bipolar scale. These seventeen satisfaction items and the subjects' organizational level were then factor analyzed, from which four factors were identified: higher- and lower-order need satisfaction, esteem-prestige, and participation in the management process. Organizational level, which showed no relationship to the first three factors, loaded positively on the participation in management processes factor. Thus, in this sample at least, organizational level was positively associated with only a subset of satisfaction (with participation) and was not associated with any other satisfaction items or with overall job satisfaction.

Herman and Hulin (1973) measured job satisfaction with both the PNSQ and the JDI. Subjects were four levels of managerial personnel from one large organization. The organizational level/job satisfaction hypothesis was examined via discriminant analysis on the JDI and the PNSQ separately. Analysis of the JDI resulted in a significant overall solution and two significant linear functions discriminating between the four organizational levels. Scale loadings in this analysis indicated that group differences were based primarily on satisfaction with work and pay. In these functions, satisfaction generally tended to increase with each organizational level. The results of the discriminant analysis of the PNSQ, however, did not result in a significant overall solution.

A final group of studies to be reviewed in this section has many or all of the multivariate characteristics described above, and to varying degrees may be seen as partial replicates of one another in terms of methodology and/or variables studied.

Herman and Hulin (1972) utilized a number of aspects of the multivariate studies reviewed above in examining the organizational level/job satisfaction hypothesis. They measured multiple individual characteristics, organizational structural variables, and job attitudes on 307 salaried supervisory and nonsupervisory personnel in one plant of a large industrial organization. The individual variables consisted of respondents' age, education, and plant tenure. The structural variables were functional division, organizational level, and department. The attitudinal variables measured included general job attitudes, job satisfaction, and attitudes toward line/staff relations and leadership. Component analysis (Eckert and Young, 1936) was used to reduce this set of items to seventeen attitude variables. Discriminant analysis was then performed on these dependent variables for each of the three structural variables and the three personal variables. In addition, a multivariate analog of omega squared was used to estimate the power of the solutions and to compare the six independent variables. This analysis indicated that the three structural variables accounted for more attitude variance (department = .82; function = .60; level = .43) than did the individual variables (tenure = .40; age = .39; education level = .37). With respect to organizational level, the four classifications of respondents were first-level supervisors, second- and third-level supervisors, nonsupervisory personnel, and staff assistants who were not engaged in supervision. Herman and Hulin reported that differences between hierarchical groups related primarily to line/staff relation and co-worker satisfaction, and to evaluation of both superior/ subordinate relationships and supportive services. Thus, changes in organizational level were related to only a subset of total job satisfaction. While finding an overall increase in satisfaction with increasing organizational level, Herman and Hulin (1972) did not find a consistent increase in satisfaction at each higher level. Even disregarding the nonsupervisory staff group, respondents at the middle organizational level were more satisfied than their higher- or lower-level counterparts.

Herman, Dunham, and Hulin (1975) investigated both structural properties and participant demographic characteristics as these related to a number of attitudinal dependent variables. Herman et al. studied 392 employees in a printing plant. Organizational level and position (i.e., membership in an organizationally defined department and work shift) were examined. Using canonical and part canonical analysis, it was found that: (1) 22 percent of the total response variance across all dependent variables was accounted for by the total set of demographic and structural varibles, (2) the demographic variables accounted for 9 percent of the response variance, (3) the structural variables accounted for 19 percent of the response variance, (4) controlling for the covariation of demographic and structural variables, structural indices accounted for 13 percent of the response variance while demographic indices accounted for only 3 percent of the response variance. In terms of specific structural relations, persons in higher job levels expressed greater satisfaction with work and pay and experienced greater work-related motivation. In general, persons in nonproduction departments (primarily office and maintenance) expressed greater satisfaction and motivation than those in production departments.

In a related study, Adams, Laker, and Hulin (1977) have found a significant main effect ($p < .01$) of job level (three levels) and functional specialty (five classifications) on four of the JDI dimensions (excluding satisfaction with promotional opportunities) on a sample of 1,313 employees of a printing company. In addition, the job level × functional specialty interaction was significant ($p \leqslant .01$) and was due to the univariate F for satisfaction with work. Job level was determined by ratings of jobs done by the investigators across factors like training time or education required, responsibility, and authority. The ratings were done on 20-point scales applied to fifteen job categories. The resulting rating distribution was trichotomized into unskilled, skilled, and professional-supervisory categories. Job satisfaction increased on each of the four JDI dimensions across the unskilled, skilled, and professional-supervisory categories respectively.

In a sample of 710 employees from all levels and departments of one insurance company, Newman (1975) related personal characteristics (age, sex, education, tenure, and number of dependents) and organizational structure variables (hierarchical level, functional division, department, and work group) to job attitudes. The latter included job satisfaction (JDI) and a measure of personal work involvement (Lodahl and Kejner, 1965). Analysis was carried out via discriminant analysis separately on each personal and organizational characteristic and included the multivariate analogue of omega squared.

Results indicated that the greatest amount of attitude variance was associated with the personal variables of age (.30) and tenure (.28) and the structural variables hierarchical level (.31), department (.22), and work group (.34). Hierarchical level was positively related to satisfaction with the work itself, pay, supervision, co-workers, and to job involvement.

O'Reilly and Roberts (1975) also related individual characteristics (ability, personality traits, and motivational traits all measured by Ghiselli's [1971] Self-Description Inventory) and structural variables (organizational level, organizational tenure, and job tenure) to job satisfaction as measured by the JDI. The sample consisted of 578 officers and enlisted men in one "high technology" naval aviation unit. Analysis was conducted with canonical correlation. With structural variables partialed out, there were no significant relationships between personality variables and job satisfaction. However, with personality variables controlled, there were significant relationships between structure and job satisfaction. While the authors do not present specific data, they suggest that organizational level is positively related to satisfaction with pay and promotion.

Szilagyi, Sims, and Keller (1976) found satisfaction with work, supervision, and co-workers as measured by the JDI to increase with occupational level. They sampled 931 hospital employees at five occupational levels. In the hospital studies, occupational and organizational level appear to be highly and positively related. In addition, role ambiguity (versus role conflict) was more highly related to satisfaction at higher occupational levels while role conflict was more highly related to satisfaction at lower occupational levels. Thus, occupational level was both directly related to satisfaction indices and also moderated the relationship betwen two role dynamics variables and these satisfaction indices. In a related study, Sims and Szilagyi (1976) found that occupational level moderated the relation between perceptions of *task* characteristics (measured via the *JDS*) and satisfaction with work (measured via the *JDI*). For example, feedback and satisfaction with work were more highly related at higher than lower occupational levels while variety and satisfaction with work were more highly related at lower than higher occupational levels. Thus, occupational and organizational level may not only impact satisfaction directly, but may also influence the relation between the nature of the immediate work and affective reactions to that work.

While these studies are to be lauded for their analytical techniques and inclusion of personal variables, there are several weaknesses which should be pointed out. First, Herman and Hulin (1972), Herman et al. (1975), Newman (1975), and Szilagyi, Sims, and Keller (1976) all include department, work group, and/or occupational group as structural variables. These variables were measured by the organization's designation of department or work group. Thus, these are nominal variables which have meaning only within the context of the *particular* organization. Failure to specify what this variable means in terms of a more generic structural definition precludes generalization to other organizations and situations, and in several cases raises the danger of confounding several generic structural variables (e.g., line/staff position, specialization) and/or individual level variables (e.g., selection, supervision) encompassed within the department variable. The conclusion that structural variables account for more attitude variance than do personal variables must be regarded with caution, at least at this stage of the analysis.[3]

A second problem involves the classification of variables concerning the interface of the individual and the organization. For example, Herman et al. (1975) grouped tenure, participation in an apprentice program, and work shift as structural variables. O'Reilly and Roberts also classified organizational tenure and job tenure as structural variables. On the other hand, Herman and Hulin (1972) and Newman (1975) include tenure as a personal variable. Given that these variables are in fact meaningful only with respect to a specific organization-individual interface, attempting to proportion variance in attitudes between structural variables and personal or demographic variables is inappropriate.

To summarize, studies relating organizational level to one or more aspects of job satisfaction have been presented. These studies are summarized in Table 1. As can be seen from this table, samples, measures, and analytical techniques have varied widely, as have the results. Military samples have provided only limited support for the organizational level/ job satisfaction hypothesis as stated in the Porter and Lawler (1965) review. Samples of government workers (Rhinehart et al., 1969; Lichtman, 1970) and national union officials (Miller, 1966a) did show support for the hypothesis.

Eleven studies have tested the hypothesis in business or industrial samples. While there is considerable method variance across the studies, and while job satisfaction does not appear to increase *consistently* with each organizational level, there is a relatively stable pattern between studies which indicates an overall positive association of organization level and job satisfaction.

Summary of Studies of the Job Satisfaction-Organizational Level Hypothesis

Study	Sample (n)	Number levels	Satisfaction measure	Multivariate aspects	Statistical procedure	Significant overall relationship	Direction of relationship	Consistent with each level
Miller, 1966	Union officials (171)	2	PNSQ	Craft/industrial	Signed-rank tests	Yes	Positive	Yes
Porter and Mitchell, 1967	Military (1,297)	6	PNSQ	None	Signed-rank tests	Yes	Positive	No
Johnson and Marcum, 1968	Military (504)	3	PNSQ	None	Kruskal-Wallis	Yes	Positive	No
Mitchell, 1970	Military (675)	3	PNSQ	Line-staff controlled	Signed-rank tests	No	Zero	No
Rhinehart et al., 1969	Government (2,026)	4	PNSQ	None	Signed-rank tests	Yes	Positive	Yes
Lichtman, 1970	Government (94)	3	Harris (1949)	None	ANOVA	Yes	Positive	Yes
Slocum, 1971	Business (210)	2	PNSQ	None	t-tests	Yes	Positive	Yes
Locke and Whiting, 1974	Business (911)	5	Locke and Whiting (1974)	Occupation controlled demographic	F-tests	Yes	Positive	No
Lawler and Porter, 1966	Business (1,916)	4	PNSQ (pay)	Structural individual	Multiple correlation	Yes	Negative	Yes
Schwab and Wallace, 1974	Business (273)	Not identified	MSQ; JDI (pay)	Individual	Multiple correlation	Yes	Negative	Yes
Cummings and El Salmi, 1970 / El Salmi and Cummings, 1968	Business (425)	3	PNSQ	None	Signed-rank tests (see text)	No	Zero / Interactive	No
Waters and Roach, 1973	Business (101)	4	PNSQ (and 4 other items)	Analysis	Factor analysis	Yes (one of 4 factors)	Positive	Yes
Herman and Hulin, 1973	Business (121) (158)	4	PNSQ / JDI	Analysis Analysis	Discriminant analysis	No / Yes	Zero / Positive	No / Yes
Herman and Hulin, 1972	Business (307)	4	Multiple items (see text)	Structural; individual; analysis	Component analysis; discriminant analysis	Yes	Positive	No
Herman, Dunham, and Hulin, 1975	Business (392)	Not identified	JDI	Structural; individual; analysis	Component analysis; canonical correlation	Yes	Positive	See text

Summary of Studies of the Job Satisfaction-Organizational Level Hypothesis (*continued*)

Study	Sample (n)	Number levels	Satisfaction measure	Multivariate aspects	Statistical procedure	Significant overall relationship	Direction of relationship	Consistent with each level
Newman, 1975	Business (710)	All levels	JDI	Structural; individual; analysis	Discriminant analysis	Yes	Positive	See text
O'Reilly and Roberts, 1975	U.S. Navy (578)	Not identified	JDI	Structural; individual; analysis	Canonical correlation	Yes	Positive	See text
Szilagyi, Sims, and Keller, 1976	Hospital (931) Business (192)	5 3	JDI	Analysis	Zero- and first-order correlations and nonparametric procedures	Yes	Positive	No
MacEachron, 1977	Hospital Nurses (70)	3	JDI	Structural; individual	Pearson product moment correlation; moderator analysis via subgrouping	No for 4 JDI dimensions; yes for 1 dimension	Positive for pay satisfaction	Not tested
Adams, Laker, and Hulin, 1977	Business (1,313)	3	JDI	Structural; analysis	MANOVA and discriminant analysis	Yes	Positive	Yes

182

Sources of job satisfaction   Three studies have examined the components and sources of job satisfaction as a function of the organizational level and context of the subjects. Starcevich (1972) examined the importance of eighteen job components in contributing to job satisfaction and dissatisfaction across three organizational levels. He found that organizational level (first-line managers, middle managers, and professional employees) did not affect the judged importance of the job components in determining satisfaction or dissatisfaction. Satisfaction means by organizational level were not reported. In general, however, job content factors (versus job context factors) were judged to be more important for both job satisfaction and dissatisfaction.

Harris and Locke (1974) found that white-collar employees (generally in higher-level positions) tended to derive satisfaction and dissatisfaction from "motivator" factors while blue-collar employees tended to derive satisfaction from "hygiene" factors. Most of the difference was attributable to the white-collar employees mentioning achievement-related events while blue-collar employees attributed satisfaction-dissatisfaction to monetary events.

Locke and Whiting (1974) have examined both environmental events and agents as sources of satisfaction and dissatisfaction in a sample of blue- and white-collar male employees in the solid waste management industry from both the private and public sectors. Events identified as sources of satisfaction for white-collar employees were promotions, feelings of achievement, and smoothness of work flow while for blue-collar employees sources of satisfaction included money, amount of work, working conditions, and interpersonal atmosphere. The pattern for sources of dissatisfaction was parallel. Agents serving as sources of satisfaction for white-collar employees were self and subordinates while blue-collars attributed satisfaction primarily to the organization. Dissatisfaction was attributed to subordinates and the union by the white-collars while blue-collars blamed supervisors and co-workers.

Other attitudes   Several studies have related organizational level to job attitudes other than satisfaction. For example, Mitchell and Porter (1967) found that commissioned Air Force officers ranked inner-directed traits as increasingly more important, and other-directed traits as increasingly less important with increasing organizational level. A similar trend was reported for noncommissioned officers, although these two groups did not form a consistent pattern over the six organizational level surveyed.

Miller (1966b) also examined differences in the perceived importance of inner- versus other-directed traits of national union officials Within the industrial union hierarchy, higher-level officials placed significantly more emphasis on inner-directed traits than did lower-level officials. Lower-level industrial union officials placed somewhat greater, but insignificant, emphasis on other-directed traits than did higher-level officials. Within the craft unions, the differences were in the same direction, but were not statistically significant.

Lichtman (1970) related organizational level to other job attitudes and individual characteristics. He reported differences in internal and external control (Rotter, 1966), job-related tension or role strain (Indik, Seashore, and Slesinger, 1964) and need achievement (French, 1948) as a function of organizational level. Analysis of variance indicated no significant differences in need achievement or internal control across organizational level. Job-related tension or role strain was inversely and significantly related to organizational level.

In an experimental study, Dierterly and Schneider (1974) examined the effects of three heirarchical levels (two supervisory and one clerical in a single chain of command) on subjects' perceptions of personal power within a simulated organization and on perceptions of the organization's climate. Position level exerted no main effects on either dependent variable. Bedrosian (1964) found top-level managers (presidents, vice presidents, general managers, department managers) to exhibit higher socioeconomic vocational interests than middle level managers (section heads, second-level supervisors). No differences were found between line and staff managers in level or patterning of vocational interests.

Bernardin and Alvares (1975) found the organization level of employees to be related to ratings of the perceived effectiveness of three methods of conflict resolution. Higher-level employees (versus lower-level employees) rated forcing behaviors to be more effective methods. Conversely, lower-level employees (versus higher-level employees) rated confrontation behaviors to be more effective. These differences in the perceived effectiveness of conflict-resolution methods were found to be related to differences in rated leadership effectiveness by level.

Finally, Downey, Hellriegel, Phelps, and Slocum (1974) reported that the organizational level of their respondents (104 management personnel of a specialty steel firm in central Pennsylvania) moderated the correlations between five satisfactions and six perceived organizational climate measures.

These seven studies, then, indicate that (as in the earlier Porter and Lawler review) organizational level is related to differences in work attitudes other than job satisfaction and to other individual characteristics.

Summary

Taken superficially, the research as reviewed might lead one to believe that organizational level is indeed related to job attitudes as Porter and Lawler (1965) concluded. It appears that most higher-level personnel are more satisfied with most job aspects. However, research since the Porter and Lawler review also suggests that simple univariate conceptualizations and analyses are inadequate. The relationships between organizational level and job attitudes and behavior appear to vary by sample, and to some degree by method. Moreover, a number of studies have explicated differences in job attitudes by individual characteristics as well as by hierarchical level. It also appears that many of the structural variables are interrelated, and that their effects are interactive. When this is the case, univariate analyses are clearly not appropriate, and one's confidence in the results based on that type of analysis is necessarily lessened.

SUBORGANIZATIONAL: LINE AND STAFF POSITIONS

The distinction between line and staff personnel has typically been drawn on the basis of task function. Those individuals involved in the primary output of the organization are termed line, while those whose function only indirectly involve primary output are staff. The latter are often involved in coordination, control, and support of line positions (Price, 1972).

Following a similar definition, the Porter and Lawler (1965) review concluded that:

> Staff managers derive less satisfaction from their jobs, feel they have to be more other-directed, and exhibit different patterns of behavior [than line managers] (p. 33).

However, Porter and Lawler (1965) also suggested that line and staff each form separate hierarchies of authority, and thus an individual could vary by function as well as organizational level. While most research to be reviewed below has controlled for this in some manner, few have considered this problem to the extent of estimating net and joint effects of level and function in the same analysis.

Relation to Attitudes

Two of the studies involving the organizational level/job satisfaction hypothesis in military samples also included line/staff distinctions. In the Johnson and Marcum (1968) study, officers alternated between line and staff positions during their career. Respondents were asked to indicate their need satisfaction concerning recognition or "credit for accomplishment" in line and staff positions they had occupied. No significant differences were found.

Mitchell (1970) reported need satisfaction separately for command and staff personnel within each military rank. Among the highest-ranking officers, commanders were significantly more satisfied over all need categories than were staff commanders in four of five possible comparisons. No significant differences in need satisfaction were found in any other intralevel line/staff comparisons.

Turning now to studies sampling individuals in business and industrial organizations, Lawler and Porter (1966) found essentially no relationship between managers' satisfaction with their pay and line, staff, or combined line/staff position in either zero-order or partial correlational analyses. A finding of no relationship between line, staff, or combined line/staff positions and several need-related variables (fulfillment, importance, satisfaction, and possibility of fulfillment) was reported by Cummings and El Salmi (1970) using univariate analysis, and by El Salmi and Cummings (1968) using multivariate analysis.

Herman and Hulin (1972) also examined the relation of line/staff position to job satisfaction. This multivariate study (described in more detail in the preceding section) related line/staff position to seventeen attitude variables using discriminant function analysis. The multivariate analog of the omega squared measure of association indicated that the function of line/staff position was the second most powerful predictor of job attitudes ($\omega^2 = .60$).

One additional study pertaining to attitudes other than job satisfaction as a function of line/staff position has been reported by Lifter, Bass, and Nussbaum (1971). This study sought to relate the line and staff position of 122 first-line supervisors to differential levels of the perceived importance of effort expenditure in determining salary increases. The results indicated that neither supervisors' perceived actual importance of effort nor their perceptions of normative importance of effort differed significantly as a function of line or staff position. Unfortunately, the line and staff supervisor's average span of control also differed significantly (line managers supervised an average of 8.8 more subordinates than did staff managers; $p < .001$). This confounding was not controlled for, and hence the results are totally ambiguous.

Summary

The seven studies reviewed above have not provided strong support for the existence of differential need satisfaction associated with line and staff positions. While Johnson and Marcum (1968) found no differences in recognition or accomplishment, Mitchell's (1970) military data suggests the possibility of an organizational level by line/staff function interaction. The Lawler and Porter (1966), El Salmi and Cummings (1968), and Cummings and El Salmi (1970) studies, although designed to test this hypothesis, did not find evidence of an interactive relationship. It is important to note that these studies (although varied in several respects) all had the common feature of measuring need satisfaction as a dependent variable. The Herman and Hulin (1972) study not only used instruments of different format, but of more specificity, which might be expected to yield specific attitudes affected by line/staff differences. Their data suggest, however, that such functional differences may relate to attitudinal differences concerned with *specific* interrelationships between line and staff groups, rather than need satisfactions in general.[4]

SUBORGANIZATIONAL: SPAN OF CONTROL

Porter and Lawler (1965) after noting that there had been considerable prescriptive debate on the subject, concluded that little research had been published on the effects of span of control, making it impossible to draw conclusions concerning span's impact on attitudes and behavior. They also noted that the effects of span of control may covary with other organizational variables such as hierarchical level and technology. This lack of empirical support and/or the proposed complexity of the phenomena may have bewildered, but not motivated, future researchers, inasmuch as only two studies pertaining to span of control have been reported during the period of this review.[5]

McDonald and Gunderson (1974) found a positive relation between number of persons supervised and superior job satisfaction in a sample of 5,851 Navy enlisted men. Span of control significantly and positively entered a multiple-regression equation in a cross-validation sample. In a longitudinal study of 151 engineers, Farris (1969) found that each of four performance measures (number of patents, number of technical reports, and two supervisory ratings) were positively related to span of control in two time periods. While these relationships were generally statistically significant, their magnitudes were generally small to moderate.

SUBORGANIZATIONAL: SUBUNIT SIZE

Subunit size, defined as any clearly delineated intraorganizational work unit, also has been decreasingly studied since 1965. Porter and Steers (1973) note that no post-1965 studies were found relating subunit size to absenteeism and turnover, and the present review found only one study relating subunit size to individual attitudes and one relating it to individual behavior.

Cummings and El Salmi (1970) found that subunit size was not significantly related to need fulfillment or need importance. Need satisfaction and the possibility of need fulfillment both increased with subunit size, although the former did not reach traditionally accepted significance levels. These results stand in contradiction to research reviewed by Porter and Lawler (1965), who found that need and job satisfaction *decreased* with increasing subunit size. Cummings and El Salmi suggest that differences between their results and previous research may be due to sample characteristics (managerial versus nonmanagerial). Unfortunately, with the paucity of other studies since 1965 on either blue-collar or managerial samples, this is difficult to evaluate.

Cummins and King (1973) examined the interaction between group (subunit) size and task structure as they impact group member performance in a large manufacturing plant. Subunit size was found to be positively related to productivity ratings only for highly structured, routine tasks. In subunits working on unstructured, ambiguous tasks, size and productivity were negatively, though not significantly, related.

TOTAL ORGANIZATION: SIZE

Total organization size is one area in which previous research was not particularly abundant. Much of the pre-1965 research had focused on subunit size to the neglect of total size, an outcome which led Porter and Lawler to conclude in a very tentative manner that job satisfaction tended to decrease with organization size. They also noted the possibility that organization size and hierarchical position might interact as evidenced by differential effects of size on the job satisfaction of blue-collar workers versus executives.

Relation to Attitudes

Several studies relating organization size to aspects of job and/or need satisfaction have been reported in the last decade. Lawler and Porter (1966) found that organization size had a small but statistically significant zero-order and partial correlational relationship to managerial pay satisfaction. Cummings and El Salmi (1970) found that while total size was not related to need satisfaction or need importance, need fulfillment was greater in medium-sized companies than in either large or small firms. In a multivariate analysis of this data, El Salmi and Cummings (1968) found significant interaction effects of organization level and total size on need fulfillment. In small organizations, the highest-level executives experienced more need fulfillment than did their counterparts in large organizations. However, lower levels of managers in large firms reported more fulfillment than did those in small firms.

In a sample of sixty chapters of an undergraduate business fraternity, Osborn and Hunt (1975) found a significant, positive correlation between organizational size and satisfaction with work as well as with overall satisfaction as measured by the JDI. In addition, they found that size interacted with leader consideration and initiating structure in relation to several JDI satisfaction dimensions. Multiple linear regression analysis indicated that size contributed little explained variance to satisfaction beyond that contributed by leader behaviors.

One study relating attitudes other than need or job satisfaction to total organization size has been reported by England and Lee (1973). Organization size and perceived organizational goals were examined in an international sample (U.S.A., Japan, Korea). Perceptions of organizational goals were measured by eight dimensions of England's Personal Value Questionnaire (England, 1967). In general, the concern with organizational goals increased with size of the organization, and in particular three specific goals (concern with high productivity, profit maximization, and organizational growth) showed significant linear increases with organization size. There were no cross-cultural differences.

Relation to Behavior

Only one study relating total organization size to job behaviors at the individual level of analysis has been found.[6] Ingham (1970) related organizational size (as measured by number of employees) to turnover and absenteeism of industrial workers in England. While there was no relationship between size and turnover, absenteeism showed a significant positive linear relationship with total organization size.

Summary

While the number of researchers studying attitudinal and behavioral differences associated with varying organization size has increased since Porter and Lawler's (1965) review, there has not been enough data reported to justify strong conclusions. Available evidence suggests that size may interact with other structural variables (such as hierarchical level and subunit size) in determining such differences. It also may be that the effects of total organization size operate through intervening constructs at the group and individual levels of analysis. Certainly, organizational size can be expected to influence information processing, communication systems, and control systems. These, in turn, may exert direct impacts on individual attitudes and behaviors as well as indirect impacts through intra- and intergroup processes.

TOTAL ORGANIZATION: VERTICAL AND HORIZONTAL COMPLEXITY

Tall versus Flat Shape and Specialization

Price (1972) has defined complexity as "the degree of structural differentiation within a social system" (p. 70). Encompassed within this definition are two subdimensions. The first of these is vertical complexity or tall versus flat shape. This dimension is generally defined and measured as the number of hierarchical levels relative to the organization's size. The second subdimension is horizontal complexity, and focuses on specialization or "role differentiation." Research relating to both of these dimensions will be reviewed.

Relation to Attitudes

Five studies have related vertical complexity and job attitudes. El Salmi and Cummings (1968) found that vertical complexity interacted with hierarchical level in its effect on need satisfaction. At the highest managerial levels, complex (tall) structures were associated with more need satisfaction than were intermediate or flat structures. At lower managerial levels, the relationship was reversed, with flat and intermediate structures being associated with more need satisfaction.

Ghiselli and Johnson (1970) found that tall or flat shape moderated the relationship between managerial need satisfaction and organizational success. Tall or flat shape was measured by the total number of hierarchical levels in the organization within a given total size category. Using a "slightly shortened version" of the PNSQ and Ghiselli's (1964) index of managerial success, the correlation between need satisfaction and success was determined for 413 managers from a diverse group of organizations. These relationships were then compared between tall and flat organizations over each need category. Only for higher-order needs (esteem, autonomy, and self-actualization) was need satisfaction significantly more positively related to managerial success in tall than in flat organizations.

In a similar study, Ghiselli and Siegel (1972) investigated the relationship between managerial success and attitudes toward authoritarian-democratic leadership styles as moderated by tall or flat shape. Four dimensions of managers' leadership attitudes were measured by an instrument developed by Haire, Ghiselli, and Porter (1966) and each of these was correlated with managerial success. These correlations, which were generally quite low (range, $-.14$ to $.29$), re then compared between tall and flat organizations. There were no ...ferences in the leadership attitude-success relationship on the "faith in others" or "participation" dimensions. The "sharing information" (with subordinates) dimension was negatively related to success in flat organizations ($r = -.14$)

while this relationship was positive ($r = .10$) in tall organizations. With respect to "internal group control," managers holding authoritarian views were more successful in flat organizations ($r = .29$) than in tall organizations ($r = .02$).

Gannon and Paine (1974) related the vertical complexity measure of unity of command to several attitudinal dependent variables in a sample of 304 General Services Administration employees, 181 of whom reported only to one superior while 123 reported to two or more superiors. Those reporting to a single boss reported that (1) personnel selections were more likely to be based on ability, (2) they experienced less role conflict and job pressure, (3) they experienced less need for coordination, and (4) that the responsibility and autonomy associated with their jobs were more adequate.

In a study of 295 trade salesmen in three organizations, Ivancevich and Donnelly (1975) found that salesmen in flat (versus tall and medium) organizations expressed greater satisfaction of self-actualization and autonomy needs and expressed less anxiety and psychological stress. Unfortunately, the three organizations differed on characteristics other than organizational shape, making it risky to attribute any of the differences to shape.

Relation to Behavior and Performance

A study relating vertical complexity to behavior was reported by Carzo and Yanouzas (1969), who tested the effects of tall and flat structures on performance (as measured by decision time, profits, and rate of return) in a laboratory experiment. Results indicated that decision time was not affected by the structural variable, while profits and rates of return were somewhat greater in tall than in flat structures. However, the main effects of the structural variable were not statistically significant for any of the three dependent variables.

Ivancevich and Donnelly (1975) reported that while salesmen in flat, medium, and tall structures did not differ on absenteeism and a route-coverage index, the salesmen from the flat organization were rated by their superiors as more efficient (total orders received by a salesman divided by the total number of retail outlets visited).

Horizontal Complexity and Behavior

Hage, Aiken, and Marrett (1971) related horizontal complexity (as measured by the "number of occupational specialties and degree of professional activity;" p. 866) to work-related verbal communications in sixteen health and welfare organizations. Communications were measured both as scheduled and unscheduled exchanges. The frequency of meetings and proportion of staff members involved also were measured. The occupational specialty measure was significantly related to three of eleven communication measures. The degree of professional activities measure was significantly associated with two of the eleven communication measures.

Summary

The studies grouped under the general category of organizational complexity have been diverse in subject population, design, analysis, and dependent variable measured. Clearly, it appears that complexity (particularly in the vertical dimension) affects organizational members' attitudes and behaviors. Results must be interpreted with caution, however, since few replications have been attempted.

TOTAL ORGANIZATION: CENTRALIZED OR DECENTRALIZED SHAPE

Noting that the literature through 1964 focused on a variety of definitions and a plenum of prescriptive formulations as well as an absence of confirmatory empirical evidence, Porter and Lawler (1965) concluded that:

> The studies reviewed offer no clear-cut support for the proposition that decentralization can produce either improved job attitudes or performance (p. 46).

Research on individual attitudinal and/or behavioral correlates of centralization *since* 1964 has not been abundant. The four studies to be reviewed in this section all focus on aspects of member participation in decision making as the operationalization of decentralization.

Relation to Attitudes

Aiken and Hage (1966) focused on alienation from work and from expressive relationships as a function of centralization in health and welfare organizations. Six interview questions concerning satisfaction with respondents' work and two questions concerning satisfaction with coworkers constituted the measures of alienation from work and expressive relationship respectively. Centralization was measured by indices of hierarchy of authority and participation in decision making. Formalization, as indicated by rule observation and job codification, also was measured. Participation in decision making was negatively related, while job codification and rule observation were positively related to alienation from work.

Bachman, Smith, and Slesinger (1966) measured centralization in terms of interpersonal and task control, and five bases of superiors' power or influence (referent, expert, reward, coercive, and legitimate power) over their subordinates. These, in turn, were related to respondents' satisfaction with their office manager. Data analysis was based on zero-order correlations. Satisfaction was referenced to ". . . the way your office manager is doing his job" (p. 130). The results indicated a positive relationship with several task and interpersonal control measures. With respect to power, satisfaction was negatively related to reward, coercive, and legitimate power, and positively related to referent and expert power.

In an experimental study, O'Connell, Cummings, and Huber (1976) found that male student subjects experienced less of three types of tension when making decisions within centralized, hierarchically structured groups. This was particularly the case when the subjects were required to assimilate highly specific information about their decision environment.

Relation to Behavior

The Bachman et al. (1966) study also related the power and control indices of centralization to a standardized measure of the salesmen's performance. Although generally of lesser magnitude, the pattern of results was the same for performance as it was for satisfaction.

In a study related to the one reported by Aiken and Hage (1966) and described above, Hage, Aiken, and Marrett (1971) related the participation in decision-making index of centralization to various measures of scheduled and unscheduled verbal work-related communication. Five of eleven types of scheduled and unscheduled communications were significantly and positively associated with decentralization in zero-order correlational analyses.

Summary

It appears, then, that there is some limited evidence that positive job attitudes increase, and the frequency of verbal communciations increase, as decentralization increases. However, as Porter and Lawler (1965) cautioned, there are likely to be many other variables affecting the desirability of decentralized decision making. Other structural, personal, and task-related variables are likely to be of considerable importance. Aiken and Hage's data suggest that satisfaction with work is affected only by one index of centralization (i.e., participation). Further, the data analyzed by Aiken and Hage (1966) and by Hage et al. (1971) is directly relevant only to professionals and semiprofessionals in service organizations. Generalization beyond these samples requires further evidence.

TENTATIVE CONCLUSIONS

It is tempting to conclude that the research reviewed here has established consistent relationships between structural properties of organizations and members' attitudes and behaviors. One could make such conclusions, particularly in areas such as organizational level where much research has been reported. However, for a number of reasons such conclusions are unwarranted, or warranted only in very limited terms. Problems such as interrelated structural variables, inappropriate designs and analyses, breaches in levels of analysis, variance between measurement techniques, and the lack of coherent conceptual and theoretical frameworks all plague our ability to draw conclusions concerning the research reviewed above.

METHODOLOGICAL CONSIDERATIONS

In order to interpret the observed relationships between organizational and individual variables which have been reviewed above, one must be able to eliminate competing hypotheses concerning research methods as sources of variance. Such hypotheses may derive from the definition and measurement of variables, research designs, and/or analytic procedures. In this section, these methodological problems are considered, and existing alternatives are noted where appropriate.

DEFINITION AND MEASUREMENT OF VARIABLES

Structural Variables

A fundamental prerequisite for comparing studies purportedly dealing with the same topic is agreement on the conceptual definitions of the variables involved. As one author has pointed out, however, ". . . agreement on naming of variables does not necessarily imply conceptual and/or operational agreement" (Pennings, 1973, p. 688). And, as noted in the review section above, such agreement often does not exist.

While lack of comparability is a serious drawback in itself, the problem is compounded in a number of ways when the measures are operationalized. First, as Freeman and Kronenfeld (1973) and Meyer (1971) have pointed out, the unintended redundancy in conceptual definitions leads to biased empirical estimation in the form of tautologies. Such self-fulfilling prophecies stem from "definitional dependencies" at the conceptual level, and lead to positive empirical results in which the observed relationship may be more illusory than real (Freeman and Kronenfeld, 1973).

A second problem involving the interface of conceptual and operational procedures involves the method of gathering information from the organization. The "institutional" or "objective" approach relies on the organization's records and information gathered from a few "key informants" who are generally near the top of the organizational hierarchy and thus are assumed to possess broad knowledge of the organization. An alternative is the "survey" or "subjective" approach, which relies on questionnaires and interviews in attempting to sample a broader spectrum of organizational members. Even when applied to the same construct, in the same sample, these techniques may lead to differing estimates of a given variable. Pennings (1973) and Azumi and McMillan (1973) found very little convergent and discriminant validity between the subjective and objective approaches concerning centralization and formalization. Similar problems may exist with other measures of structural variables. For example, such a "clean" variable as organization size may present difficulties. In the literature reviewed above, size has typically been measured by the number of employees in the organization. However, as Price (1972) points out, this measure confounds differing degrees of labor intensity, and he suggests that "scale of operations" may be a more meaningful construct.

A third problem with the operational measures used in many of the studies reviewed above concerns categorization of potentially continuous variables. For example, studies which trichotomized size (large, medium, or small) or dichotomized vertical complexity (tall versus flat shape) or centralization all lost considerable empirical information. Statistical analyses (to be discussed below) appropriate to categorical variables typically have less power than their counterparts involving continuous variables. Furthermore, estimates of the magnitude and form of main effects and estimation of both the direction and magnitude of interaction effects are less efficient when potentially continuous variables are categorized.

A final problem concerns the control of empirically related structural variables. As many of the studies reviewed here have noted, the dimensions of organization structure are interrelated and may interact in their effects on members' attitudes and behaviors. Multivariate statistical techniques designed to estimate main and interactive effects of nonorthogonal variables are available (e.g., Harris, 1975; Morrison, 1967; Tatsuoka, 1971). However, since survey designs cannot dictate independence of the explanatory variables, one *must* rely on conceptual analysis to assure the appropriateness of the empirical technique and to guide interpretation results (Draper and Smith, 1966, Chapter 8; Gordon, 1968).

Thus theoretical/conceptual analysis is necessary at all stages of the research process, and *particularly* at the stage of definition and measurement of variables if we are to have structural measures that (1) are valid, (2) can serve as the basis for choosing appropriate methods of analysis, and (3) provide a meaningful interpretation of results.

Dependent Variables

In comparison to the structural variables discussed above, some of the individual level dependent variables are less prone to measurement ambiguity. In particular, the DJI (Smith et al., 1969) and the MSQ (Weiss et al., 1967) as measures of job satisfaction are well developed and documented. On the other hand, measures of job behaviors have been almost nonexistent while attitudinal measures other than job satisfaction (e.g., "inner-other directedness") are less well understood. While these shortcomings are regrettable in that they lead to the same problems as noted above concerning structural variables, they will not be discussed explicitly here. Rather, one attitudinal measure which has been widely used in the research reviewed above and which poses some particular methodological problems will be discussed in detail.

PNSQ purports to measure need satisfaction as based on Maslow's need hierarchy theory of motivation. Five need categories (security, social, esteem, autonomy, and self-actualization) are measured by a total of thirteen items. The format of each of these items consists of asking the respondent *(a)* how much of an opportunity there "is now" to satisfy a particular need, *(b)* how much of an opportunity there "should be," and *(c)* how important the need is to the individual. Need fulfillment is measured by the subject's response to the "is now" question while need fulfillment deficiencies are measured as the differences between the "should be" and "is now" responses.

Recently, there have been a number of investigations concerning the characteristics of this instrument. Three studies have been designed to interpret the factor structure of the PNSQ. The results of studies by Herman and Hulin (1973), Roberts, Walter, and Miles (1971), and Waters and Roach (1973) have all shown that the PNSQ does not factor into the five a priori Maslow need categories. Further, the results of the factor analysis do not show a consistent pattern of factors. At best, there seems to be two semi-independent factors representing, perhaps, higher- and lower-order needs.

Two other studies have examined the measurement characteristics of need fulfillment deficiency scores ("d-scores") generated by the PNSQ. Wallace and Berger (1973) found that the d-scores did not meet accepted standards of reliability in any of two managerial and two nonmanagerial samples. Imparato (1972) investigated the scale characteristics of the PNSQ as a function of the size and location of d-scores. In his sample of professional and technical workers, Imparato found that while the size of the d-score was not related to job satisfaction (as measured by the JDI) the location of the d-scores (above and below the midpoint) was positively related to JDI measures of satisfaction.

Finally, Herman and Hulin (1973) and Imparato (1972) investigated the construct validity of the PNSQ via convergent and discriminant validity with the JDI. Both studies failed to find evidence of convergent or discriminant validity.

The results of these studies suggest that:

1. The PNSQ does not represent the five Maslow need categories and does not factor into any consistent pattern.

2. The scale scores (in either the form of the thirteen individual items or the aggregation of these items into the five need categories) are not independent.

3. The PNSQ does not show construct validity with well developed and validated measures of job satisfaction such as the JDI.

4. The d-scores (representing need satisfaction) generated by the PNSQ do not appear to be reliable or to represent interval scales.

Unfortunately, these problems with the PNSQ affect many of the studies reviewed above. In fact, well over one-third of the studies reviewed have used the PNSQ to measure need fulfillment and need satisfaction. In the most heavily researched area (hierarchical level), over half of the studies reviewed have used the PNSQ. The general lack of reliability and validity evidence on the PNSQ, combined with the more consistent results found with better-developed measures of satisfaction (e.g., Herman and Hulin, 1973) suggest that the most parsimonious explanation of the inconsistencies between structural variables and need satisfaction may simply be measurement error.

Recent research has suggested that alternative formulations and operationalizations of a need category concept may provide a firmer base for relating structural constructs to need importance and satisfaction (Alderfer, 1969; Hall and Nougaim, 1968; Lawler and Suttle, 1972; Mitchell and Moudgill, 1976; Schneider and Alderfer, 1973; Wahba and Bridwell, 1974).

RESEARCH DESIGNS

A second general area of methodological concern in interpreting the studies reviewed above focuses on research designs. The first problem here is the lack of longitudinal designs and almost a *complete* reliance on cross-sectional surveys. This approach leaves a number of important questions unanswered. Obviously, no estimates of change or causality can be made with single point-in-time survey designs. Even more important than the potential for oversimplification is the danger of erroneous conclusions based on cross-sectional data. Holdaway and Blowers (1971), for example, have shown that while larger organizational size tended to be associated with small administrative ratios in cross-sectional data, the relationship did not hold up in longitudinal analysis. Farris (1969), on the other hand, did find considerable stability between organizational factors and individual performance over a 6-year period. However, his research also indicated that the assumption of simultaneous measurement (i.e., zero time lag) implicitly made in cross-sectional research may be tenuous.

A second major problem with cross-sectional designs (as used in the research reviewed here) is that they highlight the relationship between structural variables and only a very limited set of variables describing organizational members. While job satisfaction and job performance are clearly of major importance as organizational outcomes, they may be an inadequate sample of the range of the phenomena to be explained.

This problem is compounded by research designs which jump directly from the macro (organizational) level of analysis to the micro (individual) level with simple bivariate and cross-sectional survey designs. For example, in the data reviewed above suggesting that job satisfaction and job performance are positive functions of organizational level, it may be that individuals further up in the hierarchy consistently are more satisfied and/or perform better partially because (1) of personal attributes (such as ability or education), (2) they are selectively promoted to higher positions based on their performance, (3) they have higher expectations concerning salient job outcomes, and/or (4) they receive more organizational rewards.

The use of longitudinal designs, combined with a richer sampling of personal attributes, may well be one method of increasing our understanding of the relationships between structural variables and members' attitudes and behaviors. In fact, a more inclusive model of individual attitudes and behavior may be a first step toward bridging the gap between macro and micro levels of analysis.

A second design consideration involves the use of survey versus experimental or quasi-experimental techniques. The vast majority of the studies reviewed above did not involve manipulation of any structural variable. Such reliance on survey data, while perhaps justifiable on the grounds of access or cost considerations, highlights our inability to make causal arguments. The conclusions of the present review in this respect are the same as Porter and Lawler who, in 1965, warned that:

. . . experimental "proof" of cause-effect relationships between structure and employee attitudes and behavior is elusive and almost nonexistent (p. 47).

The inability to delineate the effects of organizational variables through experimental procedures, combined with problems of interrelated structural variables highlights again the need for sound conceptual analysis, rigorous definition, measurement and control procedures as well as the need for longitudinal designs and a more complete description of organization members' attitudes, behaviors, and personal attributes.

STATISTICAL ANALYSIS

The problems outlined above concerning measurement and design have important bearing on the appropriateness of analytical techniques, and the conclusions derivable from those analyses. With respect to the literature reviewed above, there are two pressing problems: the first centers on the use of inappropriate analytical techniques; the second concerns the evaluation of research using inappropriate analyses. Comments on the second issue are offered in the next section.

In deciding on an appropriate technique for an analysis, one must consider the nature of the phenomena under investigation. It should be clear from numerous studies of organizations, and from much of the literature reviewed above, that organizational structure variables are likely to be interrelated, are often interactive in their effects on members' attitudes and behaviors, and that these effects occur over time. Indeed, Porter and Lawler warned of this 10 years ago.

Organizations appear to be much too complex for a given variable to have a consistent unidirectional effect across a wide variety of types of conditions. . . . there has been a tendency to oversimplify vastly the effects of particular structural variables (Porter and Lawler, 1965, p. 48).

What, then, are the appropriate techniques for dealing with this problem? Clearly, method variance due to definitional dependencies, measurement problems, etc., must be eliminated. But even with these problems resolved, the multivariate nature of the phenomena requires multivariate design and analysis. The need is compounded by an inability to beable to disentangle the effects of structural variables experimentally. What is needed are clear a priori conceptual frameworks matched by appropriate multivariate analyses.

A number of problems arise when univariate or inadequate multivariate models are applied to such complex phenomena as the effects of organizational structure on individual attitudes and behavior. In terms of the general linear regression model (which underlies many of the statistical techniques employed in the research reviewed above) when a relevant explanatory variable which is correlated with the included explanatory variables is omitted, the estimated effects of the included explanatory variables will be biased and inconsistent (Kmenta, 1971; Theil, 1971).

An additional problem arises when multiple univariate statistical tests are used to test correlated hypotheses. Such is the case, for example, when the PNSQ is used to measure need fulfillment and need satisfaction across each of several levels of an organizational hierarchy. As has been pointed out by Herman and Hulin:

> Whenever multiple comparisons are made on correlated dependent variables, the [type-I error rate] for each hypothesis is unknown. Trend analysis on the means of the hierarchical groups, first on the 13 Porter need deficiency items, then on the five derived scale scores . . . [is typically performed]. To the extent that the dependent variables covary, the significance of mean differences will be highly overstated by the assumption of independence made in multiple significance tests (Herman and Hulin, 1973, pp. 118–119).

This is of particular importance when a large number of hypotheses are tested, as has been done in the vast majority of studies using the PNSQ. Further, the tendency for nonsignificant results to be hidden and/or not reported (Dunnette, 1966; Bakan, 1966) leads to an overemphasis of what may be chance results relative to nonsignificant results and/or failures to replicate (Walster and Cleary, 1970).

A final problem arises in the use of the PNSQ, the scales of which ·apparently are not interval. Statistical techniques, then, must be nonparametric, and the use of these techniques limits the testing of interaction hypotheses (although, see El Salmi and Cummings, 1968, as an exception). A final limitation with nonparametric techniques is the lack of measures of the magnitude of effects.

DISCUSSION AND CONCLUSIONS

The limiting factors discussed above clearly must qualify and attenuate the conclusions which may be drawn from the research reviewed. Yet, with those limitations in mind, the following summary statements may be made.

1. The bulk of the research has focused on attitudinal rather than on behavioral differences associated with structural variations. Future research clearly should be directed toward behavioral as well as attitudinal variables. There is little evidence that we will learn much about behavior in organizations by studying attitudinal dependent variables. Furthermore, it is likely that a broader sampling of organizational members' attitudes, behaviors, and personal attributes will increase our understanding.

2. In terms of specific attitudes investigated, job and need satisfaction clearly have attracted the most attention. Need fulfillment and need importance (as measured by the PNSQ) have not shown consistent results. This may, as noted above, be an artifact of that particular instrument. Further use of the PNSQ should be preceded by additional investigation and modification of the instrument.

3. In terms of the structural variables investigated, the most frequently investigated characteristic has been hierarchical level. Given the large number of studies reviewed on this topic, it seems clear that organizational level is positively related to job and need satisfaction. However, it is not clear that this relationship is linear or even that need satisfaction *consistently* increases over ascending organizational levels. Results of level variations also appear to be modified by the type of social system studied (e.g., military versus business) and by interactions with other structural variables. Further, the relationship between organizational level and satisfaction may have been overestimated by a number of univariate studies testing multiple correlated hypotheses. Finally, to the extent that longitudinal and/or field-experimental studies have been absent, we know only that the relationship is associative. Future research should be directed at these questions, and should include more extensive conceptualizations involving other organizational and personal variables.

4. Other structural variables have been researched less than hierarchical level. Only two studies investigating the effects of span of control and two studies concerning subunit size have been found within the time perspective of this review. With respect to line-staff differences, attitudinal measures focusing on specific line-staff outcomes found significant differences while studies using more general measures of need satisfaction found mixed results. Research on total organization properties (total size, complexity, centralization) has been sparse and difficult to compare. While there appear to be some relationships with individuals' attitudes, the available evidence is mixed and suggests that these variables may interact with other structural variables in their effect on organizational members.

It is distressing to note the many problems plaguing our ability to draw strong conclusions from the literature reviewed above. Even more distressing is the fact that Porter and Lawler (1965) pointed out many of the problems (e.g., interactive effects of structural variables and the need for longitudinal and experimental designs) well before most of the present research was done. If, however, we are to benefit by this past research, then we should become aware of and *act on* the problems and potentials facing future researchers.

As a minimal first step, then, we need to focus upon the recurrent problems of measurement, design, hypothesis testing, and analysis. Solutions for these methodological problems are available and should be applied. If this is done, the difficulties of interpreting future results should be lessened. However, even if methodological problems were to be lightened, there would still be difficulty in interpreting results. It appears that we are at a point at which inductive empiricism cannot advance the state of our knowledge much further without the accompaniment of sound conceptual analysis. Simply the explicit awareness that many studies leap from the aggregate macro-organizational level (as independent variables) directly to the individual level of analysis (as dependent variables) should warn us that causality cannot be direct and simple cross-sectional, bivariate relationships between these two discrepant levels of analysis will not provide adequate explanation.

Surely, organizational processes such as selection and promotion of individuals within organizational ranks, the use of valued skills and abilities of individuals in various positions, and organizational controls on individual behavior, to name only a few, must intervene between structural and individual variables. In addition, the psychological processes through which structural characteristics impact individual attitudes and behavior are largely ignored in the studies reviewed. Inductive empiricism alone cannot solve these problems for several reasons. First, the sheer volume of research needed to obtain the empirical information necessary to eliminate rival hypotheses involving such a large set of variables would be overwhelming. Second, and perhaps more important, the interpretation of large amounts of empirical data without sound conceptual analysis is extremely inefficient, if not meaningless. This is not to argue that we should focus solely on theoretical analyses. Rather, rigorous conceptual *and* empirical approaches should be combined to complement one another if this area of inquiry is to progress further.

FOOTNOTES

1. The authors wish to acknowledge helpful comments on this manuscript by Randall Dunham, Alan Filley, Jeanne Herman, and Donald Schwab.
2. Organizational level was operationalized through a company-developed point system

of job evaluation. The most significant input into this system was the skill level required by the job (personal communication, D. P. Schwab).

3. Additional analyses of this variable are being performed by several of the authors concerned with this problem (personal communications, C. L. Hulin, June, 1975; R. B. Dunham, January, 1976).

4. See footnote 3 and associated discussion in the text.

5. As noted in the previous section, the Lifter, Bass, and Nussbaum (1971) study could be construed as pertaining to span of control equally as well as line-staff function, since these two variables were uncontrolled covariates.

6. A number of studies have been found which relate subunit size and/or total size to aggregated behavioral data (e.g., department or plant) beyond the individual level of analysis (e.g., Mahoney, Frost, Crandall, and Weitzel, 1972; Hrebiniak and Alutto, 1973; Ronan and Prien, 1973). These studies have been excluded as being beyond the scope of the present review in that they do not deal with behaviors at the *individual* level of analysis.

REFERENCES

1. Adams, E. F., D. R. Laker, and C. L. Hulin (1977) "An investigation of the influence of job level and functional specialty on job attitudes and perceptions," *Journal of Applied Psychology 62*, 335–343.

2. Aiken, M., and J. Hage (1966) "Organizational alienation: A comparative analysis," *American Sociological Review 31*, 497–507.

3. Alderfer, C. P. (1969) "An empirical test of a new theory of human needs," *Organizational Behavior and Human Performance 4*, 142–175.

4. Azumi, K., and C. J. McMillan (1973) "Subjective and objective measures of organization structure: A preliminary analysis," paper presented at the American Sociological Association meetings, New York.

5. Bachman, J. G., C. G. Smith, and J. A. Slesinger (1966) "Control, performance, and satisfaction: An analysis of structural and individual effects," *Journal of Personality and Social Psychology 4*, 127–136.

6. Bakan, D. (1966) "The test of significance in psychological research," *Psychological Bulletin 66*, 423–437.

7. Bedrosian, H. (1964) "An analysis of vocational interests at two levels of management," *Journal of Applied Psychology 48*, 325–328.

8. Bernardin, H. J., and K. M. Alvares (1975) "The effects of organizational level on perceptions of role conflict resolution strategy," *Organizational Behavior and Human Performance 14*, 1–9.

9. Bidwell, C. E., and J. D. Kasarda (1975) "School district organization and student achievement," *American Sociological Review 40*, 55–70.

10. Campbell, D. T., and J. C. Stanley (1966) *Experimental and Quasi-Experimental Designs for Research,* Chicago: Rand McNally & Company.

11. Carrell, M. R., and N. F. Elberg (1974) "Some personal and organizational determinants of job satisfaction of postal clerks," *Academy of Management Jounral 17*, 368–373.

12. Carzo, R., Jr., and J. N. Yanouzas (1969) "Effects of flat and tall organization structure," *Administrative Science Quarterly 14*, 178–191.

13. Crozier, M. (1972) "The relationship between micro and macrosociology," *Human Relations 25*, 239–251.

14. Cummings, L. L., and A. M. El Salmi (1968) "Empirical research on the bases and correlates of managerial motivation: A review of the literature," *Psychological Bulletin 70*, 127–144.

15. ———, and ———. (1970) "The impact of role diversity, job level, and organizational size on managerial satisfaction," *Administrative Science Quarterly 15*, 1–12.

16. Cummins, R. C., and D. C. King (1973) "The interaction of group size and task structure in an industrial organization," *Personnel Psychology 26*, 87–94.

17. Dieterly, D. L., and B. Schneider (1974) "The effect of organizational environment on perceived power and climate: A laboratory study," *Organizational Behavior and Human Performance 11*, 316–337.

18. Downey, H. K., D. Hellriegel, M. Phelps, and J. W. Slocum, Jr. (1974) "Organizational climate and job satisfaction: A comparative analysis," *Journal of Business Research 2*, 233–248.

19. Draper, N. R., and H. Smith (1966) *Applied Regression Analysis*, New York: John Wiley & Sons, Inc.

20. Dunnette, M. D. (1966) "Fads, fashion, and folderol in psychology," *American Psychologist 21*, 343–352.

21. Eckert, C., and G. Young (1936) "The approximation of one matrix by another of a lower rank," *Psychometrika 1*, 211–218.

22. El Salmi, A. M., and L. L. Cummings (1968) "Managers' perceptions of needs and need satisfactions as a function of interactions among organizational variables," *Personnel Psychology 21*, 465–477.

23. England, G. W. (1967) "Organizational goals and expected behavior of American managers," *Academy of Management Journal 10*, 107–117.

24. ———, and R. Lee (1973) "Organizational size as an influence on perceived organizational goals: A comparative study among American, Japanese, and Korean managers," *Organizational Behavior and Human Performance 9*, 48–58.

25. Farris, G. H. (1969) "Organizational factors and individual performance: A longitudinal study," *Journal of Applied Psychology 53*, 87–92.

26. Freeman, J. H., and J. E. Kronenfeld (1973) "Problems of definitional dependency: The case of administrative intensity," *Social Forces 52*, 108–121.

27. French, E. (1958) "Development of a measure of complex motivation," in J. W. Atkinson (ed.), *Motives in Fantasy, Action, and Society*, Princeton: D. Van Nostrnad.

28. Gannon, M. J., and F. T. Paine (1974) "Unity of command and job attitudes of managers in a bureaucratic organization," *Journal of Applied Psychology 59*, 392–394.

29. Gavin, J. F. (1975) "Organizational climate as a function of personal and organizational variables," *Journal of Applied Psychology 60*, 135–139.

30. Ghiselli, E. E. (1964) *Theory of Psychological Measurement*, New York: McGraw-Hill.

31. ———. (1971) *Explorations in Management Talent*, Pacific Palisades, Calif.: Goodyear.

32. ———, and D. A. Johnson, (1970) "Need satisfaction, managerial success, and organizational structure," *Personnel Psychology 23*, 569–576.

33. ———, and J. Siegel, (1972) "Leadership and managerial success in tall and flat organization structures," *Personnel Psychology 25*, 617–624.

34. Goldberger, A. S., and O. D. Duncan (1973) *Structural Equation Models in the Social Sciences*, New York: Seminar Press, Inc.

35. Gordon, R. A. (1968) "Issues in multiple regression," *American Journal of Sociology 74*, 592–616.

36. Hage, J., M. Aiken, and C. B. Marrett (1971) "Organization structure and communications," *American Sociological Review 36*, 860–871.

37. Haire, M., E. E. Ghiselli, and L. W. Porter (1966) *Managerial Thinking*, New York: John Wiley & Sons, Inc.

38. Hall, D. T., and K. E. Nougaim (1968) "An examination of Maslow's need hierarchy in an organizational setting," *Organizational Behavior and Human Performance 3*, 12–35.

39. Harris, C. W. (1963) *Problems in Measuring Change*, Madison, Wis.: University of Wisconsin Press.
40. Harris, F. J. (1949) "The quantification of an industrial employee survey. I. Method," *Journal of Applied Psychology 33*, 103–111.
41. Harris, R. J. (1975) *A Primer of Multivariate Statistics*, New York: Academic Press, Inc.
42. Harris, T. C., and E. A. Locke (1974) "Replication of white-collar, blue-collar differences in sources of satisfaction and dissatisfaction," *Journal of Applied Psychology 59*, 369–370.
43. Hellriegel, D., and J. W. Slocum, Jr. (1974) "Organizational climate: Measures, research, and contingencies," *Academy of Management Journal 17*, 225–280.
44. Herman, J. B., R. B. Dunham, and C. L. Hulin (1975) "Organizational structure, demographic characteristics, and employee responses," *Organizational Behavior and Human Performance 13*, 206–232.
45. ———, and C. L. Hulin (1972) "Studying organizational attitudes from individual and organizational frames of reference," *Organizational Behavior and Human Performance 8*, 84–108.
46. ———, and ———. (1973) "Managerial satisfactions and organizational roles: An investigation of Porter's need deficiency scales," *Journal of Applied Psychology 57*, 118–124.
47. Holdaway, E. A., and T. A. Blowers, (1971) "Administrative ratios and organization size: A longitudinal analysis," *American Sociological Review 36*, 278–286.
48. Hrebiniak, L. G., and J. A. Alutto (1973) "A comparative organizational study of performance and size correlates in inpatient psychiatric departments," *Administrative Science Quarterly 18*, 365–382.
49. Imparato, N. (1972) "Relationship between Porter's need satisfaction questionnaire and the Job Descriptive Index," *Journal of Applied Psychology 56*, 397–405.
50. Indik, B., S. E. Seashore, and J. Slesinger (1964) "Demographic correlates of psychological strain," *Journal of Abnormal and Social Psychology 69*, 26–38.
51. Ingham, G. (1970) *Size of Industrial Organization and Worker Behavior*, Cambridge, England: Cambridge University Press.
52. Ivancevich, J. M., and J. H. Donnelly, Jr. (1975) "Relation of organizational structure to job satisfaction, anxiety-stress, and performance," *Administrative Science Quarterly 20*, 272–280.
53. James, L. R., and A. P. Jones (1974) "Organizational climate: A review of theory and research," *Psychological Bulletin 81*, 1096–1112.
54. ———, and ———. (1976) "Organizational Structure: A review of structural dimensions and their conceptual relationships with individual attitudes and behavior," *Organizational Behavior and Human Performance 16*, 74–113.
55. Johannesson, R. E. (1973) "Some problems in the measurement of organizational climate," *Organizational Behavior and Human Performance 10*, 118–144.
56. Johnson, P. V., and R. H. Marcum (1968) "Perceived deficiencies in individual need fulfillment of career army officers," *Journal of Applied Psychology 52*, 457–461.
57. Kmenta, J. (1971) *Elements of Econometrics*, New York: Macmillan, Inc.
58. Krichner, W. K., and N. B. Mousley (1963) "A note on job performance: Differences between respondent and nonrespondent salesmen in an attitude survey," *Journal of Applied Psychology 47*, 223–224.
59. Lawler, E. E., III, D. T. Hall, and G. R. Oldham (1974) "Organizational climate: Relationship to organizational structure, process, and performance," *Organizational Behavior and Human Performance 11*, 139–155.

60. ———, and L. W. Porter (1966) "Predicting managers' pay and their satisfaction with their pay," *Personnel Psychology 19*, 363–373.
61. ———, and J. L. Suttle (1972) "A causal correlational test of the need hierarchy concept," *Organizational Behavior and Human Performance 7*, 265–287.
62. Lichtman, C. M. (1970) "Some intrapersonal response correlates of organizational rank," *Journal of Applied Psychology 54*, 77–80.
63. Lifter, M. L., H. R. Bass, and H. Nussbaum (1971) "Effort expenditure and job performance of line and staff personnel," *Organizational Behaivor and Human Performance 6*, 501–515.
64. Locke, E. A., and R. J. Whiting (1974) "Sources of satisfaction and dissatisfaction among solid waste management employees," *Journal of Applied Psychology 59*, 145–156.
65. Lodahl. T., and M. Kejner (1965) "The definition and measurement of job involvement," *Journal of Applied Psychology 49*, 24–33.
66. MacEachron, A. E. (1977) "Two interactive perspectives on the relationship between job level and job satisfaction," *Organizational Behavior and Human Performance 19*, 226–246.
67. Mahoney, T. A., P. Frost, N. F. Crandall, and W. Weitzel (1972) "The conditioning influence of organizational size upon managerial practice," *Organizational Behavior and Human Performance 8*, 230–241.
68. Maslow, A. H. (1954) *Motivation and Personality*, New York: Harper & Row, Publishers.
69. McDonald, B. W., and E. E. E. Gunderson (1974) "Correlates of job satisfaction in naval environments," *Journal of Applied Psychology 59*, 371–373.
70. Meyer, M. W. (1971) "Some constraints in analyzing data on organizational structures: A comment on Blau's paper," *American Sociological Review 36*, 294–297.
71. ———. (1972) "Size and the structure of organizations: A causal analysis," *American Sociological Review 37*, 434–440.
72. Miller, E. L. (1966a) "Job satisfaction of national union officials," *Personnel Psychology 19*, 261–274.
73. ———. (1966b) "Job attitudes of national union officials: Perceptions of the importance of certain personality traits as a function of job level and union organization structure," *Personnel Psychology 19*, 395–410.
74. Mitchell, V. F. (1970) "Need satisfactions of military commanders and staff," *Journal of Applied Psychology 54*, 282–287.
75. ———, and P. Moudgill (1976) "Measurement of Maslow's need hierarchy," *Organizational Behavior and Human Performance 16*, 334–349.
76. ———, and L. W. Porter (1967) "Comparative managerial role perceptions in military and business hierarchies," *Journal of Applied Psychology 51*, 449–452.
77. Morrison, D. F. (1967) *Multivariate Statistical Methods*, New York: McGraw-Hill.
78. Newman, J. E. (1975) "Understanding the organizational structure-job attitude relationship through perceptions of the work environment," *Organizational Behavior and Human Performance 14*, 371–397.
79. O'Connell, M. J., L. L. Cummings, and G. P. Huber (1976) "The effects of environmental information and decision unit structure on felt tension," *Journal of Applied Psychology 61*, 493–500.
80. Oltman, P. K. (1968) "A portable rod-and-frame test," *Perceptual Motor Skills 26*, 503–506.
81. O'Reilly, C. A., and K. H. Roberts (1975) "Individual differences in personality, position in the organization, and job satisfaction," *Organizational Behavior and Human Performance 14*, 144–150.

82. Osborn, R. N., and J. B. Hunt (1975) "Relations between leadership, size, and subordinate satisfaction in a voluntary organization," *Journal of Applied Psychology 60*, 730–735.

83. Payne, R. (1970) "Factor analysis of a Maslow-type need satisfaction questionnaire," *Personnel Psychology 23*, 251–268.

84. Pennings, J. (1973) "Measures of organizational structure: A methodological note," *American Journal of Sociology 79*, 686–704.

85. Porter, L. W. (1962) "Job attitudes in management: I. Perceived deficiencies in need fulfillment as a function of job level," *Journal of Applied Psychology 46*, 375–384.

86. ———. (1964) *Organizational Patterns of Managerial Job Attitudes*, New York: American Foundation for Management Research.

87. ———, and M. M. Henry (1964) "Job attitudes in management: V. Perceptions of the importance of certain personality traits as a function of job level," *Journal of Applied Psychology 48*, 31–36.

88. ———, and E. E. Lawler III (1965) "Properties of organization structure in relation to job attitudes and job behavior," *Psychological Bulletin 64*, 23–51.

89. ———, and V. F Mitchell (1967) "Comparative study of need satisfaction in military and business hierarchies," *Journal of Applied Psychology 51*, 139–144.

90. ———, and Steers, R. M. (1973) "Organizational, work, and personal factors in employee turnover and absenteeism," *Psychological Bulletin 80*, 151–176.

91. Price, J. L. (1972) *Handbook of Organizational Measurement*, Lexington, Mass.: D. C. Heath & Company.

92. Pritchard, R. D., and B. W. Karasick (1973) "The effects of organizational climate on managerial job performance and job satisfaction," *Organizational Behavior and Human Performance 9*, 126–146.

93. Rhinehart, J. B., R. P. Barell, A. S. De Wolfe, J. E. Griffin, and F. E. Spaner (1969) "Comparative study of need satisfactions in government and business hierarchies," *Journal of Applied Psychology 53*, 230–235.

94. Rice, L. E., and T. R. Mitchell (1973) "Structural determinants of individual behavior in organizations," *Administrative Science Quarterly 18*, 56–70.

95. Roberts, K. H., G. A. Walter, and R. E. Miles (1971) "A factor analytic study of job satisfaction items designated to measure Maslow need categories," *Personnel Psychology 24*, 205–220.

96. Ronan, W. W., and E. P. Prien (1973) "An analysis of organizational behavior and organizational performance," *Organizational Behavior and Human Performance 9*, 78–99.

97. Rotter, J. B. (1966) "Generalized expectancies for internal versus external control of reinforcement," *Psychological Monographs 80* (1, Whole No. 609).

98. Runkel, P. J., and H. E. McGrath (1972) *Research on Human Behavior*, New York: Holt, Rinehart and Winston, Inc.

99. Schneider, B., and C. P. Alderfer (1973) "Three studies of measures of need satisfaction in organizations," *Administrative Science Quarterly 18*, 489–505.

100. Schwab, D. P., and M. J. Wallace, Jr. (1974) "Correlates of employee satisfaction with pay," *Industrial Relations 13*, 78–89.

101. Selvin, H. C., and A. Stuart (1966) "Data dredging procedures in survey analysis," *American Statistician 20*, 20–23.

102. Siegel, S. (1956) *Nonparametric Statistics*, New York: McGraw-Hill.

103. Sims, H. P., Jr., and A. D. Szilagyi (1976) "Job characteristic relationships: Individual and structural moderators," *Organizational Behavior and Human Performance 17*, 211–230.

104. Slocum, J. W., Jr. (1971) "Motivation in managerial levels: Relationship of need satisfaction to job performance," *Journal of Applied Psychology 55*, 312–316.
105. Smith, P. C., L. M. Kendall, and C. L. Hulin (1969) *The Measurement of Satisfaction in Work and Retirement*, Chicago: Rand McNally & Company.
106. Starcevich, M. M. (1972) "Job factor importance for job satisfaction and dissatisfaction across different occupational levels," *Journal of Applied Psychology 56*, 467–471.
107. Szilagyi, A. D., H. P. Sims, Jr., and R. T. Keller (1964) "Role dynamics, locus of control, and employee attitudes and behavior," *Academy of Management Journal 19*, 259–276.
108. Tatsuoka, M. M. (1971) *Multivariate Analysis*, New York: John Wiley & Sons, Inc.
109. Theil, H. (1971) *Principles of Econometrics*, New York: John Wiley & Sons.
110. Wahba, M. A., and L. G. Bridwell (1974) "Maslow reconsidered: A review of research on the need hierarchy theory," *Academy of Management Proceedings* pp. 514–520.
111. Wallace, M. J., and P. K. Berger (1973) "The reliability of difference scores: A preliminary investigation of a need deficiency satisfaction scale," *Academy of Management Proceedings* pp. 421–427.
112. Walster, G. W., and T. A. Cleary (1970) "A proposal for a new editorial policy in the social sciences," *American Statistician 24*, 16–19.
113. Waters, L. K., and D. Roach (1973) "A factor analysis of need-fulfillment items designed to measure Maslow need categories," *Personnel Psychology 26*, 185–190.
114. Weiss, D. J., R. V. Dawis, G. W. England, and L. H. Lofquist (1967) *Manual for the Minnesota Satisfaction Questionnaire*, Minneapolis: University of Minnesota, Industrial Relations Center.
115. Weitzel, W., P. R. Pinto, R. V. Dawis, and P. A. Jury (1973) "The impact of the organization on the structure of job satisfaction: Some factor analytic findings," *Personnel Psychology 26*, 545–557.

TOWARD A THEORY OF ORGANIZATIONAL SOCIALIZATION

John Van Maanen, MASSACHUSETTS INSTITUTE OF TECHNOLOGY

Edgar H. Schein, MASSACHUSETTS INSTITUTE OF TECHNOLOGY

ABSTRACT

The process of organizational socialization is examined in terms of the strategic forms it typically assumes. Attention is also directed to those specific organizational boundaries crossed by persons when acquiring a new work role. A set of propositions is then derived which attempts to link a particular organizational socialization tactic to the behavioral responses of individuals subjected to that tactic during a boundary passage. An underlying theme of the essay is simply that *what* people learn about their work roles in organizations is often a direct result of *how* they learn it.

Research in Organizational Behavior, Vol. 1, pp. 209–264.
Copyright © 1979 by JAI Press, Inc.
All rights of reproduction in any form reserved.
ISBN 0-89232-045-1

I. ORGANIZATIONAL SOCIALIZATION

Introduction

Work organizations offer a person far more than merely a job. Indeed, from the time individuals first enter a workplace to the time they leave their membership behind, they experience and often commit themselves to a distinct way of life complete with its own rhythms, rewards, relationships, demands, and potentials. To be sure, the differences to be found within and between organizations range from the barely discernible to the starkly dramatic. But, social research has yet to discover a work setting which leaves people unmarked by their participation.

By and large, studies of work behavior have, to date, focused primarily upon the ahistorical or "here and now" behavior and attitudes assumed by individual members of an organization that are associated with various institutional, group, interactional, and situational attributes. Relatively less attention has given to the manner in which these responses are thought to arise. In particular, the question of how it is that only certain patterns of thought and action are passed from one generation of organizational members to the next has been neglected. Since such a process of socialization necessarily involves the transmission of information and values, it is fundamentally a cultural matter.[1]

Any organizational culture consists broadly of long-standing rules of thumb, a somewhat special language, an ideology that helps edit a member's everyday experience, shared standards of relevance as to the critical aspects of the work that is being accomplished, matter-of-fact prejudices, models for social etiquette and demeanor, certain customs and rituals suggestive of how members are to relate to colleagues, subordinates, superiors, and outsiders, and a sort of residual category of some rather plain "horse sense" regarding what is appropriate and "smart" behavior within the organization and what is not. All of these cultural modes of thinking, feeling, and doing are, of course, fragmented to some degree, giving rise within large organizations to various "subcultures" or "organizational segments."[2]

Such cultural forms are so rooted in the recurrent problems and common experiences of the membership in an organizational segment that once learned they become viewed by insiders as perfectly "natural" responses to the world of work they inhabit. This is merely to say that organizational cultures arise and are maintained as a way of coping with and making sense of a given problematic environment. That organizations survive the lifetimes of their founders suggests that the culture established by the original membership displays at least some stability through time. Metaphorically, just as biologists sometimes argue that "gene pools" exploit individuals in the interest of their own survival, organizations, as sociocultural forms, do the same. Thus, the devout believer is the Church's way of ensuring the survival of the Church; the loyal citizen is the State's way of ensuring the survival of the State; the scientific apprentice is Physics' way of ensuring the survival of Physics; and the productive employee is the Corporation's way of ensuring the survival of the Corporation.

This is not to say, however, that the transfer of a particular work culture from generation to generation of organizational participants occurs smoothly, quickly, and without evolutionary difficulty. New members always bring with them at least the potential for change. They may, for example, question old assumptions about how the work is to be performed, be ignorant of some rather sacred interpersonal conventions that define authority relationships within the workplace, or fail to properly appreciate the work ideology or organizational mandate shared by the more experienced members present on the scene. Novices bring with them different backgrounds, faulty preconceptions of the jobs to be performed within the setting, including their own, and perhaps values and ends that are at odds with those of the working membership.

The more experienced members must therefore find ways to insure that the newcomer does not disrupt the ongoing activity on the scene, embarrass or cast a disparaging light on others, or question too many of the established cultural solutions worked out previously. Put bluntly, new members must be taught to see the organizational world as do their more experienced colleagues if the traditions of the organization are to survive. The manner in which this teaching/learning occurs is referred to here as the *organizational socialization process.*

What Is Organizational Socialization?

At heart, organizational socialization is a jejune phrase used by social scientists to refer to the process by which one is taught and learns "the ropes" of a particular organizational role. In its most general sense, organizational socialization is the process by which an individual acquires the social knowledge and skills necessary to assume an organizational role. Across the roles, the process may appear in many forms, ranging from a relatively quick, self-guided, trial-and-error process to a far more elaborate one requiring a lengthy preparation period of education and training followed by an equally drawn-out period of official apprenticeship.[3] In fact, if one takes seriously the notion that learning itself is a continuous and life-long process, the entire organizational career of an individual can be characterized as a socialization process (Schein,1971a; Van Maanen, 1977a). At any rate, given a particular role, organizational socialization refers minimally, though, as we shall see, not maximally, to the fashion in which an individual is taught and learns what behaviors and perspectives are customary and desirable within the work setting as well as what ones are not.

Insofar as the individual is concerned, the results of an organizational socialization process include, for instance, a readiness to select certain events for attention over others, a stylized stance toward one's routine activities, some ideas as to how one's various behavioral responses to recurrent situations are viewed by others, and so forth. In short, socialization entails the learning of a cultural perspective that can be brought to bear on both commonplace and unusual matters going on in the workplace. To come to know an organizational situation and act within it implies that a person has developed some commonsensical beliefs, principles, and understandings, or in shorthand notation, a *perspective* for interpreting one's experiences in a given sphere of the work world. As Shibutani (1962) suggests, it provides the individual with an ordered view of the work life that runs ahead and guides experience, orders and shapes personal relationships in the work setting, and provides the ground rules under which everyday conduct is to be managed. Once developed, a perspective provides a person with the conventional wisdom that governs a particular context as to the typical features of everyday life.

To illustrate this highly contingent and contextual process, consider the following hypothetical but completely plausible exchange between an experienced patrolman and a colleague in a police department. When asked about what happened to him on a given shift, the veteran officer might well respond by saying, "We didn't do any police work, just wrote a couple of movers and brought in a body, a stand-up you know." The raw recruit could hardly know of such things, for the description given clearly presumes a special kind of knowledge shared by experienced organizational members as to the typical features of their work and how such knowledge is used when going about and talking about their job. The rookie must learn of these understandings and eventually come to make use of them in an entirely matter-of-fact way if he is to continue as a member of the organization. At root, this is the cultural material with which organizational socialization is concerned.

At this point, however, it is important to note that not all organizational socialization can be assumed to be functional for either the individual or the organization. Organizations are created and sustained by people often for other people and are also embedded deeply within a larger and continually changing environment. They invent as well as provide the means by which individual and collective needs are fulfilled. Whereas learning the orgnaizational culture may always be immediately *adjustive* for an individual in that such learning will reduce the tension associated with entering an unfamiliar situation, such learning, in the long run, may not always be adaptive, since certain cultural forms may persist long after they have ceased to be of individual value. Consider, for example, the pervasive practice in many relatively stable orgnaizations of encouraging most lower and middle managerial employees to aspire to high position within the organization despite the fact that there will be very few positions open at these levels. Perhaps the discontent of the so-called "plateaued manager" can then be seen as a result of a socialization practice that has outlived its usefulness.

Consider also that what may be adjustive for the individual may not be adaptive for the organization.[4] Situations in which the careless assignment of an eager and talented newcomer to an indifferent, disgruntled, or abrasively cantankerous supervisor may represent such a case wherein the adjustive solution seized upon by the new member is to leave the organization as soon as employment elsewhere has been secured. Socialization practices must not therefore be taken for granted or, worse, ignored on the basis that all cultural learning is fundamentally functional. The sieve that is history operates in often capricious and accidental ways and there is little reason to believe that all aspects of a culture that are manufactured and passed on by members of an organization to other incoming members are necessarily useful at either the individual or collective levels.

We must note also that the problems of organizational socialization refer to any and all passages undergone by members of an organization. From beginning to end, a person's career within an organization represents a potential series of transitions from one position to another (Van Maanen, 1977b; Glaser, 1968; Hall, 1976; Schein, 1971a). These transitions may be few in number or many, they may entail upward, downward, or lateral movement, and demand relatively mild to severe adjustments on the part of the individual. Of course, the intensity, importance, and visibility of a given passage will vary across a person's career. It is probably most obvious (both to the individual and to others on the scene) when a person first enters the organization—the outsider to insider passage. It is perhaps least obvious when an experienced member of an organization undergoes a simple change of assignment, shift, or job location. Nevertheless, a period of socialization accompanies each passage. From this standpoint, organizational socialization is ubiquitous, persistent, and forever problematic.

II. BACKGROUND AND UNDERLYING ASSUMPTIONS

With few exceptions, observers of organizations have failed to give systematic attention to the problem of how specific bits of culture are transmitted within an organization. The empirical materials that do exist are scattered widely across all disciplines found in the social sciences and hence do not share a common focus or a set of similar concepts.[5] Even within sociology and anthropology, the disciplines most concerned with cultural matters, the published studies devoted to socialization practices of groups, organizations, subcultures, societies, tribes, and so forth tend to be more often than not anecdotal, noncomparative, and based upon retrospective informant accounts of the process rather than the observation of the process *in situ*. Indeed, then, general statements about the process, content, agents, and targets of organizational socialization are grossly impressionistic. In other words, a total conceptual scheme for attacking the problem may be said to be presently nonexistent.

In this and the sections to follow, we offer the beginnings of a *descriptive* conceptual scheme which we feel will be useful in guiding some much-needed research in this crucial area. Our efforts are directed toward building a sound theoretical base for the study of organizational socialization and not toward proffering any normative theory as to the "effectiveness" or "ineffectiveness" of any given organizational form. We are interested consequently in generating a set of interrelated theoretical propositions about the structure and outcome of organizational socialization processes. Such a theory, to be analytically sound, must accomplish at least three things. *First,* it must tell us where to look within an organization to observe socialization in its most salient and critical forms. *Second,* such a thoery must describe, in a fashion generally applicable to a large number of organizational contexts, the various cultural forms organizational socialization can take. And, *third,* the theory must offer some explanation as to why a particular form of a socialization occurring at a given location within an organization tends to result in certain kinds of individual or collective outcomes rather than others. Only in this fashion will it be possible to build a testable theory to direct research in the area.[6]

Some Assumptions

There are, of course, many assumptions that undergird our theory building efforts in this regard. *First,* and perhaps of most importance, is the well-grounded assumption that individuals undergoing any organizational transition are in an anxiety-producing situation. In the main, they are more or less motivated to reduce this anxiety by learning the functional and social requirements of their newly assumed role as quickly as possible. The sources of this anxiety are many. To wit, psychological tensions are promoted no doubt by the feelings of loneliness and isolation that are associated initially with a new location in an organization as well as the performance anxieties a person may have when assuming new duties. Gone also is the learned social situation with its established and comfortable routines for handling interaction and predicting the responses of others to oneself. Thus, stress is likely because newcomers to a particular organizational role will initially feel a lack of identification with the various activities observed to be going on about them. Needless to say, different kinds of transitions will invoke different levels of anxiety, but any passage from the familiar to the less familiar will create some difficulties for the person moving on.

Second, organizational socialization and the learning that is associated with it does not occur in a social vacuum strictly on the basis of the official and available versions of the new role requirements. Any person crossing into a new organizational region is vulnerable to clues on how to proceed that originate within the interactional zone that immediately surrounds him. Colleagues, superiors, subordinates, clients, and other associates support and guide the individual in learning the new role. Indeed, they help to interpret the events one experiences such that one can eventually take action in one's altered situation. Ultimately, they provide the individual with a sense of accomplishment and competence (or failure and incompetence).

Third, the stability and productivity of any organization depends in large measure upon the ways newcomers to various positions come eventually to carry out their tasks. When the passing of positions from generation to generation of incumbents is accomplished smoothly with a minimum of disruption, the continuity of the organization's mission is maintained, the predictability of the organization's performance is left intact. And, assuming the organizational environment remains reasonably stable, the survival of the organization is assured—at least in the short run. It could be said that the various socialization processes carried out within an organization represent the glue which holds together the various interlocking parts of an ongoing social concern.

Fourth, the way in which individuals adjust to novel circumstances is remarkably similar though there is no doubt great variation in the particular content and type of adjustments achieved (or not achieved). In some cases, a shift into a new work situation may result in a sharply altered organizational and personal identity for an individual, as often occurs when a factory worker becomes a foreman or a staff analyst becomes a line manager. In other cases, the shift may result in only minor and insignificant changes in a person's organizational and personal identity, as perhaps is the case when a craftsman is rotated to a new department or a fireman changes from working the hook-and-ladder to a rescue squad. Yet, in any of these shifts there is still likely to be at least some surprise or what Hughes (1958) calls "reality shock" in store for the individual involved when he first encounters the new working context. When persons undergo a transition, regardless of the information they already possess about the new role, their a priori understandings of that role will undoubtedly change.[7] In short rarely, if ever, can such learning be complete until a newcomer has endured a period of initiation within the new role. As Barnard (1938) noted with characteristic clarity, "There is no instant replacement, there is always a period of adjustment."

Fifth, the analysis that follows makes no so-called functional assumptions about the necessity of organizations to socialize individuals to particular kinds of roles. Indeed, we reject any implicit or explicit notions that certain organizationally relevant rules, values, or motivations must be internalized by people as "blueprints for behavior" if they are to participate and contribute to the organization's continued survival. Such a view leaves little room for individual uniqueness and ignores the always problematic contextual nature of the various ways organizational roles can be filled. While, there are no doubt reasons why certain socialization tactics are used more frequently by one organization than another, these reasons are to be located at the human level of analysis, not at the structural or functional levels. From this perspective, we are very much committed to a symbolic interactionist view of social life, one that suggests that individuals, not organizations, create and sustain beliefs and what is and is not functional (Strauss, 1959). And, as in all matters individual, what is functional for one actor may be dysfunctional for another.

Sixth, and finally, we assume here that a theory of organizational socialization must not allow itself to become too preoccupied with individual characteristics (age, background, personality characteristics, etc.), specific organizations (public, private, voluntary, coercive, etc.), or particular occupational roles (doctor, lawyer, crook, banker, etc.). To be of value to researchers and laymen alike, the theory must transcend the particular and peculiar and aim for the general and typical. At least at this stage in the construction of a theory, there are, as we will show, some rather recognizable and pervasive socialization processes used across virtually all organizational settings and all kinds of individuals that can be understood far more quickly and directly if we do not bog ourselves down in the examination of every dimension that conceivably could influence the outcome of a given process. In other words, the theory we sketch out below does not seek to specify its own applications or uniqueness. What we attempt to accomplish here is the identification of the likely effects upon individuals who have been processed into a general organizational location through certain identified means. Our concern is therefore with the effects of what can be called *"people processing"* devices. The frequency and substantive outcome of the use of these devices across particular types of people, organizations, and occupations are then peripheral to our analytic concern and properly lie beyond the scope of this paper, for these are questions best handled by detailed empirical study.

Plan of This Paper

Given this rather lengthy presentation of introductory matters, the following section, Part III, provides a model of the general setting in which organizational socialization takes place. As such, it is a theoretical depiction of an organization within which certain boundaries exist and therefore demark particular transition points where socialization can be expected to occur. In Part IV, several types of individual responses or outcomes to the socialization process are described in terms we believe to be both organizationally and theoretically revevant. That is, these outcomes are potential effects of a given socialization process and are considered largely in terms of how an individual actually behaves in the new organizational role, not in terms of how an individual may or may not feel toward the new role. It is, therefore, the performance or action of a person that concerns us in this section and not attitudes, motives, beliefs, or values that may or may not be associated with an individual's handling of a given organizational role. In Part V, we present the basic propositions which comprise the core analytic materials of this paper and specify a set of strategic or tactical means by which organizational socialization is typically accomplished. Each strategy or tactic is discussed generally and then related systematically to its probable absence or presence at a given boundary as well as its probable effect upon individuals who are crossing a particular boundary. Part VI concludes the paper with a brief overview and guide to future research in this area.

III. THE ORGANIZATIONAL SETTING: SEGMENTS AND BOUNDARIES

Perhaps the best way to view an organization follows the anthropological line suggesting that any group of people who interact regularly over an extended period of time will develop a sort of unexplicated or tacit mandate concerning what is correct and proper for a member of the group to undertake as well as what is the correct and proper way to go about such an undertaking. At a high level of abstraction, then, members of ongoing business organizations, for example, orient their efforts toward "making money" in socially prescribed ways just as members of governmental agencies orient their efforts toward "doing public service" in socially prescribed ways. More concretely, however, organizations are made up of people each following ends that are to some degree unique. But, since these people interact with one another and share information, purposes, and approaches to the various everyday problems they face, organizations can be viewed as arenas in which an almost infinite series of negotiated situations arise over who will do what, when, where, and in what fashion. Over time, these negotiations result in an emerging set of *organizationally defined roles* for people to fill (Manning, 1970). These roles may or may not be formalized and fully sanctioned throughout the organization, yet they nonetheless appear to have some rather stable properties associated with them which tend to be passed on from role taker to role taker. Of course, these organizationally defined roles hardly coerce each role taker to perform in identical ways. Certainly, whenever a novel problem arises, people come together acting within their roles to confront and make sense of the shared event. Such events, if serious enough, give rise to altered definitions of both the organizational role and the organizational situation in which the role is carried out. From this standpoint, an organization is little more than a situated activity space in which various individuals come together and base their efforts upon a somewhat shared, but continually problematic, version of what it is they are to do, both collectively and individually.[8]

The problem we face here concerns the manner in which these versions of what people are to do—organizationally defined roles—are passed on and interpreted from one role occupant to the next. To do so, however, requires a model of the organization such that members can be distinguished from one another and from outsiders on the basis of as few organizational variables as possible. Furthermore, we need a model that is flexible enough to allow for as much descriptive validity as possible across a wide variety of organizational contexts.

Schein (1971a) has developed a model of the organization that provides a quite useful description of an organizationally defined role in terms of three dimensions that are discernible empirically. The first dimension is a *functional* one and refers to the various tasks performed by members of an organization. Thus, most organizations have departmental structures, which for enterprises located in the business sector of the economy might include the functions of marketing, finance, production, administrative staff, personnel, research and development, and so forth. In the public sector, an organization such as a police department might have functional divisions corresponding to patrol, investigations, communications, planning, records, custody, and the like. Visually, we can map the functional domains of an organization along departmental and subdepartmental or program lines as if each function and subfunction occupied a part of a circle or pie-shaped figure. Each function then covers a particular portion of the circumference of the circle depending upon its proportionate size within the organization. Consider, for example, the XYZ Widget Company as depicted in Figure 1.

Each slice in the figurative representation is a functional division with relatively distinct boundaries such that most persons in the organization

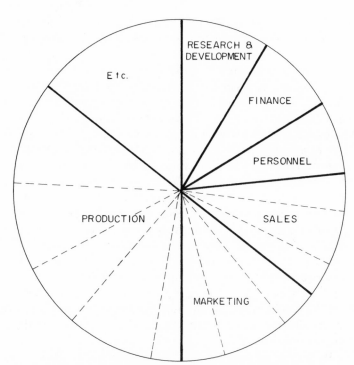

Figure 1. Functional domains of organizations.

could easily locate themselves and others within a slice of the circle. Clearly, no two organizations would be precisely the same, because even if the department and subdepartment structures were identical, the numbers of people contained within each slice would no doubt differ.

The second dimension identified by Schein concerns the *hierarchical* distribution of rank within an organization. This is essentially a matter of who, on paper, is responsible for the actions of whom. It reflects the official lines of supervisory authority within an organization, but does not presume that such authority carries with it the power to direct the behavior of underlings.

According to the model, very decentralized organizations will have, for example, relatively few hierarchical distinctions, whereas very centralized organizations will have many. Mapping this dimension on paper, it would typically make a triangular shape (the traditional organizational pyramid) wherein the highest ranks are held by relatively few people located at the apex. For example, Figure 2 illustrates the hierarchical dimension in five hypothetical, but possible, organizations.

As Figure 2 suggests, a vast number of hierarchical possibilities exist. The XYZ Widget Company (2-A) is perhaps the most typical in that it fits textbook models of a management structure wherein increasing rank is assumed by decreasing numbers of people in a relatively smooth way. The Metropolitan Police Department (2-B) is representative of a large number of service bureaucracies. These agencies have been tagged "street level" organizations, because, in part, most of their membership occupies positions that carry low rank. To wit, over 75 percent of the employees in most police organizations work as patrolmen or investigators, the lowest-ranked positions in these organizations (Van Maanen, 1974; Lipsky, 1971). Zipper Sales, Inc. (2-C) illustrates an organization with a very steep authority structure within which each rank supervises relatively few people but there are many ranks. Pyramid sales organizations and peacetime armies are good examples in this regard. The Zero Research Institute (2-D) displays what a relatively flat hierarchical structure looks like in this scheme. Here there are few ranks for members to seek to ascend. Finally, the Stuffed Mattress Corporation (2-E) is included here to demonstrate something of the range of possibilities available to describe the hierarchical spread which potentially can characterize an organization. As can be seen, the Stuffed Mattress Company has a bulging number of middle managers. In fact, there are more managers than workers in this hypothetical firm.

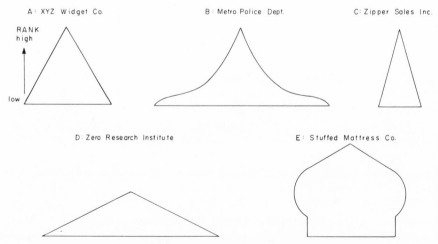

Figure 2. Hierarchical domains of organizations.

The third dimension in Schein's model is the most difficult to conceptualize and concerns the social fabric or interpersonal domain of organizational life. This is fundamentally an interactional dimension and refers to a person's *inclusion* within the organization. It can be depicted as if it were a radial dimension extending from the membership edge of a slice of organizational members in toward the middle of the functional circle. As Figure 3 indicates, movement along this dimension implies that a member's relationship with others in some segment of the organization changes. One moves toward the "center of things" or away toward the "periphery." When examining this dimension, the question must be

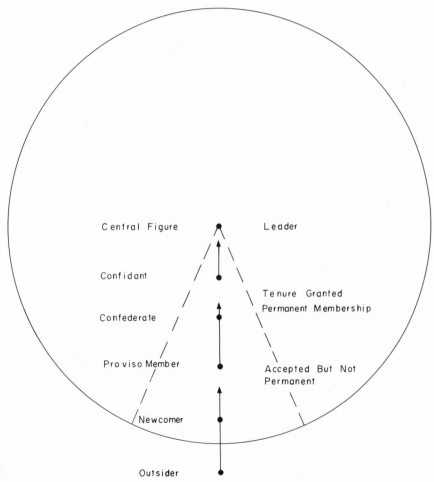

Figure 3.   Inclusionary domains of organizations.

asked how important to others on the immediate sense is a given member's role in the workings of a particular group, department, or organization? Thus, this radial dimension must involve the social rules, norms, and values through which a person's worthiness to a group is judged by members of that group. It concerns in part, then, the shared notions of what the "realwork" of any organizational segment is at any given time. To move along this dimension is to become accepted by others as a central and working member of the particular organizational segment and this can normally not be accomplished unless the member-in-transition demonstrates that he or she too shares the same assumptions as others in the setting as to what is organizationally important and what is not.

Newcomers to most hierarchical levels and functional areas in virtually all organizations inevitably remain "on the edge" of organizational affairs for some time after entrance for a host of reasons. They may not yet be deemed trustworthy by other members on the scene. They may not yet have had time to develop and present the sort of affable, cynical, easy going, or hard-driving front maintained and expected by critical others in the setting which marks membership in the particular segment of the organization to which the newcomer has been assigned. Or, quite typically, newcomers must first be tested either informally or formally as to their abilities, motives, and values before being granted inclusionary rights which then permit them: (1) to share organizational secrets, (2) to separate the *presentational* rhetoric used on outsiders to speak of what goes on in the setting from the *operational* rhetoric used by insiders to communicate with one another as to the matters-at-hand, and/or (3) to understand the unofficial yet recognized norms associated with the actual work going on and the moral conduct expected of people in the particular organizational segment.

In other words, movement along the inclusionary dimension is analogous to the entrance of a stranger to any group. If things go well, the stranger is granted more say in the group's activities and is given more opportunity to display his or her particular skills, thus becoming in the process more central and perhaps valuable to the group as a whole. In short, to cross inclusionary boundaries means that one becomes an *insider* with all the rights and privileges that go with such a position. To illustrate, given a particular function and hierarchical level, passing along the inclusionary dimension can be characterized as going from an outsider to a marginally accepted novice group member, to a confederate of sorts who assists other members on certain selected matters, to a confidant or intimate of others who fully shares in all the social, cultural, and task-related affairs of the group. In certain educational institutions, the granting of university tenure represents the formal recognition of crossing a major inclusionary boundary, as well as the more obvious hierarchical passage.

When the three dimensions—functional, hierarchical, and inclusionary—are combined, the model of the organization becomes analytically most useful and interesting. From a Weberian, ideal-type perspective, organizations are conical in shape and contain within them three generic types of boundaries across which a member may pass (see Figure 4-A). And, as Schein suggests, these boundaries will differ within and between organizations as to both their number and permeability (i.e., the ease or difficulty associated with a boundary passage). Relatively tall organizations (4-B) may have, for example, many hierarchical boundaries yet relatively few functional and inclusionary ones. By implication, members moving up or down in such organizations must orient themselves more to rank and level distinctions among the membership than to the distinctions which result from either functional specialization or social status within a given rank. Military organizations and the elaborate pageantry that surrounds the hierarchical realms within them are unusually good examples of this type. On the other hand, flat organizations (4-C) such as some consulting firms have few hierarchical boundaries but many functional and inclusionary ones. Indeed, in such firms, turnover is high and few members are allowed (or necessarily desire) to pass across the relatively stringent radial dimensions to become central and permanent fixtures within the organization. Prestigious universities represent

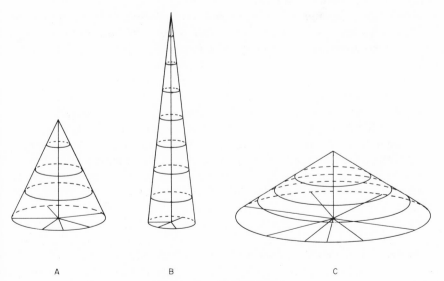

A   B   C

Figure 4. Variations I.

another good illustration in which functional boundaries are exceedingly difficult to rotate through and inclusionary boundaries are guarded by the most rigorous of tenuring policies.

Organizations also differ in the sorts of filtering processes they use to screen, select, and process those members who pass across particular boundaries. Hierarchical boundaries crossed by persons moving upward are associated usually with filtering processes carrying notions of merit, potential, and judged past performance, although age and length of service are often utilized as surrogate measures of "readiness" to move upward in an organization. Functional boundaries usually filter people on the basis of their demonstrated skill or assumed aptitude to handle a particular task. However, when functional boundaries are relatively permeable, as they often are, the filtering process may operate on the premise that there are people in the organization who "need" or "wish" to broaden their work experiences. Finally, inclusionary filters, in the main, represent evaluations made by others on the scene as to another's "fitness" for membership. Of course, such evaluations may be formal, informal, or both. Consider the new patrolman in a large urban police department who must not only serve out a period of official probation successfully, but also must pass a number of unofficial colleague-initiated tests on the street before others in the department will view him as a desirable member of the patrol division (and assigned squad within that division) within the organization (Van Maanen, 1973; Rubenstein, 1973).[9]

Given this model, some key postulates about the socialization process in organizations can be stated.

First, socialization, although continuous throughout one's career within an organization, is no doubt more intense and problematic for a member (and others) just before and just after a particular boundary passage. That is, an individual's anxiety and hence vulnerability to organizational influence are likely to be highest during the anticipatory and initiation phases of an organizational boundary passage. Similarly, the more boundaries that are crossed by a person at any one time, the more profound the experience is likely to be for the person. This is one reason why the outsider-to-insider passage in which an individual crosses over all three organizational boundaries at once is so often marked by dramatic changes in a person, changes of a sort that are rarely matched again during other internal passages of the individual's career (Van Maanen, 1976; Glaser, 1968; Becker et al., 1961; Hughes, 1958).

Second, a person is likely to have the most impact upon others in the organization, what Porter, Lawler, and Hackman (1975) call the "individualization" process and what Schein (1971a) refers to as the "innovation" process, at points furthest from any boundary crossing. In other words, the influence of the organization upon the individual peaks during passage, whereas the individual's influence upon the organization peaks well after and well before any further movement.

Third, because of the conical shape typically displayed by organizations, socialization along the inclusionary dimension is likely to be more critical to lower-placed members than higher-placed members since, according to the model, to move up in the organization indicates that some, perhaps considerable movement has already occurred inward. This presumes, however, an ideal-type, symmetrically shaped organization in which central members from the top to the bottom of the organization all share roughly the same norms and values. In fact, as Figure 5-A shows, organizations may be nonsymmetrically skewed, thus, hierarchically favoring the movement up of only those persons coming from a particular functional or inclusionary location. Consider, for example, those business concerns whose top executives invariably come from only certain functional areas of the organization. Similarly, organizations may also be tipped radically to the side (Figure 5-B). In such cases, certain inclusionary prerequisites for career movements and their associated boundary passages have been more or less altered because "insiders" at one level are "outsiders" at another. Nor are "insiders" in a favorable position to move upward in the organization, as might be the case in more symmetrically shaped firms where certain key values are shared by all "insiders" regardless of level. To take an example, certain organizations headed by reform-minded top officials may make "mountain climbers" out of some members who literally scale the vertical dimension of the organization from an outsider's or noninclusionary position. Yet, it is probably also true that during such a climb the climber has little effect upon any of the various groups in which he or she may have claimed membership, since the climber will never have developed a persuasive or influential position within these organizational segments.[10]

We have now reached the stage where it makes sense to return to the

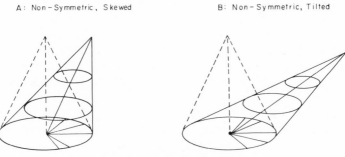

A: Non-Symmetric, Skewed   B: Non-Symmetric, Tilted

Figure 5. Variations II.

individual level of analysis for a time and consider the ways in which people can respond to an organizational socialization process. And, after considering this problem briefly, we can then proceed to the central matters of our concern, the examination of various processes through which organizationally defined roles (consisting of hierarchical, functional, and inclusionary properties) are passed on from generation to generation of organizational members.

IV. INDIVIDUAL RESPONSES TO ORGANIZATIONAL SOCIALIZATION ROLE COMPONENTS—KNOWLEDGE, STRATEGY, AND MISSION

Any organizationally defined role includes what Hughes (1958) called a "bundle of tasks." Whether one is a lathe operator, dentist, beauty operator, or computer programmer, each role includes many specific actions and tasks to be performed, ranging from perhaps sweeping the floor to mediating disputes between colleagues, or, from filling in for an absent coworker to utilizing one's own somewhat special and unique skills in the performance of a given task. In general, then, a role is merely the set of often diverse behaviors that are more or less expected of persons who occupy a certain defined position within a particular social system, in this case, an organization (Parsons, 1951; Newcomb, 1952; Biddle and Thomas, 1966). Moreover, it usually follows that if these expectations are met or exceeded, certain organizational rights and rewards are passed on to the person performing the role. If not, however, is usually follows that certain remedial actions are taken or punishments meted out.

All roles which are created, sustained, and transmitted by people include both content characteristics (i.e., what it is people should do) and process characteristics (i.e., how it is they should do it). The content of a particular role can be depicted both in terms of a general, almost ideological mandate that goes with it and in terms of the general set of mandate-fulfilling actions that are supposed to be performed by the role occupant. Thus, doctors are thought to "heal the sick" by prescribing available "cures" to be found somewhere within the vast catalogue of "medical" knowledge." Similarly, the process associated with the performance of a role also has associated with it general strategies and specific practices. The doctor "does diagnosis" by taking a patient's blood pressure, eliciting a history, reading an X-ray, and so forth. Finally, linked to all these concerns are social norms and rules which suggest, for example, the appropriate mannerisms, attitudes, and social rituals to be displayed when performing various parts of the "bundle of tasks" called a role. Doctors, to continue our illustration, have "bedside manners," often assume a pose of distance or remoteness toward certain emotionally trying events in the lives of their patients, and take a characteristically "all knowing" stance toward most of the nursing personnel with whom they come into contact.

Putting these conceptual matters together, organizationally defined roles can be seen to possess, *first,* a *content or knowledge base* which, if accepted by the role occupant, indicates the range of existing solutions to the given problems encountered regularly on the job. Engineers know, for instance, the heat limits to which certain metals can be exposed before the molecules of the materials rearrange themselves. *Second,* an organizationally defined role includes a *strategic base,* which suggests the ground rules for the choosing of particular solutions. Hence, the engineers may be out to "cut costs" or "beat the competition" in some organizations when designing a particular product or piece of machinery. *Third,* organizationally defined roles are invested historically with something of an *explicit and implicit mission, purpose, or mandate* which is, in part, traceable to the knowledge and strategy bases of the roles, but also is grounded in the total organizational mission and in the relationships that a particular role has with other roles within and outside the organization. Engineering roles, to wit, are defined and supported by other managerial, technical support, and sales roles in both organizational and client contexts, and hence are influenced by their relative position in the overall scheme of things. While the professionalization of a particular occupational role can be viewed as an attempt to reduce such dependencies through the claim made by role practitioners to have an autonomous and special knowledge base, such professionalization in an orgnaizational society such as our is very incomplete.[11] At any rate, the missions associated with organizationally defined roles serve to legitimate, justify, and define the ends pursued by role occupants and, thus, support to some degree the various strategies and norms followed by those presently performing the role.

These three features of an organizationally defined role—knowledge base, strategy, and mission—and the norms that surround them are, of course, highly intertwined. A change in the knowledge base of a given role may alter the means and ends followed by practitioners. Indeed, the recognized failure to achieve a given end may provoke the development of new knowledge. Strategic failures are not unknown either and may lead to disenchantment and change in the mission and knowledge bases of a particular role. Nevertheless, given the situation in which a newcomer is asked to take on an organizationally defined role, that newcomer must respond in some fashion to these three elements.

Responses to Socialization

Custodianship, content innovation, and role innovation   Perhaps the easiest or most expedient response of a newcomer to a given role is to assume a custodial or caretaker stance toward the knowledge, strategies, and missions associated with the role (Schein, 1971b). Taking such a stance, the newcomer does not question but accepts the status quo. Certainly, there are powerful reasons for adopting such a custodial or conforming orientation. First and foremost among them is the plain fact that the inherited past assumed by the newcomer may have much to recommend it in terms of functional achievement. If the enterprise has been successful, why "rock the boat"? One simply learns the substantive requirements of the job and the customary strategies that have been developed to meet these requirements (and the norms of use that surround them) and the successful accomplishment of the mission is assured.

On the other hand, as a newcomer one may feel for a variety of reasons somewhat impatient with or uneasy about the knowledge base of a particular organizational role that is transmitted and, hence, be unwilling to limit oneself to the use of such knowledge in the performance of the role. A newly promoted marketing manager may, for instance, take issue with the quality of some of the regional reports used by his predecessor to inform his decision making. The new manager may then aggressively seek out other information on which to base his decisions. As a result, new strategies and perhaps even new objectives may eventually develop in this department. Similarly, tactical alternatives as to the means to certain ends may be sought out by individuals after assuming a new role. The new marketing manager may decide to involve more salesmen and engineers in group meetings devoted to developing new product lines instead of relying only on his or her own marketing people.

Schein (1971b) refers to this response as "content innovation." It is marked by the development of substantive improvements or changes in the knowledge base or strategic practices of a particular role. The "reformer" in public service agencies, for example, rarely seeks to change the stated objectives of the agency mission, but rather seeks to improve, make more efficient or less corrupt the existing practices by which given ends are collectively sought. In such cases, traditional ends and norms of practice are accepted by the newcomer, but the person is troubled by the existing strategies or technologies-in-use for the achievement of these ends and perhaps is troubled too by the degree to which the traditional norms are circumvented in practice.

Pushing the analysis one more step, an individual may seek to redefine the entire role by attacking and attempting to change the *mission* associated traditionally with that role. This response is characterized by a complete rejection of most of the norms governing the conduct and performance of a particular role. The "Rebel" or "Guerrilla" or "Insurgent" are popular tags we attach to and associate with such responses. Take, for example, Ralph Nader's attempts within certain communities of lawyers who work for the federal government to create and sustain an organizationally defined role of consumer advocate, industrial safety proponent, or even whistleblower. Also note the recent questioning raised by health care officials as to the appropriate aims of medical practice. Some doctors have in fact argued vigorously in both words and deeds for professional roles that are proactive and preventive-centered rather than the historically fixed reactive and treatment-centered roles. Schein (1971b) has called this response "role innovation in that a genuine attempt is made by a role holder to redefine the ends to which the role functions.

Thus, there are two poles toward which a newcomer's response to an organizationally defined role can gravitate. At one extreme is the caretaking response, marked by an acceptance of the role as presented and traditionally practiced by role occupants. We will label this response and the various forms it can take as "custodial." At the other extreme, we can group responses of an "innovative" nature. Perhaps most extreme are those responses which display a rejection and redefinition of the major premises concerning missions and strategies followed by the majority of the role occupants to both practice and justify their present role—what we label here "role innovation." Less extreme, but perhaps equally as innovative in some cases are those responses indicative of an effort to locate new knowledge on which to base the organizationally defined role or improved means to perform it—what we label here "content innovation." Of course, such new knowledge, if discovered, may lead only to a further rationalization of the present practices and goals, but, nevertheless, the search itself, to differing degrees, represents something of an innovative response. For our purposes, then, those individuals who after assuming a given role seek actively to alter its knowledge base, strategic practices, or historically established ends display a generic response type we will label "innovative," which can be further broken down into role innovation and content innovation.

The central but still nagging question remains, of course, as to the reasons that provoke one or the other response types. Certainly, individuals vary in their backgrounds, value systems, and predispositions to calmly accept things as they are or to vigorously strive to alter them. It is true too that changes in the larger environment within which organizationally defined roles are played out may force certain changes upon role occupants despite perhaps vehement resistance or whatever particular backgrounds, values, or predispositions define those who presently perform a given role. But, these factors go well beyond our interests here, for they essentially lie outside an organizational analysis. *The causal mechanism we seek to examine here is the organizational socialization process itself.* Therefore, the argument to follow suggests that there are particular forms of socialization that can enhance or retard the likelihood of an innovative or custodial response to an organizationally defined role no matter what the attributes of the people being processed or how the particular environment is characterized within which the process occurs. What these forms are and how they work is the topic now to be addressed.

V. PEOPLE PROCESSING—THE TACTICAL DIMENSIONS OF ORGANIZATIONAL SOCIALIZATION AND THEIR EFFECTS

The phrase "tactics of organizational socialization" refers to the ways in which the experiences of individuals in transition from one role to another are structured for them by others in the organization (Van Maanen, 1978). These tactics may be selected consciously by the management of an organization, such as the requirement that all newcomers attend a formal training session or orientation program of some kind before assuming the duties of a particular role. Or they may be selected "unconsciously" by management, representing merely precedents established in the dim past of an organization's history, such as the proverbial "sink or swim" method of socialization used on certain jobs by which individuals must learn how to perform the new role on their own. From the perspective of those learning the role, the selection of teaching methods is often made by persons not of their own situation but rather by those long gone. Yet, these choices may still bind contemporary members of the organization. Or, teaching methods may arise simply from certain latent and unexamined premises or assumptions underlying present practices. However, regardless of the manner of choice, any given tactic represents a distinguishable set of events which influence the individual in transition and which may make innovative responses from that individual more likely than custodial (or vice-versa). It is possible, therefore, to denote the various tactics used by organizations and then to explore the differential results of their own use upon the people to whom they are directed.

The analysis presented in this section explores these tactics from primarily a structural standpoint. That is, we are interested in describing various forms and results of socialization as they occur when persons move across hierarchical, functional, and inclusionary boundaries. The main focus is consequently upon the external or structural properties peculiar to a specified tactic. The tactics are essentially process variables akin to but more specific than such general transitional processes as education, training, apprenticeship, or sponsorship. Furthermore, the process variables themselves are not tied to any particular type of organization. Theoretically, at least, they can be used in virtually any setting in which individual careers are played out, be they business careers, school careers, political careers, service careers, blue-, white-, or pink-collar careers, civil service careers, and so on. The analysis follows, then, the most fundamental premise that people respond to particular organizationally defined roles differently not only because people and organizations differ, but also because socialization processes differ. And, like a sculptor's mold, certain forms of socialization can produce remarkably similar outcomes no matter what individual ingredients are used to fill the mold or no matter where the mold is typically set down.

Each tactic we discuss below operates in a way that somewhat uniquely organizes the learning experiences of a newcomer to a particular role. Although much of the evidence presented here on the effects of a given strategy comes from studies conducted on the outsider-to-insider passage wherein a person first becomes a member of the organization, the analysis seems to go beyond these transitions by examining the effects of each tactic across the three organizational boundaries separately, reinforcing, at the same time, the proposition that socialization occurs periodically throughout the organizational careers of individuals.

The various tactics we will describe are not mutually exclusive. Indeed, they are usually combined in sundry and sometimes inventive ways. The effects of the tactics upon people are consequently cumulative. Except for a short summary section, we discuss each tactic in relative isolation. However, the reader should be aware that any recruit to an organizational position often encounters all the listed tactics simultaneously. Additionally, each tactic is discussed along with its counterpart or opposing tactic. In other words, each can be thought of as existing on a continuum where there is a considerable range between the two poles.

The term "tactic" is used here to describe each of the listed processes, because the degree to which any one tactic is used by an organization is not in any sense a "natural" or prerequisite condition necessary for socialization to occur. In other words, socialization itself always takes place at boundary transitions by some means or other. And, whether the tactics used are selected by design or accident, they are at least theoretically subject to rapid and complete change at the direction of the management of an organization. In other words, the relative use of a particular tactic upon persons crossing given organizational boundaries can be, by and large, a choice made by organizational decision makers on functional, economic, technical, humanistic, expedient, traditional, or perhaps purely arbitrary grounds. This is an important point, for it suggests that we can be far more self-conscious about employing certain "people processing techniques" than we have been in the past. In fact, a major purpose of this paper is to heighten and cultivate a broader awareness of what it is we do to people under the guise of "breaking them in" to an organizationally defined role. Presumably, if we gain a greater understanding and appreciation for the sometimes unintended consequences of a particular tactic, we can alter the strategy for the betterment of both the individual and the organization.

Van Maanen (1978) has identified at least six major tactical dimensions which characterize the structural side of organizational socialization. These dimensions or processes were deduced logically from empirical observations and from accounts found in the social science literature. We do not assert here that this list is exhaustive or that the processes are presented in any order or relevance to a particular organization or occupation. These are fundamentally empirical questions that can be answered only by further research. We do assert, however, and attempt to demonstrate that these tactics are quite common to a given boundary passage and of substantial consequence to people in the organization in that they partially determine the degree to which the response of the newcomer will be custodial or innovative.

Lastly, we should note that there is seemingly no logical or conclusive end to a list of organizational socialization tactics. The list may well be infinite, for these are essentially cultural forms that are continually subject to invention and modification as well as stabilization and continuity. At least at this juncture in the development of theory, questions concerning the use of and change in the various tactics of socialization are just beginning to be answered by carefully designed research. Our reasons for choosing these particular tactics are simply the visible presence (or omnipresence) of a tactic across what appears to be a wide variety of organizations as well as the seeming importance and power of that tactic on persons who are subjected to it.

The six dimensions we will analyze are:
1. Collective vs. individual socialization processes.
2. Formal vs. informal socialization processes.
3. Sequential vs. variable socialization processes.
4. Fixed vs. variable socialization processes.
5. Serial vs. disjunctive socialization processes.
6. Investiture vs. diversiture socialization processes.

Collective vs. Individual Socialization Processes

Definition Collective socialization refers to the tactic of taking a group of recruits who are facing a given boundary passage and putting them through a common set of experiences together. A number of good examples of this process is readily available: basic training or boot camp in military organizations, pledging in fraternal orders, education in graduate schools for the scholarly and professional trades, intensive group training for salesmen in business firms, management training courses to which groups of prospective or practicing managers are sent for an extended period of common education, and so forth.

At the other extreme, socialization in the *individual* mode refers to the tactic of processing recruits singly and in isolation from one another through a more or less unique set of experiences, Apprenticeship programs, specific intern or trainee assignments, and a plain "on-the-job training" wherein a recruit is expected to learn a given organizationally defined role on his or her own accord are typical examples. As Wheeler (1977) notes, the difference between the two tactical forms is analogous to the batch versus unit styles of production. In the batch- or mass-production case, recruits are bunched together at the outset and channeled through an identical set of events with the results being relatively uniform. In the unit or made-to-order case, recruits are processed individually through a rather different series of events, with the results being relatively variable.

As Becker (1964) and others have argued quite persuasively, when individuals experience a socialization program collectively, the thoughts, feelings, and actions of those in the recruit group almost always reflect an "in the same boat" consciousness. Individual changes in perspective are therefore built upon an understanding of the problems faced by all group members. In Becker's words, "As the group shares problems, various members experiment with possible solutions and report back to the group. In the course of collective discussions, the members arrive at a definition of their situation and develop a consensus."

Collective socialization processes often promote and intensify the demands of the socialization agents. Indeed, army recruits socialize each other in ways the army could never do, or, for that matter, would not officially be permitted to do. Similarly, graduate students are often said to learn more from one another than from the faculty. And, while the socialization agents may have the power to define the nature of the collective problem, the recruits often have more resources available to them to define the solution—time, expereince, motivation, expertise, and patience (or lack thereof). In many cases, collective tactics result in formation of an almost separate subworld within the organization comprised solely of recruits, complete with its own argot, areas of discourse, and unique understandings. A cultural perspective is developed that can be brought to bear upon common problems faced by the group.[13] Dornbush (1955) suggested, for example, that a "union of sympathy" developed among recruits in a Coast Guard Academy as a result of the enforced regimentation associated with the training program. Sharing similar difficulties and working out collective solutions clearly dramatized to the recruits the worth and usefulness of colleagial relationships.

Individual strategies also induce personal change. But the views adopted by people processed individually are likely to be far less homogeneous than the views of those processed collectively. In psychoanalytic training, for example, the vocabulary of motives a recruit-patient develops to interpret his or her situation is quite personal and specific compared to the vocabulary that develops in group therapy (Laing, 1960). Of course, such socialization can result in deep individual changes—what Burke (1950) refers to as "secular conversion"—but they are lonely changes and are dependent solely upon the particular relationship which exists between agent and recruit.

Apprenticeship modes of work socialization are sometimes quite similar to the therapist-patient relationship. If the responsibility for transforming an individual to a given status within the organization is delegated to only one person, an intense, value-oriented process is most likely to follow. This practice is common whenever a role incumbent is viewed by others in the organization as being the only member capable of shaping the recruit. Caplow (1964) notes the prevalence of this practice in the upper levels of bureaucratic organizations. Since the responsibility is given to only one organizational member, the person so designated often becomes a role model whose thoughts and actions the recruit emulates. Police departments, craftlike trades, and architectural firms all make extensive use of the individual socialization strategy. Outcomes in these one-on-one efforts are dependent primarily upon the affective relationships which may or may not develop between the apprentice and master. In cases of high affect, the new member is liable to quickly and fully appreciate and accept the skills, beliefs, and values of his or her mentor and the process works relatively well. However, when there are few affective bonds, the socialization process may break down and the hoped-for transition will not take place.

From this standpoint, individual socialization processes are most likely to be associated with complex roles. Further, such modes are most frequently followed when there are relatively few incumbents compared to many aspirants for a given role and when a collective identity among recruits is viewed as less important than the recruit's learning of the operational specifics of the given role.

On the other hand, collective socialization programs are usually found in organizations where there are a large number of recruits to be processed into the same organizationally defined role; where the content of this role can be fairly clearly specified; and, where the organization desires to build a collective sense of identity, solidarity, and loyalty within the cohort group being socialized. Overall, the individual processes are expensive both in time and money. Failures cannot be easily recycled or rescued by reassignment. And, with growing bureaucratic structures, collective socialization tactics, because of their economy, ease, efficiency, and predictability, have tended to replace the more traditional individual modes of specialization such as apprenticeship and "on-the-job training" in the modern organization (Salaman, 1973; Perrow, 1972; Blau and Schoenherr, 1971).

Given these considerations, we can now derive some propositions about the relationship of this socialization dimension to boundary passages and recruit responses.

Propositions

1. Collective socialization is most likely to be associated with functional boundaries (where new skills of a technical or functional nature have to be learned) or with the external—nonmember to member—inclusionary boundary of a given organizational segment (where some period of orientation or training is required before it is felt recruits are capable of entering into even the simplest of role relations associated with the new role).

2. Individual socialization is most likely to be associated with hierarchical boundaries where preparation for promotion requires the complex learning of skills, attitudes, and values, and where specific judgments of a given individual must be made by certain others in the organization as to the person's "fitness" for promotion (or demotion). Similarly, one would expect individual socialization to precede passage through the innermost inclusionary boundaries within an organizational segment. To be granted tenure or a very central position in any organizational segment implies that the individual has been evaluated by others on the scene as to his or her trustworthiness and readiness to defend the common interests of other "insiders." Clearly, such delicate evaluations can only be accomplished on a relatively personal and case-by-case basis.

3. Whatever the boundary being passed, collective socialization is most likely to produce a custodial (or, at best, a content innovative) orientation among newcomers. It is least likely to produce role-innovative outcomes, because the group perspective which develops as a result of collective socialization acts as a constraint upon the individual.[14] The likelihood of rebellion must be mentioned, however, because the consensual character of the solutions to the boundary passage problems worked out by the group may allow the members to collectively deviate more from the standards set by the agents than is possible under the individual mode of socialization. Collective processes provide a potential base for recruit resistance.

Classic illustrations of the dilemma raised by the use of the collective strategy can be found in both educational and work environments. In educational settings, the faculty may beseech a student to study hard while the student's compatriots exhort him to relax and have a good time. In many work settings, supervisors attempt to insure that each employee works up to his level of competence while the worker's peers try to impress hupon him that he must not be a "rate buster." To the degree that recruits are backed into the proverbial corner and can not satisfy both demands at the same time, they will typically follow the dicta of those with whom they spend most of their time. Thus, in collective modes, the congruence between agent objectives and the actual perspectives and practices adopted by the group is always problematic. "Beating the system" by selectively ignoring or disobeying certain agent demands is far more likely to occur in a collective socialization process than in an individualized one wherein agent surveillance is closer at hand to correct whatever "wrongs" the newcomer may be learning.

4. Individual socialization is most likely to produce the specific outcomes desired by the socialization agent(s). Because of the relatively greater control an agent has over a recruit in the individual mode, these outcomes can be custodial, content innovative, or role innovative.

The implication here is simply that if one is attempting to train for content or role innovation (i.e., set up socialization situations which will maximize the likelihood of innovative responses), it is probably essential to minimize as much as possible any collective processes, thus avoiding the formation of recruit group norms based on a common or shared fate. More so than individual norms, group norms are likely to be both traditional and custodial in orientation (often reflected by the popular idioms, "the path of least resistance" or the "lowest common denominator"), which serves to severely limit the newcomers' potential responses to their novel work situation.

Formal vs. Informal Socialization Processes

Definition Formal socialization refers to those processes in which a newcomer is more or less segregated from regular organizational members while being put through a set of experiences tailored explicitly for the newcomer. Formal processes, then, leave no doubt as to the recruit's "special" role of the scheme of things organizational (Wheeler, 1966), These processes are illustrated by such socialization programs as police academies, professional schools, various sorts of internships, and apprenticeships in which the activities that are to be engaged in by the apprentice are prescribed officially and clearly.

Informal socialization processes, in contrast, do not distinguish the newcomer's role specifically, nor is there an effort made in such programs to rigidly differentiate the recruit from the other more experienced organizational members. As such, informal tactics provide a sort of laissez-faire socialization for recruits whereby new roles are learned, it is said, through trial and error. Examples here include those proverbial "on-the-job-training" assignments, apprenticeship programs in which the apprentice's role is not tightly specified, and, more generally, any situation where the newcomer is accepted from the outset as at least a provisional member of a work group and not officially placed into a recruit role by the use of specific labels, uniforms, assignments, or other symbolic devices designed to distinguish newcomers from others on the scene.

This dimension is related closely to the collective-individual dimension, but it is, in principle, different. While most collective socialization processes are also formal ones, there are some which are informal. To wit, there are those situations where a cohort of new employees is brought into an organization together, where meetings are held periodically to assess how the group is collectively getting along, but where the work assignments of each member of the cohort are to different departments within which each member of the cohort is trained through informal means. On the other hand, one can also imagine a very formal socialization program existing for an individual which entails the labeling of the person as a recruit and also specifies quite minutely a series of activities that must be performed as part of the training regime. Would-be partners in law firms are often subject to such socialization tactics whereby they must first handle the "dirty work" of the firm for some period. Certainly this sort of "pledge class of one" is not that uncommon in many occupational spheres.

Formal socialization processes are typically found in organizations where specific preparation for new status is involved and where it is deemed important that a newcomer learn the "correct" attitudes, values, and protocol associated with the new role. To put the matter bluntly, the more formal the process, the more concern there is likely to be shown for the recruit's absorption of the appropriate demeanor and stance associated with the target role—that one begins to think and feel like a United States Marine, an I.B.M. executive, or a Catholic priest.

The greater the separation of the recruit from the day-to-day reality of the work setting, the less the newcomer will be able to carry over and generalize any abilities or skills learned in the socialization setting (Bidwell, 1962; Schein and Bennis, 1965). Formal processes concentrate, therefore, more upon attitude than act. Such results may be implicit or unintended, however. Consider, for example, the research which suggests that police recruits, student nurses, and sales trainees commonly denounce their formal training as irrelevant, abstract, and dull. Paradoxically, these newcomers are also expressing in their attitude precisely those components of the valued subcultural ethos that characterizes their particular occupation—autonomy, pragmatism, and the concern for personal style (Van Maanen, 1974; Shafer, 1975; Olesen and Whittaker, 1968).

It is important to note too that formal periods of socialization not only serve to prepare recruits to assume particular statuses in an organizational world, they also serve to provide an intensive period in which others in the organization can rather closely judge the newcomer's commitment and deference to the critical values of the occupation. Recruits in police academies are, for example, assessed quite thoroughly by staff members as to their loyalty not only to the organization, but to their fellow recruits as well. And, those who do not adhere to particular norms thought crucial to the trade (e.g., the "no rat rule") are ushered as unceremoniously out of police departments as they were rushed ceremoniously in (Manning and Van Maanen, 1978). It is true, of course, that merely passing through a rigorous formal process serves also as a test of the recruit's willingness to assume the new role. Often, simply the sacrifice and hard work it takes a recruit to complete a very long formal process serves effectively to fuse the newcomer to the prepared-for role. Thus, given a lengthy and demanding formal process, it is unlikely that one will later wish to jeopardize the practical value of such a course by quitting or appearing to forget occupational lessons once learned.

Learning through experience in the informal socialization mode is an entirely different matter. First, such tactics place recruits in the position where they must select their own socialization agents. The value of this mode to the newcomer is then determined largely by the relevant knowledge possessed by an agent and, of course, the agent's ability to transfer such knowledge. The freedom of choice afforded recruits in the more informal processes has therefore a price: They must force others in the setting to teach them.[15] Second, mistakes or errors made by recruits in an informal socialization process must be regarded as more costly and serious than mistakes occurring in formal processes. Because real work is interfered with, a recruit who makes a mistake may create considerable trouble for both himself and others. The rookie patrolman who "freezes" while he and his partner strive to settle a tavern brawl on the street rather than in an academy role-playing exercise may find himself ostracized from the inner circle of his squad. The forgetful novice beautician who provokes a customer by dyeing her hair the wrong color may be forced to look elsewhere for an organization in which to complete the mandatory licensing requirement of the trade. Experienced organizational members know full well that "mistakes happen," but a recruit is under a special pressure to perform well during an informal initiation period—or to at least ask before acting.

With these considerations in mind, the following general propositions can now be stated. '

Propositions

1. Formal socialization is most likely to be associated with hierarchical and inclusionary boundary passages wherein a newcomer is expected to assume a new *status* or *rank* in the organization (complete with the values, attitudes, and demeanor that go with such new status). Informal socialization, on the other hand, is most likely to be associated with functional boundary passages wherein the newcomer must learn new skills, methods, or practical abilities. If, however, the new skills to be learned also require a new knowledge base, a formal training period dealing specifically with such knowledge and its use may precede the boundary passage. Since the teaching of such knowledge is likely to occur in idealized or "theoretical" situations in a formal process, an informal process of socialization dealing with the applications of the knowledge will still be required upon the recruit's entrance into the new role.[16]

In effect, this proposition alerts us to the apparent functional necessity for the use of formal socialization tactics when there exists a cultural gap between the organizational segments to be traversed by the individual. For example, a company sending an American manager to head an overseas subsidiary should probably allow for a formal period of socialization, including perhaps language training, briefings on the new culture, guided tours of the key areas, and so forth. All of this must occur under the formal tutelage of someone who knows what sorts of culture shocks are likely to be encouraged during the transition. Such movements are not limited theoretically to hierarchical or inclusionary boundaries, but rather reflect the size of the cultural differences that exist at any boundary. In some organizations, a move from engineering to sales may involve as much culture shock for an individual as a promotion from project leader to group supervisor or a transfer from staff analyst to line manager.

2. Formal socialization tactics are most likely to be found where the nature of the work and/or the values surrounding the work to be performed in the target role are seen to involve high levels of risk for the newcomer, colleagues of the newcomer, the organization itself, and/or clients of the organization.

Thus the training of doctors, professional pickpockets, lawyers, and airline pilots involves long periods of formal socialization largely because the work involved in all these cases is complex, difficult, and usually entails a very high penalty for making a mistake. Formal training for electricians, soldiers, and machinists is also predicated on the need to minimize the minimizable risks—human or otherwise—such as damaging expensive equipment.Where the cost of a mistake is relatively low, informal socialization processes are more likely to be found.

3. Whatever the boundary passage, formal socialization is most likely to produce a custodial orientation.

As implied above, formal tactics tend to emphasize the "proper" or "accepted" ways to accomplish things in an organization. Even the fact that the target role can be presented in isolation from its everyday performance implies that there are available various traditional means of accomplishing the task. However, a caveat is appropriate here, for it is often the case that once recruits have begun to perform the role in an official capacity, they "unlearn" much of what they learned in the formal process and begin to substitute "practical" or "smart" ways of doing things for the "proper" or "standard" strategies they were once taught. From this standpoint, formal socialization processes represent frequently only the "first wave" of socialization and are followed by a "second wave" of informal socialization once the newcomer is located in a particular organizational slot and begins to discover the actual practices that go on there (Inkeles, 1966).[17] Whereas the first wave stresses a broad stance toward the job, the second wave emphasizes specific actions, unique applications of the general rules, and the odd nuances thought necessary by others on the scene to perform the role in the work setting. When the gap separating the two sorts of learning is rather large, disillusionment with the first wave may set in, causing the newcomer to disregard virtually everything learned in the formal socialization process. Thus, while formal processes tend to produce custodial orientations among recruits, these orientations may not be all too stable unless the lessons of the formal process are reasonably congruent with those of the informal process which may follow.

4. Informal socialization, like individual socialization, carries with it the potential for producing more extreme responses in either the custodial or innovative directions than formal socialization.

If, for example, a recruit is assigned in the informal mode to a work group or a boss characterized by an "organization man" orientation, he or she is likely to become very custodial in orientation—at least in the short run. On the other hand, if that same recruit is assigned to a work group or boss characterized by an innovative orientation, he or she might then become quite innovative too. What we are saying, in effect, is that individual and informal socialization are potentially more powerful techniques of shaping work behavior than formal and collective modes, because they involve on-the-job contingencies as well as teaching by people who are clearly doing the work. In contrast, formal processes artificially divide up concerns that must be approached simultaneously on the job and are often under the control of instructors (agents) whose credibility is lacking. It would appear then that if formal and collective processes are to "succeed" from an agent's perspective, first, they must be long enough to almost force recruits to learn their lessons well and perhaps practice them too and, second, they must be run by persons who have considerable legitimacy in the eyes of the recruits.

Sequential vs. Random Steps in the Socialization Process

Definition The degree of formality and the degree to which the process of socialization is collective are, as indicated, associated with major boundary passages, with basic orientation activities, and, most often, with the initial entry of a recruit into the organization. However, for some roles in an organization, the socialization process may cover a broad spectrum of assignments and experiences, taking sometimes many years of preparation. The person wishing to become a medical specialist has, for instance, to go through an undergraduate premed program, medical school, internship, and residency before becoming eligible to simply take the specialist board examinations. Similarly, a person being groomed for a general manager position may have to rotate through several staff positions as a junior analyst, through various functional divisions in order to learn the "areas of the business," and through various supervisory levels to build up experiences and a so-called "good track record" which would then warrant the ultimate "goal job" (Gordon, 1977).

Sequential socialization refers to the degree to which the organization or occupation specifies a given sequence of discrete and identifiable steps leading to the target role. *Random socialization* occurs when the sequence of steps leading to the target role is unknown, ambiguous, or continually changing. In the case of most professional training such as medicine, we have a very sequential process in that the steps leading to the professional role must be negotiated in a specific order. In the case of the general manager, however, we have a sequential process only with respect to supervisory or rank levels, but the sequence of rotating through functional positions and divisions is often unspecified and, in some organizations, left more or less to "random" events. Thus, in random processes, while there may be a number of steps or stages leading to the taking of certain organizational roles, there is no necessary order specified in terms of the steps that are to be taken.

When examining sequential strategies, it is crucial to note the degree to which each stage builds or expands upon the preceding stage. For example, the courses in most technical training programs are arranged in what is thought to be a simple-to-complex progression. On the other hand, some sequential processes seem to follow no internal logic. Management education is, for instance, quite often disjointed, with the curriculum jumping from topic to topic with little integration across stages. In such cases, what is learned by a recruit in the program is dependent simply upon what is liked best in the sequence. If, however, the flow of topics or courses is harmonious and connected functionally or logically in some fashion, what may seem like minor alterations required of an individual at each sequential stage will accumulate so that at the end persons will "discover" themselves to be considerably different than when they began (e.g., in training for a specific skill). One sees this effect most clearly in the acquisition of complex skill or in the complete "professional" perspective or in the value systems built up after many years of graduate study.[18]

Relatedly, if several agents handle various portions of a sequential process, the degree to which the aims of the agents are common is very important. For example, in some officer's training schools of peacetime military organizations, the agents responsible for physical and weapons training tend to have very different perspectives toward their jobs and toward the recruits than those agents who are in charge of classroom instruction (Wamsley, 1972). Recruits quickly spot such conflicts when they exist and sometimes exploit them, playing agents off against one another. Such incongruities often lead to a more relaxed situation for the recruits, one in which they enjoy watching their instructors pay more attention to each other than they pay to the training program.

We should note that many of these concerns apply to random processes as well. In both random and sequential arrangements, agents may be unknown to one another, they may be quite far apart spatially, and may have thoroughly different images of their respective tasks. Both Merton (1957) and Glaser (1964) have remarked upon the difficulty many scientists apparently have when moving from a university to an industrial setting to practice their trade (a random socialization process). The pattern is seemingly quite disconcerting for many scientists when they discover that their academic training emphasized a far different set of skills, interests, and values than is required in the corporate environment. As Avery (1968) observed, to become a "good" industrial scientist, the individual has to learn the painful lesson that to be able to sell an idea is at least as important as having one in the first place. From this standpoint, empathy must certainly be extended to the so-called juvenile delinquent who receives "guidance" at various times from the police, probation officers, judges, social workers, psychiatrists, and correctional officials. Such a process, sometimes sequential but typically random, evocatively suggests that the person may well learn to be only whatever the immediate situation demands.

In a sequential process, there is likely to be a strong *bias* in the presentation by each agent to make the next stage appear benign. Thus, a recruit is told that if he will just "buckle down and apply himself" in Stage A, Stages B, C, D. and E will be easy. Agents usually mask, unwittingly or wittingly, the true nature of the stage to follow, for, if recruits feel that the future is bright, rewarding, and assured, they will be most cooperative at the stage they are in, not wishing to risk the future they think awaits them. To wit, note the tactics of high school mathematics teachers who tell their students that if they will just work hard in algebra, geometry will be a "cinch." An extreme case of this sequential "betrayal" occurs in state executions, where condemned persons are usually told by their "coaches" on the scene that their demise will be quick, painless, and likely to speed them on their way to a "better place" (Eshelman, 1962).

Given these sensitizing definitions and the qualifications that apply to this socialization tactic, some theoretical propositions can now be stated.

Propositions

1. Sequential socialization is most likely to be associated with hierarchical boundaries.

Hierarchies are typically organized from the outset on the assumption that higher-level positions cannot be fulfilled adequately until lower-level ones have first been fulfilled. Such an assumption is not built into functional or inclusion boundaries where a person can demonstrate a readiness for passage at any given time. At least in part, hierarchies preserve sequential socialization processes in order to maintain the image that the hierarchy itself is a valid base for the distribution of authority. If one could skip levels, the whole concept of authority, it is thought, would be undermined. Of course, in some executive promotions, skipping is accomplished for all practical purposes through the extremely rapid advancement of someone viewed as unusually talented, "fast tracked," fortuitously connected, just plain lucky, or all of these attributes together.

To pass inclusion boundaries may take a long time while one is proving oneself to be trustworthy to many different people, but the process typically does not specify a sequence in which such a test can or must be passed. In the case of functional boundaries, there may be many specific steps associated with the education or training activities involved in preparing to cross the boundary, but, sometimes at least, one may be given a job on the basis of education received at a much earlier time or on the basis of certain experiences which are seen as "equivalent" to education or training. Inclusionary and functional boundary passages are, therefore, associated more with various sorts of random socialization processes.

2. Sequential socialization is more likely to produce custodial orientations among recruits than innovative orientations because the recruits remain "locked in," as it were, to the conforming demands of others in the organization for a long period of time before the target role is achieved. Even the ability of the organization to specify a sequence implies a set of fairly clear norms about what is required to perform the target role. And, the clearer the role, the more likely it is that the training for that role will produce custodial response.

On the other hand, recruits who encounter various socialization experiences in a *random* fashion may find themselves exposed to a wide and diverse variety of views and perceptions of the target role which would make it more likely than is true of sequential socialization to lead to innovative orientations. It would seem therefore that a company who wishes to groom innovative general managers would do well to avoid sequential prcesses and encourage more *ad hoc* decision-making procedures in the organization concerning managerial job moves and training experiences.

Fixed vs. Variable Socialization Processes

Definition   This dimension refers to the degree to which the steps involved in a socialization process have a *timetable* associated with them that is both adhered to by the organization and communicated to the recruit. *Fixed socialization* processes provide a recruit with the precise knowledge of the time it will take to complete a given passage (Roth, 1963). Thus, while organizations may specify various career paths having different timetables, all of these paths may be more or less fixed in terms of the degree to which the recruit must follow the determined timetable. Some management trainees, for instance, are put on so-called "fast tracks" and required to accept new rotational assignments every year or so despite their own wishes. Similarly, others said to be on "slow" or "regular" tracks may be forewarned not to expect an assignment shift for at least 4 or 5 years. Consider also that promotional policies in most universities explicitly specify the number of years a person can be appointed to a given rank. They also spell out precisely when a tenure decision must be reached on a given individual. The process can sometimes be speeded up.

Variable socialization processes give a recruit few clues as to when to expect a given boundary passage. Thus, both the prisoner of war who is told by his captors that he will be released only when he has "learned the truth" and the patient in a psychiatric hospital who cannot return home until he is again judged "normal" are in pure versions of the variable process. On a more mundane level, most upwardly mobile careers in business organizations are marked by variable socialization processes rather than fixed ones, because many uncontrolled factors such as the state of the economy and the turnover rates in the upper echelons of management may partially determine whether and when any given person will be promoted to the next higher level.

Furthermore, what may be true for one person is not true for another in variable socialization processes. Such a situation requires a recruit to search out clues as to the future. To wit, apprenticeship programs often specify only the minimum number of years a person must remain in the apprentice role and leave open the time a person can be expected to be advanced into the journeyman classification. However, since the rates of passage across any organizational boundary are a matter of some concern to people, transitional timetables may be developed by recruits anyway on the most flimsy and fragmentary information. Rumors and innuendos about who is going where and when they are going characterize situations marked by the presence of the variable strategy of socialization. Indeed, the would-be general manager often pushes quite hard to discover the signs of a coming promotion (or demotion). The individual listens closely to stories concerning the time it takes one to advance in the organization, observes as carefully as possible the experiences of others, and, in general, develops an age consciousness delineating the range of appropriate ages for given positions. And, whether or not this age consciousness is accurate, the individual will measure his or her progress against such beliefs.

Relatedly, Roth (1963) suggests that a special category of "chronic sidetrack" may be created for certain types of role failures. Thus, in the fixed socialization processes of public schools, the retarded are shunted off to distinct classes where the notion of progress does not exist. Similarly, in some police agencies, recruits unable to meet certain agent demands, particularly during that portion of the socialization process which is fixed and takes recruits typically through the academy to the patrol division, are provided long-term assignments as city jailers or traffic controllers, not as patrolmen. Such assignments serve as a signal to the recruit and others in the organization that the individual has left the normal career path. To the extent that such organizational "Siberias" exist and can be identified with certainty by those in the setting, chronic sidetracking from which there is rarely a return is a distinct possibility in fixed socialization processes. On the other hand, sidetracking is usually more subtle and problematic to a recruit operating in a variable socialization track. Indeed, many people who are working in the middle levels of management are often unable to judge just where they are, where they are going, or how they are doing. Consequently, variable processes are likely to create much anxiety and perhaps frustration for individuals who are unable to construct reasonably valid timetables to inform them of the appropriateness of their movement (or lack of movement) in the organization.

It also should be noted that variable processes are a very powerful antidote in the formation of group solidarity among potential recruits to certain organizationally defined roles. The movement of people at different rates and according to different patterns makes it virtually impossible for a cohort group to remain cohesive and loyal to one another. Indeed, in highly competitive situations, recruits being processed in a variable mode tend to differentiate themselves, both socially and psychologically, from each other. Furthermore, they often are obsequious to authority, suspicious of colleagues, and, more generally, adopt strategies of passage that minimize risk. Therefore, if, from the organization's point of view, peer group solidarity in a recruit is desirable, care should be taken to use only fixed timetables in the socialization processes.

We now look to certain propositions which arise on the basis of this discussion of fixed and variable socialization tactic.

Propositions

1. Fixed timetables for socialization processes are most likely to be associated with hierarchical boundary passages and least likely to be used with inclusionary boundary passages; functional boundaries present a mixed case.

Thus, in some organizations, one can almost guarantee that after a certain number of years to the day, one will be promoted to a higher rank. Consider here the military and certain civil service bureaucracies. To the contrary, one cannot guarantee that after a certain length of time a person will have learned what is necessary to make a functional move or will have acquired the trust and support required to move closer to the core of the organization. Those latter moves are more likely to be made on the basis of situational or *in situ* assessments and can involve very long or very short periods of time.

2. Fixed socialization processes are most likely to produce innovative responses; *variable socialization* processes are most likely to produce custodial responses.

The logic behind this proposition is simply that a variable situation leads to maximum anxiety and this anxiety operates as a strong motivator toward conformity. Intuitively, most managers utilize this principle when they attempt, for example, to control their most rebellious or difficult subordinates by telling them that their next career move "may or may not happen" within a given time frame. Doctors too use this tactic to induce patients to "get well" by refusing to provide them with any kind of timetable for their release from the hospital. And, of course, interrogators in police organizations and prison camps use the vagueness that surrounds one's expected length of sentence to pressure prisoners to make confessions and change attitudes (Schein, 1961; Goffman, 1961).

Variable socialization processes keep a recruit maximally off balance and at the mercy of socialization agents. In effect, the agent says to a recruit, "I will pass you along to the next stage when you are ready, but I will decide when you are ready." In fixed processes, such as a 4-year medical school program, a 3-month boot-camp, a 1-year apprenticeship, a set 2-year tour of duty to another geographical district of a business firm, persons can usually gear themselves to the situation better than in the variable case and therefore can plan innovative activities to fit the time-table. It should also be noted, however, that a fixed process may undermine the power of the innovator vis-à-vis the group of which he is a part. This is particularly the case near the end of a given stage, since others in the organization typically also know that the innovator is now in a "lame duck" period. Consequently, from the point of view of the innovator in certain roles, it is desirable to be in a position to know one's own time-table but to conceal this knowledge from others.

Serial vs. Disjunctive Socialization Processes

Definition A *serial socialization* process is one in which experienced members of the organization groom newcomers who are about to assume similar kinds of positions in the organization. In effect, these experienced members serve as role models for recruits. In the police world, for exam-ple, the serial mode—whereby rookies are assigned only older veteran officers as their first working partners on patrol—is virtually taken for granted, and some observers have suggested that it is this aspect of polic-ing that accounts for the remarkable intergenerational stability of patrol-men behavior patterns (Westley, 1970; Rubenstein, 1973; Manning and Van Maanen, 1978). Serial modes create something analogous to Mead's (1956) notion of a postfigurative culture. Just as children in stable societies are able to gain a sure sense of the future that awaits them by seeing in their parents and grandparents an image of themselves grown older, employees in organizations can gain a surer sense of the future by seeing in their more experienced elders an image of themselves further along in the organization. A danger exists, of course, that this image will be neither flattering nor desirable from the perspective of the recruits and many newcomers may leave the organization rather than face what ap-pears to be an agonizing future. In industrial settings where worker morale is low and turnover is high, a serial pattern of initiating newcomers into the organization would maintain and perhaps amplify an already poor situation.

When newcomers are not following the footsteps of immediate or re-cent predecessors, and when no role models are available to recruits to inform them as to how they are to proceed in the new role, the socializa-tion process is a *disjunctive* one. Many examples can be cited. Take, for instance, the case of a black firefighter entering a previously all-white engine company or a woman entering managerial ranks in a firm in which such ranks had previously been occupied only by males. In these cases, there are few, if any, persons on the scene who have shared the unique problems faced by the recruit. Certainly such situations make things extremely difficult and anxiety-provoking for the newcomer. An interesting illustration is also provided by the "heroic myth" to be found in many cultures and presented by Campbell (1956). In most versions of this saga, a young man is deliberately sent away from his homeland and "suffers" through a series of trials and tribulations in order to discover new ways of thinking about and doing things. Typically, after some most disjunctive adventures and misadventures, the hero is given some sort of magic gift and brings it back to his home society as a way of revitalizing it. Such disjunctive themes are also central ones in western fairy tales (Bettelheim 1976).[19]

The analytic distinction between serial and disjunctive socialization processes is sometimes brought into sharp focus when an organization undertakes a "house cleaning" whereby old members are swept out of the back door and new members brought in the front door to replace them. In extreme cases, an entire organization can be thrown into a disjunctive mode of socialization with the result that the organization will no longer resemble its former self. It is also true that occasionally the person who is presumably being socialized by another organizational member has more experience and knowledge than the one doing the socializing. To wit, in colleges where faculty members are constantly entering and exiting, long-term students exert much control over the institution. Certainly, in other organizations such as prisons and mental hospitals, recruit turnover is often considerably smaller than staff turnover. It should not be surprising then that these organizations are often literally run by the inmates.

Sometimes, what appears to be a serial process is actually disjunctive. In many work organizations, it is the case that if someone is exceptionally good and is promoted to project leader by age 25, that same person must be exceptionally mediocre to be in that same position at age 50 or 55. Because of such circumstances, the age-graded stereotype of the youthful, naive, and passive junior member of the firm being coached wisely by a mature, informed, and active mentor is frequently false. The process may have been designed as a serial one, but, to the recruit, the process may be disjunctive if he or she is unwilling to take the mentor seriously. Roth (1963) labels this problem "gapping" and it appears to be a serious one associated with serial strategies. Gapping refers to the historical, social, or ideological distance between recruits and agents. And, when the past experiences, reference groups, or values of the agents are quite removed from those of the recruits, good intentions aside, the serial process may become a disjunctive one.

In summary, it is generally true that recruits representing the first class will set the tone for the classes to follow. It is not suggested that those who follow are paginated *seriatim,* but simply that for those to come, it is easier to learn from others already on hand than it is to learn on their own as originators. As long as there are others available in the socialization setting whom the newcomers consider to be "like them," these others will act as guides, passing on consensual solutions to the typical problems faced by a recruit. Mental patients, for example, often report that they were only able to survive and gain their release because other more experienced patients "set them wise" as to what the psychiatric staff deemed appropriate behavior and indicative of improvement (Stanton and Schwartz, 1959; Goffman, 1961).

We can now state some propositions which relate these above considerations to the theoretical variables of interest.

Propositions

1. Serial socialization is most likely to be associated with inclusionary boundary passages.

This association results because to become a central member of any organizational segment normally requires that others consider one to be affable, trustworthy, and, of course, central as well. This is unlikely to occur unless these others perceive the newcomer to be, in most respects, similar to themselves. Recruits must at least seem to be taking those with whom they work seriously or risk being labeled deviant in the situation and hence not allowed across inclusionary boundaries.

2. Serial socialization processes are likely to be found only at those functional or hierarchical boundary passages which are seen by those in control of the process as requiring a continuity of skills, values, and attitudes. *Disjunctive* processes are most likely to be found at those functional and hierarchical boundary passages which are seen as not requiring such continuity. In other words, there is no a priori reason why serial or disjunctive processes would be found at either of these two types of boundaries. Organizations seemingly can arrange for a serial or disjunctive process at these locations according to criteria of their own making.

3. Serial socialization processes are most likely to produce a custodial orientation; disjunctive processes are most likely to produce an innovative orientation.

Whereas the serial process risks stagnation and contamination, the disjunctive process risks complication and confusion. But, the disjunctive pattern also creates the opportunity for a recruit to be inventive and original. Certainly newcomers left to their own devices may rely on inappropriate others for definitions of their tasks. Without an old guard around to hamper the development of a fresh perspective, the conformity and lockstep pressures created by the serial mode are absent. Entrepreneurs, for example, almost automatically fall into a disjunctive process of socialization as do those who fill newly created organizational roles. In both cases, there are few role models available to the individual who have had similar experiences and could therefore coach the newcomer in light of the lessons they have learned. Consequently, if innovation is to be stimulated, for whatever reason, the socialization process should minimize the possibility of allowing incumbents to form relationships with their likely successors, for these role incumbents will typically teach the recruit the "old" ways of doing things. Instead, the process should maximize either a very broad range of role models such as might be created through the use of individual, informal, and random tactics of socialization or deliberately create situations where gaps occur between role model and recruit, or construct brand-new roles to keep the recruits "loose" in their orientation.

Investiture vs. Divestiture Socialization Processes

Definition   The final strategy to be discussed here concerns the degree to which a socialization process is constructed to either confirm or disconfirm the entering identity of the recruit. *Investiture socialization* processes ratify and document for recruits the viability and usefulness of those personal characteristics they bring with them to the organization. An investiture process says to the newcomer, "We like you just as you are." Indeed, the organization through the use of this tactic does not wish to change the recruit. Rather, it wishes to take advantage of and build upon the skills, values, and attitudes the recruit is thought to possess. From this perspective, investiture processes substantiate and perhaps enhance the newcomer's view of himself. To wit, most young business school graduates are on an investiture path, though at certain boundaries they may run into certain disconfirming expereinces. At times, positions on the bottom rungs of organizational ladders are filled by the use of this tactic wherein newcomers to these positions are handled with a great deal of concern. Investiture processes attempt to make entrance into a given organizationally defined role as smooth and trouble free as possible. Orientation programs, career counseling, relocation assistance, social functions, even a visit to the president's office with the perfunctory handshake and good wishes systematically suggest to newcomers that they are valuable to the organization.

Divestiture socialization processes, in contrast, seek to deny and strip away certain personal characteristics of a recruit. Many occupational and organizational communities almost require a recruit to sever old friendships, undergo extensive harassment from experienced members, and engage for long periods of time in doing the "dirty work" of the trade typified by its low pay, low status, low interest value, and low skill requirements. Many aspects of professional training such as the first year of medical school and the novitiate period associated with religious orders are organized explicitly to disconfirm many aspects of the recruit's entering self-image, thus beginning the process of rebuilding the individual's self-image based upon new assumptions. Often these new assumptions arise from the recruits' own discovery, gradual or dramatic, that they have an ability to do things they had not thought themselves able to do previously.

Ordinarily, the degree to which the recruit experiences the socialization process as an ordeal indicates the degree to which divestiture processes are operating. Goffman's (1961) "total institutions" are commonly thought typical in this regard in the deliberate "mortifications to self" which entry into them entails. But, even in total institutions, socialization processes will have different meaning to different recruits. Thus, the degree to which the process is one of devestiture or investiture to a recruit is, in part, a function of the recruit's entering characteristics and orientation toward the role. Perhaps Goffman and others have been overimpressed with the degree of humiliation and profanation of self that occurs in certain organizations. Even the harshest of institutional settings, some recruits will undergo a brutal divestiture process with a calculated indifference and stoic nonchalance. Some recruits too will have been through divestiture processes so frequently that new socialization attempts can be undergone rather matter-of-factly. Furthermore, "total institutions" sometimes offer a recruit a sort of home-away-from-home that more or less complements the recruit's entering self-image. Thus, for convicted robbers, becoming a member of, say, a thief subculture in a prison acts more as an investiture than a divestiture process. In such situations, one's preinstitutional identity can be sustained, if not enhanced, with ease.

Yet, the fact remains that many organizations consciously promote ordeals designed to make the recruit whatever the organization deems appropriate, what Schein has described as "up-ending" experiences (Schein, 1964). In extreme circumstances, recruits are forced to abstain from certain types of behavior, must publicly degrade themselves and others, and must follow a rigid set of rules and regulations. Furthermore, measures are often taken to isolate recruits from former associates who presumably would continue to confirm the recruit's old identity. The process, when voluntarily undergone, serves to commit and bind the person to the organization and is typically premised upon a strong desire on the part of the recruit to become an accepted member of the organization (or an organizational segment). In brief, the recruit's entrance into the role or system is aided by his or her "awe" of the institution and this "awe" then sustains the individual's motivation through subsequent ordeals of divestiture. Consider here, first-year law students at elite universities (Turow, 1977) or young women entering religious orders (Hulme, 1956).

There are many familiar illustrations of organizations in this society that require a recruit to pass robust tests in order to gain privileged access into their realms: religious cults, elite law schools, self-realization groups, professional athletic teams, many law enforcement agencies, military organizations, and so on. Even some business occupations such as certified public accounting have stiff licensing requirements which, to many recruits, appear much like a divestiture process. It should be kept in mind, however, that these stern tactics provide an identity-bestowing as well as an identity-destroying process. Coercion is not necessarily a damaging assault on the person. Indeed, it can be a device for stimulating many personal changes that are evaluated positively by the person and others. What is, of course, problematic with coercion is its nonvoluntary aspects and the possibility of misuse in the hands of irresponsible agents.

Given these concerns, some propositions can now be presented which seek to further explicate the workings of this socialization tactic in organizational settings.

Propositions

1. Divestiture processes are most likely to be found (1) at the point of initial entry into an organization or occupation, and (2) prior to the crossing of major inclusionary boundaries where a recruit must pass some basic test of worthiness for membership in an organizational segment.

Once the person has passed these initial boundaries, subsequent boundary passages are much more likely to be of an investiture nature unless movement from one segment of the organization to another involves a major change of skills, value, or self-image. For example, one can imagine the college graduate engineer going into an engineering department of a company and experiencing this process as basically an investiture one. If, at a later time, this person decides to move into line management and goes through an extensive formal or informal management training process, such training may well be experienced as a divestiture process, because it may challenge many of the individual's cherished values which were associated with and rooted in the old engineering role.

2. Divestiture processes are most likely to lead to a custodial orientation: investiture processes are most likely to lead to an innovative orientation (unless the recruit enters and is rewarded for holding a custodial orientation at the outset).

Divestiture processes, in effect, remold the person and, therefore, are powerful ways for organizations and occupations to control the values of incoming members. Such processes lie at the heart of most professional training, thus helping to explain why professionals appear to be so deeply and permanently socialized. For, once a person has successfully completed a difficult divestiture process and has constructed something of a new identity based on the role to which the divestiture process was directed, there are strong forces toward the maintenance of the new identity. The strongest of these forces is perhaps the fact that the sacrifice involved in building the new identity must be justified, consequently making any disclaimers placed on the new identity extremely difficult for the person to accomplish. Furthermore, since the person's self-esteem following the successful completion of a divestiture process comes to rest on the new self-image, the individual will organize his present and future experiences to insure that his self-esteem can be enhanced or at least maintained (Goffman, 1959; Schein, 1961; Schein and Bennis, 1965). In short, the image becomes self-fulfilling.

Interaction of the Socialization Tactics

In the preceding portions of this "people processing" discussion, we identified some of the major tactical dimensions of socialization processes. These tactics were presented as logically independent of each other. Furthermore, we examined, through a series of propositional acts, the likelihood that each tactic would be associated with certain kinds of organizational boundary passages and the likelihood that each tactic would lead to either a custodial or innovative response. On examining real organizations, it is empirically obvious that these tactical dimensions are associated with one another and that the actual impact of organizational socialization upon a recruit is a cumulative one, the result of a combination of socialization tactics which perhaps enhance and reinforce or conflict and neutralize each other. It is also obvious that awareness of these tactical dimensions makes it possible for managers to design socialization processes which maximize the probabilities of certain outcomes. In the following section, we suggest some propositions about strategic combinations of socialization tactics in relation to the critical search for the conditions under which an organization can expect to promote from its recruits custodial, content-innovative, or role-innovative responses.[20]

Propositions

1. A custodial response will be most likely to result from a socialization process which is (1) sequential, (2) variable, (3) serial, and (4) involves divestiture processes.

In other words, the conditions which stimulate a custodial orientation derive from processes which involve the recruit in a definite series of cumulative stages (sequential); without set timetables for matriculation from one stage to the next, thus implying that boundary passages will be denied the recruit unless certain criteria have been met (variable); involving role models who set the "correct" example for the recruit (serial); and processes which, through various means, involve the recruit's redefinition of self around certain recognized organizational values (divestiture).

2. Content innovation is most likely to occur through a socialization process which is (1) collective, (2) formal, (3) random, (4) fixed, and (5) disjunctive.

In other words, for content innovation to occur in a role, it is desirable to train the role recruits as a formal group in which new ideas or technologies are specifically taught such that the value of innovation is stressed. Furthermore, it is desirable to avoid training sequences which might reinforce traditional ways of doing things but also to avoid variable timetables which might induce anxiety and promote divisive competition among recruits in which the best way to succeed is to "play it safe." Finally, the more the role models are themselves innovative (or absent altogether), the more the recruit will be encouraged (or forced) to innovate.

3. Role innovation, the redefining of the mission or goals of the role itself, is the most extreme form of innovation and is most likely to occur through a socialization process which is (1) individual, (2) informal, (3) random, (4) disjunctive, and (5) involves investiture processes.

In other words, for an individual to have the motivation and strength to be a role innovator, it is necessary for that person to be reinforced individually by various other members of the organization (which must be an informal process since it implies disloyalty to the role, group, organizational segment, or total organization itself), to be free of sequential stages which might inhibit innovative efforts, to be exposed to innovative role models or none at all, and to experience an affirmation of self throughout the process. It is very difficult indeed to change norms surrounding the mission or goals of an organizationally defined role. Therefore, it will probably only occur when an individual who is innovative in orientation at the outset encounters an essentially benign socialization process which not only does not discourage role innovation, but genuinely encourages it.

VI. SUMMARY AND CONCLUSIONS

What we have presented in this paper includes: a model of the organization and its major internal boundaries; a concept of role and role learning; the notions of custodial or innovative responses to socialization experiences; and a detailed analysis of six different dimensions of the socialization process which can be thought of as distinct "tactics" which managers (agents) can employ when socializing new recruits into the organization or at various boundary passages.

We have attempted to spell out, through a series of propositions, the likelihood that any given tactic would or would not be associated with any particular kind of organizational boundary passage. Also, we have developed several propositions about the likelihood of any given tactic leading to custodial, content-innovative, or role-innovative responses. Finally, we have proposed a combination of tactics which one might hypothesize as being most likely to produce each of the specific organizational responses.

We do not consider this a completed theory in that we do not as yet have enough empirical evidence to determine in a more tightly arranged and logical scheme how the various socialization tactics can be more or less ordered in terms of their effects upon recruits being initiated into organizational roles. We do feel, however, that the six analytically distinct dimensions of the socialization process represent a first and important step in this direction. We believe, then, that we have displayed some theory which can now be tested empirically.

In any event, we feel that the specification of the dimensions themselves at least opens up—both for researchers and managers in organizations—an analytic framework for considering the actual processes by which people are brought into new roles in the workplace. Indeed, it is time to become more conscious of the choices and consequences of the ways in which we "process people." Uninspired custodianship, recalcitrance, and even organizational stagnation are often the direct result of how employees are processed into the organization. Role innovation and ultimately organizational revitalization, at the other extreme, can also be a direct result of how people were processed. From this perspective, organizational results are not simply the consequences of the work accomplished by people brought into the organization; rather, they are the consequences of the work these people accomplish after the organization itself has completed its work on them.

FOOTNOTES

1. The view of social action taken in this paper is based essentially upon Meadian social psychology and is expressed most succinctly by the symbolic interactionists (see, for example, the work of Mead, 1930; Goffman, 1959; Blumer, 1969; Hughes, 1971; Becker, 1970). Personal change within this framework always requires the analytic occasion of "surprise." Such surprise prompts, even if only momentarily, a kind of disengagement from the concerns of the moment and perhaps the apprehension of those affairs that the person has not hitherto noticed at all. Philosophically, the perspective is related closely to that of phenomenology. For some groundings here, see Schutz, 1970; Lyman and Scott, 1970; Psathas, 1972; and, especially, Zaner, 1970.

2. We use the phrase "organizational segment" quite broadly in this paper. We mean by the phrase simply the joining of actions undertaken by different organizational members in the pursuit of certain ends. Departments are, therefore, organizational segments, as are work groups or project teams. Vertical and horizontal cliques, cabals, and conspiracies also fall under this rubric, for their existence implies an unofficial, though nonetheless real, merging of individual efforts. See Manning (1977) and Burns (1955, 1958, 1961) for a more elaborate use of this concept.

3. In general, any form of adult socialization, including the organizational variety, is analogous to that of childhood socialization, but an adult socialization process must contend with the individual's "culture of orientation," which may stand in the way of the organization's efforts. For an introduction to the various forms of adult and organizational socialization, see, for example, Becker and Strauss, 1956; Schein, 1961, 1964, 1968; Becker, 1964; Caplow, 1964; Brim and Wheeler, 1966; Roth, 1963; Moore, 1969; Inkeles, 1966; Manning, 1970; and Van Maanen, 1976. For an earlier statement of some of the ideas in this paper, see Van Maanen and Schein, 1977. Another introduction to the topic can be located in Porter, Lawler, and Hackman (1975) under the partially misleading chapter title "Adaptation Processes."

4. To some extent, those adjustments that turn out to be nonadaptive fall under the classification of what Platt (1972) calls a "social trap." In brief, such traps may involve, first, a time delay before the ill effects of a particular adjustment are felt as is the case with smoking and lung cancer or industrial pollution and environmental decay. Second, social traps also describe situations wherein strong individual incentives (or disincentives) seemingly prohibit people from acting in their collective best interest as exemplified by the infamous Kitty Genovese slaying in New York City or in game situations marked by the "Prisoner's Dilemma."

5. To wit, psychologists of a developmental stripe emphasize cognitive learning (e.g., Erikson, 1959, 1968; Piaget, 1962, 1969; Kroll et al., 1970; Keen, 1977) whereas psychologists more concerned with individual differences emphasize the matching of persons and setting in their socialization studies (e.g., Holland, 1966; Roe, 1957; Super et al., 1963). On the other hand, political scientists seem most concerned with how newcomers gain "control of things" (e.g., Hyman, 1959; Bell and Price, 1975; Edelman, 1967). Students of complex organizations nearly always focus on the effectiveness of the newcomer (e.g., Berlew and Hall, 1966; Feldman, 1976; Schein, 1978). Anthropologists, when they consider *adult* socialization at all, tend to be far more interested in transitions across particular societies than those occurring within a society (e.g., Taft, 1975; Kimball and Watson, 1972; Stonequist, 1937) or with those passages within a society that mark a youth's transitition into adulthood (Van Gennep, 1960; LeVine, 1973). All this is to say that these diverse studies provide some very rich descriptive materials but rarely do the theoretical accounts of the socialization process go beyond disciplinary boundaries. We have tried at least in small measure to transcend these boundaries in this paper.

6. To be sure, even if we accomplished these ends fully, our theory would still be of only the middle range (Merton, 1957). A comprehensive theory must also consider the origins and alterations in the historical patterns of organizational socialization as well as the differential effects of the process upon people of widely diverse backgrounds, cultures, and situations. The importance of a comparative and historical approach to the design of socialization studies cannot be underestimated. While we have a number of longitudinal accounts of the process as it occurs in a particular organization or occupation (e.g., Dornbush, 1955; Lieberman, 1956; Evan, 1963; Light, 1972; Van Maanen, 1973; Rosenbaum, 1976), these remain solitary case studies complete with their own idiosyncratic conceptual frameworks. Some good examples of the type of comparative and historical empirical work needed in this regard are provided by Lortie, 1975; Faulkner, 1974; and Kanter, 1968.

7. The most general process model of socialization is the Lewinian model with its three phases of "unfreezing, changing, and refreezing." Both Schein (1961a,b; 1968) and Van Maanen (1976) have relied extensively on this general formulation when describing the organizational socialization process from the individual's perspective.

8. This is, of course, taking an anthropological or cultural perspective on complex organizations, which requires the suspension of belief in formal pronouncements or inductive fiats

as to what organizations are about until detailed empirical study has been conducted into the workings of any given organization. Such an approach has much to recommend it. Indeed, the various studies which refer to the differences between the intentional and unintentional consequences, the manifest and latent goals, the theory-in-use and theory-in-practice, and the explicit and implicit objectives of an organization all would seem to point in this direction (e.g., Gouldner, 1954; Blau, 1955; Crozier, 1964; Burns, 1961; Schein, 1970; Argyris and Schön, 1974; Blankenship, 1977).

9. Looking to the functional and hierarchical boundaries, this would appear to be the case because immediately after entrance to a new position, the individual is too wrapped up in learning the requirements of the job to have much, if any, influence upon those requirements themselves. And, just before passage, the person is probably too caught up in the transition itself to have (or desire to have) much influence on the position being left behind. Across the inclusionary boundaries, the situation is similar though perhaps less clear. Immediately after entry, the person knows few people and will have developed little of the sort of interpersonal trust with others on the scene which is necessary to exert meaningful influence. But, after having achieved a central and visible position within the particular setting, it is likely that such a position is premised upon the individual's almost total acceptance of the norms and values of the group. As anthropologists are prone to say, the person may have "gone native" and has consequently lost the sort of marginality and detachment necessary to suggest critical alterations in the social scheme of things.

10. A more specific example is useful here. Police organizations come to mind, for there are some interesting case examples: Newly appointed so-called "progressive" or "reform" chiefs of police have, after purging the top administrative ranks and inserting personnel who were sympathetic to their preconceptions of what the organization should be about, tried to insure that only the "right types" (those who were also likely to share the chief's vision) would be promoted in the system. Thus, "old timers" who had very central and influential positions within their respective ranks and functions were no longer in favorable positions to rise in the organization. Policy changes around the structure of the promotion board, oral examinations, and the educational requirements for particular ranks seemingly worked in this regard. Yet, given the short-lived tenure of the instigators of these reforms and the short-lived period of the reforms themselves, moving these departments from the top down proved to be quite difficult, if not impossible. Indeed, lower-placed members in these departments were able (through a variety of inventive means) to block reform in the long run by either forcing the new chief out entirely or by "snapping" the chief back into a position where the values of organizational members once again fell more or less along a plumb line dropped from the top of the organizational cone. See Daley, 1973; Fishgrund, 1977; Beigel and Beigel, 1977 for case materials bearing on the rather remarkable resistance to change exhibited in police organizations.

11. For some further treatments of this role, position, and claims made by occupations commonly thought to be "professional," see Wilensky, 194; Goode, 1969; Vollmer and Mills, 1966; Hughes, 1958; and especially, Blankenship, 1977.

12. Aside from the strategic matters considered directly in the text, the poles of each tactical dimension represent differences in the amount of prior planning engaged in by members of the organization, differences in the level of commitment of organizational resources to a given socialization pattern, and differences in the number of agents actively involved in the process. However, situational and historical considerations unique to any given occupation or organization limit the kind of generalizations we can make on these matters. In other words, in some lines of work, the choice of an *individual* mode of socialization may require more planning, be more costly, and require more agents than the choice of a *collective* mode. In other endeavors, however, the case may be reversed. "Quality control"

may be a crucial aspect of the organization's choice of tactics wherein due to the exacting, dangerous, or consequential nature of the task to be performed by a newcomer to the field, standardized outcomes (promoted by collective processes—see following section) are, if not required, at least socially desirable, as is the case in medicine or firefighting. Needless to say, comparative studies are crucial in this regard.

13. The strength of group understandings depends, of course, upon the degree to which all members actually share the same fate. In highly competitive collective settings, group members know that their own success is increased through the failure of others; hence, the social support networks necessary to maintain cohesion in the group may break down. Consensual understandings will develop, but they will buttress individual modes of adjustment. Junior faculty members in publication-minded universities, for instance, follow group standards, although such standards nearly always stress individual scholarship, the collective standard being, as it is, an individual one.

14. A corollary to this proposition can also be suggested. Namely, the longer recruits remain together as a collective entity, the less likely role-innovative responses become. Van Maanen (1978) refers to such lengthy collective processes within which transfer rates in and out of the recruit group are low as "closed" socialization. On the other hand, "open" socialization, according to Van Maanen, also involves collective socialization, but the mode is marked by changing personnel across time within the recruit group. An interesting study in this regard is reported by Torrance (1955), who examined the decision-making abilities of Air Force flight crews who had trained together for some ten weeks. After training, some crews were scrambled (open-collective socialization), whereas the remaining crews stayed intact (closed-collective socialization). To Torrance's surprise, the scrambled crews were far superior on the performance of various task-related problems than were the intact crews. Interpreting these results, he concluded that the relative lack of power differentials and social status among the scrambled groups allowed for a more open and honest consideration of alternative solutions to the problems facing the group than would be possible when power and status were established and relatively fixed as was the case for the intact crews. Janis (1972) has recently reported some very similar findings.

15. Part of the difficulty for recruits in this matter is that they normally have very little to offer experienced organizational members in exchange for being taught the norms of a particular role. It is not the case that veteran members dislike or distrust novices (though in some instances they may), but it is merely the case that recruits have nothing substantial to contribute to the matters at hand. Thus, newcomers in the informal mode must often first behaviorally demonstrate their value to their would-be teachers by, say, performing "gofer" duties such as fetching work materials, snacks, and coffee, running little necessary but inconsequential errands, doing the "dirty work" others on the scene wish to avoid, and displaying an "eager" or "good" attitude when engaged in such tasks. In exchange for this willingness, a teaching relationship may then emerge. See Lortie, 1975; Haas, 1972; and Rubenstein, 1973 for some good examples in this regard.

16. This suggests that many socialization programs begin with universalistic concerns in which standards are taught as well as the uniform application of these standards. However, perhaps almost as many programs end with very particularistic concerns where recruits are taught that there are shifting standards which are applied uniquely to individual cases. This certainly reflects the typical content of the two socialization phases (formal and informal) mentioned in the text. Consider too that in many organizations the strict adherence to the rules (such as what is usually taught in a formal socialization process) may well reflect a sort of cultural incompetence when the recruit actually "goes to work" rather than competence, since, as all "good" members of the organization know, it is necessary to know the operating *rules about the rules* to perform adequately on the job. A further consideration of this

popular and frequent formal-to-informal socialization sequence is presented on the following pages of the text.

17. Some illustrations are perhaps useful here. Consider the fact that in many organizations employees misrepresent their overtime statements or expense allowances; budget makers pad their budgets with either fictitious expenses or exaggerated amounts for a given item; and supervisors invariably overrate the performance of their subordinates. None of these practices are likely to be conveyed during the first wave of formal socialization. Moreover, a member who strictly adheres to the formal or correct practices (the proper) rather than the social practices currently in use within the work setting (the smart) is likely to be considered by others to be an "organizational dope" until the second wave of socialization provides the recruit with the necessary learning. In other words, the "organizational dope" is one who has not been fully socialized.

18. As Professor Barry Staw (personal communication) rightly suggests, the degree to which the substantive base of a socialization process can be presented in a sequential fashion depends, in part, upon the availability to those directing the process to call upon a fully developed and shared intellectual or disciplinary paradigm. Thus, when classifying socialization processes in educational institutions, mathematics or physics are far more likely to be presented sequentially to student-recruits in those fields than are, for example, history or sociology. In work organizations, the use of sequential processes leading to a given organizationally defined role will also vary according to the degree that agents have recourse to shared knowledge about and/or experience with the target role. From this standpoint, financial anslysts or production supervisors are perhaps more likely to be socialized in a sequential manner than are organizational development specialists or new product line managers. However, we can press this analogy too far, because in work organizations, as in educational ones, pedagogical disputes over the *proper sequence of learning* are indeed quite common even when there exists a widely accepted paradigm among socialization agents.

19. Fairy tales may sometimes come true but certainly not all disjunctive socialization processes have happy endings. An informative and perhaps limiting case is provided by Klineberg and Cottle (1973). They note that first-generation rural-to-city migrants suffer a serious break between their past and present experiences. So serious is this break, in fact, that the migrant's image of a better future usually lies unconnected to any concrete activities toward which the migrant can direct his present efforts. It would seem, therefore, that extremely disjunctive experiences risk demolishing that most delicate bridge between means and ends. If this occurs, anomie and alienation are sure to result (Van Maanen, 1977a).

20. We should note that in these summary propositions we do not take a position on all socialization tactics. When a particular tactic is not explicitly mentioned in the proposition, it is because we feel that the tactic could go either way depending on more specific circumstances. In the first proposition, for example, formal-informal and collective-individual socialization tactics are not mentioned, because we feel that their use, in any combination, neither adds to nor detracts from the prediction as stated. To include these tactics would require more information—information of the sort partially spelled out in the proposition itself. To wit, formal-individual processes are potentially the most powerful, but also the most expensive and capable of producing custodial as well as innovative responses. On the average, formal-collective processes are probably likely to produce custodial orientations, but they can also facilitate the development of group perspectives which are highly innovative. Informal-collective processes are not at all common and therefore are quite hard to predict. And, while informal-individual processes are relatively common, the results of such processes are at best ambiguous without first specifying both the individual's initial orientation toward the particular role he or she is being prepared to assume and the other tactics to be associated with the process.

REFERENCES

1. Argyris, C., and D. Schön (1974 *Theory in Practice: Increasing Professional Effectiveness*, San Francisco: Jossey-Bass, Inc.
2. Avery, R. W. (1968) "Enculturation in industrial research," in B. G. Glaser (ed.), *Organizational Careers: A Source Book for Theory*, Chicago: Aldine Publishing Company. pp. 175–181.
3. Barnard, C. (1938) *The Functions of the Executive*, Cambridge: Harvard University Press.
4. Becker, H. S. (1970) *Sociological Work*, Chicago: Aldine Publishing Company.
5. ———. (1964) "Personal change in adult life," *Sociometry 27, 40–53.*
6. ———, and A. Strauss (1956) "Careers, personality, and adult socialization," *American Journal of Sociology 62,* 404–413.
7. ———, Geer, B., and Hughes E. (1968) *Making the Grade: The Academic Side of College Life*, New York: John Wiley.— Sons, Inc.
8. ———, B. Geer, E. C. Hughes, and A. Strauss (1961) *Boys in white: Student Culture in Medical School.* Chicago: University of Chicago Press.
9. Beigel, H., and A. Beigel (1977) *Beneath the Badge: A Story of Police Corruption*, New York: Harper & Row, Publishers.
10. Bell, C. G., and C. M. Price (1975) *The First Term: A Study of Legislative Socialization*, Beverly Hills, Calif.: Sage Publications, Inc.
11. Berlew, D. E., and D. T. Hall (1966) "The socialization of managers: Effects of expectations on performance," *Administrative Science Quarterly 11,* 207–223.
12. Bettleheim, B. (1976) *The Uses of Enchantment: The Meaning and Importance of Fairy Tales*, New York: Alfred A. Knopf., Inc.
13. Biddle, B. J., and E. J. Thomas (eds.). (1966) *Role Theory: Concepts and research*, New York: John Wiley & Sons, Inc.
14. Bidwell, C. W.)May, 1962) "Pre-adult socialization," paper read at the Social Science Research Council Conference on Socialization and Social Structure. New York.
15. Blankenship, R. (Ed.) (1977) *Colleagues in Organization: The Social Construction of Professional Work.* New York: John Wiley & Sons, Inc.
16. Balu, P. M. (1955) *The Dynamics of Bureaucracy*, Chicago: University of Chicao Press.
17. ———, and R. A. Schoenherr (1971) *The Structure of Organization.* New York: Basic Books, Inc., Publishers.
18. Blumer, H. (1969) *Symbolic Interactionism,* Englewood Cliffs, N.J. Prentice-Hall, Inc.
19. Brim, O. G., and S. Wheeler (1966) *Socialization After Childhood,* New York: John Wiley & Sons, Inc.
20. Burke, K. (1950) *A Rhetoric of Motives,* Englewood Cliffs, N.J.: Prentice-Hall, Inc.
21. Burns, T. (1955) "The reference of conduct in small groups," *Human Relations 8,* 467–486.
22. ———. (1958) "Forms of conduct," *American Journal of Sociology 64,* 137–151.
23. ———. (1961) "Micropolitics: Mechanisms of institutional change," *Administrative Science Quarterly 6,* 257–281.
24. Campbell, T. (1956) *The Hero With a Thousand Faces,* New York: Anchor Books.
25. Caplow, T. (1964) *Principles of Organization,* New York: Harcourt, Brace and World.
26. Chinoy, E. (1955) *Automobile Workers and the American Dream,* New York: Random House, Inc.
27. Crozier, M. (1964) *The Bureaucratic Phenomenon,* Chicago: University of Chicago Press.
28. Daley, R. (1973) *Target Blue: An Insider's View of the New York City Police Department.* New York: Delacorte Press.

29. Dornbush, S. M. (1955) "The mititary academy as an assimilating institution," *Social Forces 33*, 316–321.
30. Edelman, M. (1967) *The Symbolic Uses of Politics*, Urbana, Ill.: University of Illinois Press.
31. Erikson, E. H. (1959) "Identity and the life cycle," *Psychological Issues 1*, 1–171.
32. ———. (1968) *Identity: Youth and Crisis*, New York: W. W. Norton & Company, Inc.
33. Eshelman, B. (1962) *Death Row Chaplain*, Englewood Cliffs, N.J.: Prentice-Hall, Inc.
34. Evan, W. M. (1963) "Peer group interaction and organization," *American Sociological Review 28*, 436–440.
35. Faulkner, R. R. (1974) "Coming of age in organizations: A comparative study of career contingencies and adult socialization," *Sociology of Work and Occupations 1*, 173–191.
36. Feldman, D. C. (1976) "A contingency theory of socialization," *Administrative Science Quarterly 21*, 433–452.
37. Fishgrund, T. J. (1977) "Policy making on decentralization in a large urban police department," unpublished doctoral dissertation. MIT.
38. Glaser, B. G. (1964) *Organizational Scientists: Their Professional Careers*, Indianapolis: The Bobbs-Merrill Co., Inc.
39. ———. (Ed.) (1968) *Organizational Careers: A Source Book for Theory*, Chicago: Aldine Publishing Company.
40. Goffman, E. (1959) *The Presentation of Self in Everyday Life*, New York: Doubleday & Co., Inc.
41. ———. (1963) *Asylums*, New York: Random House, Inc.
42. Goode, W. J. (1969) "The theoretical limits of professionalization," in A. Etzioni (Ed.), *The Semi-Professions and Their Organization*, New York: The Free Press. pp. 226–314.
43. Gordon, J. C. (1977) "The congruence between the job orientation and job content of management school alumni," unpublished doctoral dissertation, MIT.
44. Gouldner, A. W. (1954) *Patterns of Industrial Bureaucracy*, New York: The Free Press.
45. Haas, J. B. (1972) "Educational control among high steel ironworkers," in B. Geer (ed.), *Learning to Work*, Beverly Hills, Calif.: Sage Publications, Inc. pp. 31–38.
46. Hall, D. T. (1976) *Career Development*, Santa Monica, Calif.: Goodyear Publishing Co. Inc.
47. Holland, J. L. (1966) *The Psychology of Vocational Choice: A Theory of Personality Types and Environmental Models*, London: Ginn.
48. Hughes, E. C. (1971) *The Sociological Eye*, Chicago: Aldine Publishing Co.
49. ———. (1958) *Men and Their Work*, Glencoe, Ill.: The Free Press.
50. Hulme, K. (1956) *The Nun's Story*, Boston: Little, Brown and Company.
51. Hyman, H.H. (1959) *Political Socialization*, Glencoe, Ill.: The Free Press.
52. Inkeles A. (1966) "Society, social structure and child socialization," in J. A. Clausen (Ed.), *Socialization and Society*, Boston: Little, Brown and Company: pp. 146–161.
53. Janis, I. (1972) *Victims of Groupthink*, Boston: Houghton Mifflin Company.
54. *Kanter, R. M. (1968) "Commitment and social organization: A study of commitment mechanisms in utopian communities," American Sociological Review 33, 409–417.*
55. Keen. P. G. W. (1977) "Cognitive style and career specialization," in J. Van Maanen (Ed.), *Organizational Careers: Some New Perspectives*, New York: John Wiley & Sons, Inc.: pp. 89–105.
56. Kimball, S. T., and J. G. Watson (1972) *Crossing Cultural Boundaries: The Anthropological Experience*, San Francisco: Chandler Publishing Company.

57. Klineberg, S., and T. J. Cottle (1973) *The Present of Things Past,* Boston: Little, Brown and Company.
58. Kroll, A. M., L. B. Dinklage, J. Lee, E. D. Morley, and E. H. Wilson (1970) *Career Development: Growth and Crisis,* New York: John Wiley & Sons, Inc.
59. Laing, R. D. (1960) *The Divided Self,* London: Tavistock.
60. LeVine, R. A. (1973) *Culture, Behavior, and Personality,* Chicago: Aldine Publishing Company.
61. Lieberman, S. (1956) "The effects of changes in roles on the attitudes of role occupants," *Human Relations 9,* 467–486.
62. Light, D. (1970) "The socialization of psychiatrists," unpublished doctoral dissertation. Northwestern University.
63. Lipsky, M. (1971) "Street level bureaucracy and the analysis of urban reform," *Urban Affairs Quarterly 6,* 122–159.
64. Lortie, D. C. (1975) *Schoolteacher,* Chicago: University of Chicago Press.
65. Lyman, L. M. and M. B. Scott (1970) *A Sociology of the Absurd,* New York: Meredith.
66. Manning, P. K. (1970) "Talking and becoming: A view of organizational socialization," in J. D. Douglas (ed.), *Understanding Everyday Life,* Chicago: Aldine Publishing Company. pp. 239–256.
67. ———. (1977) "Rules, colleagues and situationally justified actions," in R. Blankenship (Ed.), *Colleagues in Organizations: The Social Construction of Professional Work,* New York: John Wiley & Sons, Inc. pp. 263–289.
68. ———, and J. Van Maanen (Eds.) (1978) *A View From the Streets,* Santa Monica, Calif.: Goodyear Publishing Co. Inc.
69. Mead, G. H. (1930) *Mind, Self and Society,* Chicago: University of Chicago Press.
70. Mead, M. (1956) *New Lives for Old,* New York: William Morrow & Co.
71. Merton, R. K. (1957) *Social Theory and Social Structure,* New York: The Free Press.
72. Moore, W. E. (1969) "Occupational socialization," in D. A. Goslin (ed.), *Handbook of Socialization Theory and Research,* Chicago: Rand McNally & Company, pp. 1075–1088.
73. Newcombe, T. M. (1958) "Attitude development as a function of reference groups: The Bennington study," in E. E. Maccoby, T. M. Newcomb, and E. L. Hartley (eds.), *Readings in Social Psychology* (3rd ed.). New York: Holt, Rinehart, and Winston, pp. 117–129.
74. Olesen, V. L., and E. W. Whittaker (1968) *The Silent Dialogue: A Study in the Social Psychology of Professional Socialization.* San Francisco: Jossey-Bass, Inc., Publishers.
75. Parsons, T. (1951) *The Social System,* New York: The Free Press.
76. Perrow, C. (1972) *Complex Organizations: A Critical Essay,* New York: Scott, Foresman and Company.
77. Piaget, J. (1962) *The Moral Judgment of the Child,* New York: Collier Books.
78. ———. (1969) *The Child's Conception of Time.* London: Routledge and Kegan Paul.
79. Platt, E. C. (1972) "Social Traps," *Science 56,* 18–24.
80. Porter, L. W., E. E. Lawler, and J. R. Hackman (1975) *Behavior in Organizations,* New York: McGraw-Hill.
81. Prewitt, K., H. Enlou, and B. Zisk (1966) "Political socialization for political roles," *Public Opinion Quarterly 30,* 112–127.
82. Psathas, G. (ed.) (1973) *Phenomenological Sociology,* New York: John Wiley & Sons, Inc.
83. Roe, A. (1957) "Early determinants of vocational choice," *Journal of Counseling Psychology 4,* 212–217.
84. Rose, A. M. (1960) "Incomplete socialization," *Sociology and Research 44,* 241–253.

85. Rosenbaum, J. E. (1976) *Making Inequality: The Hidden Curriculum of High School Tracking.* New York: John Wiley & Sons, Inc.

86. Roth, J. (1963) *Timetables,* Indianapolis, Ind.: The Bobbs-Merrill Co. Inc.

87. Rubenstein, J. *City police.* New York: Farrar, Strauss, and Giroux, 1973. Community and occupation. London: Cambridge University Press,

88. Schein, E. H., Schneier, I. and Baruer, C. H. (1961a) *coercive Persuasion,* New York: W. W. Norton & Company, Inc.

89. ———. (1961b) "Management development as a process of influence," *Industrial Management Review 2,* 59–77.

90. ———. (1963) "Organizational socialization in the early career of industrial managers." Office of Naval Research, MIT, Contr. No. 1941(83).

91. ———. (1964) "How to break in the college graduate," *Harvard Business Review 42,* 68–76.

92. ———. (1968) "Organizational socialization and the profession of management," *Industrial Management Review 9,* 1–15.

93. ———. (1970) *Organizational Psychology,* 2nd ed., Englewood Cliffs, N.J.: Prentice-Hall, Inc.

94. ———. (1971a) "The individual, the organization, and the career: A conceptual scheme," *Journal of Applied Behavioral Science 7,* 401–426.

95. ———. (1971b) "Occupational socialization in the professions: The case of the role innovator," *Journal of Psychiatric Research 8,* 521–530.

96. ———. (1978) *Career Dynamics: Matching Individual and Organizational Needs,* Reading Mass.: Addison-Wesley.

97. ———, W. G. Bennis (1965) *Personal and Organizational Change Through Group Methods,* New York: John Wiley & Sons, Inc.

98. Schutz, A. (1970) *On Phenomenology and Social Relations,* Chicago: University of Chicago Press.

99. Shafer, "Selling "selling," unpublished Master's thesis, MIT 1975.

100. Shibutani, T. (1962) "Reference groups and social control," in A. Rose (ed.), *Human Behavior and Social Processes,* Boston: Houghton Mifflin Company, pp. 128–147.

101. Stanton, A. H., and Schwartz, M. S. (1954) *The Mental Hospital,* New York: Basic Books, Inc., Publishers.

102. Stonequist, E. V. (1937) *The Marginal Man,* New York: Charles Scribner's Sons.

103. Strauss, A. L. (1959) *Mirrors and Masks,* Glencoe, Ill.: The Free Press.

104. Super, D. E., R. Starishevsky, N. Matlin, and J. P. Jordaan (1963) *Career Development: Self-Concept Theory,* Princeton, N.J.: College Entrance Examination Board.

105. Taft, R. (1976) "Coping with unfamiliar cultures," in N. Wareer (ed.), *Studies in Cross Cultural Psychology* Vol. 1, London: Academic Press, 452–475.

106. Torrance, E. P. (1955) "Some consequences of power differences on decision making in permanent and temporary groups," in A. P. Hore, E. F. Borgatta, and R. F. Bales (eds.), *Small Groups,* New York: Alfred A. Knopf, Inc. pp. 179–196.

107. Turow, S. (1977) *One L: An Inside Account of Life in the First Year at Harvard Law School,* New York: G. P. Putnam's Sons.

108. Van Gennp, A. (1960) *The Rites of Passage,* Chicago: University of Chicago Press.

109. Van Maanen, J. (1973) "Observations on the making of policemen," *Human Organizations 32,* 407–418.

110. ———. (1974) "Working the streets: A developmental view of police behavior," in H. Jacob (ed.), *The Potential for Reform of Criminal Justice,* Beverly Hills, Calif.: Sage Publications, Inc., pp. 53–130.

111. ———. (1976) "Breaking-In: Socialization to work, in R. Dubin (ed.), *Handbook of Work, Organization, and Society,* Chicago: Rand McNally & Company, pp. 67–130.

112. ——— . (1976) "Experiencing organization: Notes on the meaning of careers and socialization, in J. Van Maanen (ed.), *Organizational Careers: Some New Perspectives,* New York: John Wiley & Sons, Inc., pp. 15–45.

113. ——— . (1977b) "Toward a theory of the career," in J. Van Maanen (ed.), *Organizational Careers: Some New Perspectives,* New York: John Wiley & Sons, Inc, pp. 161–179.

114. ——— . (1978) "People processing: Major strategies of organizational socialization and their consequences," in J. Paap (ed.), *New Directions in Human Resource Management,* Englewood Cliffs, N.J.: Prentice-Hall, Inc.

115. ——— , and E. H. Schein (1977) "Career development," in J. R. Hackman and J. L. Suttle (Eds.), *Improving life at work.* Santa Monica, Ca.: Goodyear Publishing Co. Inc. pp. 30–95.

116. Vollmer, H. M., and D. J. Mills (Eds.). (1966) *Professionalization,* Englewood Cliffs: N.J., Prentice-Hall, Inc.

117. Wamsley, G. L. (1972) "Contrasting institutions of Air Force socialization: Happenstance or bellweather?" *American Journal of Sociology, 78,* 399–417.

118. Westley, W. (1970) *Violence and the Police,* Cambridge, Mass.: MIT Press.

119. Wheeler, S. (1966) "The structure of formally organized socialization settings," in O. G. Brim and S. Wheeler (eds.), *Socialization After Childhood,* New York: John Wiley & Sons, Inc. pp. 51–116.

120. Wilensky, H. L. (1964) "The professionalization of everyone?" *American Journal of Sociology 70,* 137–158.

121. Zaner, R. M. (1970) *The Way of Phenomenology,* New York: Pegasus.

PARTICIPATION IN DECISION-MAKING: ONE MORE LOOK[1]

Edwin A. Locke, UNIVERSITY OF MARYLAND

David M. Schweiger, UNIVERSITY OF MARYLAND

ABSTRACT

The issue of worker participation in decision-making involves more ideological connotations than any issue in organizational behavior. This has led to biased research in the
U.S. and to extensive legislation in Europe. It is argued that the ideological arguments
for PDM are not logically defensible and that it is properly treated as a practical issue.
Participation is best defined as "joint decision-making" and therefore excludes delegation (job enrichment). Various types and degrees of PDM are identified. Theoretically
it has been argued that PDM increases morale because it helps the worker to get what

Research in Organizational Behavior, Vol. 1, pp. 265–339.
Copyright © 1979 by JAI Press, Inc.
All rights of reproduction in any form reserved.
ISBN 0-89232-045-1

he wants from the job. It also has been asserted that PDM increases productive efficiency through both cognitive and motivational mechanisms. An extensive review of research on PDM leads to the conclusion that: (a) a number of experimental field studies involved so many variables that no conclusions can be drawn from them regarding the effects of PDM: (b) three other categories of studies (experimental laboratory, correlational field, and univariate experimental field studies) found that: PDM usually leads to higher satisfaction but not to higher productivity than more authoritative management styles. Numerous contextual factors which determine the conditions under which PDM will lead to increased satisfaction and productivity are identified. The relationship of PDM to three other motivational techniques (money, goal setting and job enrichment) are discussed. It is concluded that subordinate knowledge is the single most important contextual factor determining the usefulness of PDM, assuming that productivity is the major goal of an organization. It is shown why productivity and not satisfaction is the proper goal of a profit making organization.

I. INTRODUCTION: PARTICIPATION AND IDEOLOGY

No issue in the field of organizational behavior and industrial relations is more loaded with ideological and moral connotations than that of worker participation in decision making (PDM). As is characteristic of ideological movements, the impetus for PDM has come from intellectuals rather than from workers (Strauss and Rosenstein, 1970). The attitude of many contemporary intellectuals with respect to PDM is illustrated by the following two incidents (witnessed by the senior author).

A college professor was lecturing a group of executives on "humanistic management" and told them the following anecdote: "I once asked one of my graduate students what he would recommend, as a management consultant, if 95 percent of the studies of participation in decision making had found that it worked no better or worked worse than nonparticipative methods. The student answered: 'I guess I would tell managers not to use it.' At that point [said the professor] I realized that we had failed in our graduate training."

The professor's point was that the student should recommend the use of participation regardless of its practical consequences.

The second incident involved a young assistant professor being interviewed for a job at a large eastern university. He was explaining, to the senior professors interviewing him, a method he had devised for organizational change which always involved the use of participation. One of the senior professors asked him how he reconciled the use of PDM with the somewhat contradictory research findings on the subject. "I don't care what the research shows," he replied curtly, "it's a moral issue."

Views similar to these have been advocated or are cited in numerous published books and articles (Blumberg, 1968; Davies, 1967; Foy and Gadon, 1976; Hespe and Wall, undated; Lischeron and Wall, 1975; Schregle, 1970; see Dachler and Wilpert, 1978, for further documentation). Systems of management which do not stress PDM have been accused of being "exploitative," "dictatorial," "ahuman," and even "neo-Nazi" (Macrae, 1977; Davies, 1967; Leavitt, 1965; Nicol, 1948) at worst, and hopelessly anachronistic and inefficient at best. As Tannenbaum (1974) puts it, "The question for many . . . is not whether participation works but rather how to *make* it work" (p. 105).

This antagonism toward nonparticipative management and the "authoritarian-materialistic" culture with which it is often associated is revealed in the following poem which was quoted in one pro-PDM article:

> Sweet Mary your production's poor,
> Just dry your tears and go,
> For speed and greed are rated high,
> But love-for-others, no.
> (Gillespie, 1971, p. 74)

Despite the moralistic tone in which PDM is often discussed, most supporters of PDM in the United States, in contrast to those in Europe, have thus far advocated the *voluntary* adoption of participative practices in industry. There are two major reasons for this. First, most Americans still value freedom, at least to some extent. Second, many American intellectuals in the social sciences are reluctant to advocate public policy on moral grounds alone. They are skeptical of systematic ideologies and demand "practical" evidence that ideas will work before applying them.[2] Thus many articles advocating PDM have taken a pragmatic approach (e.g., Bennis, 1966; Likert, 1961, 1967; McCormick, 1938; White and Lippitt, 1960). However, since the research evidence is somewhat equivocal, this approach provides little impetus for immediate legislation.

While there are groups such as the New Left who advocate the imposition of "participatory democracy" onto society as a whole by revolutionary means, their views have not gained wide acceptance in the United States—mainly because their real motive, the establishment of dictatorship, has been effectively exposed (Hessen, 1968; Rand, 1971b).

More recently, some respected social scientists have advocated government legislation to force organizations to improve the "quality of work life" (e.g., Lawler, 1976; see Locke, 1976b, for a rebuttal), but these suggestions have not yet attracted the serious attention of Congress. This may seem surprising in view of the present Welfare State climate. However, the "quality of work life" is a vague concept and therefore difficult to use as a basis for specific legislative proposals. Furthermore, the labor movement has been effectively stressing their particular brand of "participation" (i.e., collective bargaining to determine wages, seniority rights, working conditions, fringe benefits, etc.) with legislative help for decades, and have shown a strong reluctance to being "co-opted" by management-sponsored participation schemes (e.g., Gomberg, 1966).

While the advocates of PDM in the United States have not seen their pragmatic arguments translated into legislation, their ideological pre-, commitment to PDM is clearly evident even in research articles. This pro-PDM bias can be seen at every stage of the research process: in the design of experiments, in the interpretation of results, and in the reporting of findings.

Consider first experimental design. Typically those assigned the role of participative leaders in PDM experiments are considerate, polite, friendly, and rational. Such leaders are often compared with "authoritarian" leaders, who are cold, punitive, tactless, arbitrary, and often not very knowledgeable about the task being performed (Anderson, 1959). Not surprisingly, experimental subjects typically like the participative leaders better. Such package dealing diverts the reader from consideration of other possible combinations of traits, e.g., an authoritative leader who is considerate and highly knowledgeable, or a participative leader who is ignorant and unassertive.

Interpretations of research are also frequently biased. For example, one can find many reports of research projects in organizations in which anywhere from two or three to nine new policies or procedures were introduced (e.g., changes in incentives, technology, hours of work, participation, etc.), but in which PDM is arbitrarily singled out as *the* variable responsible for the positive results obtained. Or, in cases where the PDM group performed less well than the non-PDM group, it is baldly asserted that the effectiveness of the latter group *would* have deteriorated if the experiment had lasted longer. Or, in studies where only correlational analyses were performed, the results are interpreted as demonstrating the causal efficacy of PDM. (PDM advocates, however, are by no means the only group guilty of interpreting correlations as proving causality; Locke, 1969.) Another tactic is "criterion switching" in which certain experimental outcomes such as productivity are downplayed, while other outcomes more favorable to PDM, such as satisfaction, are arbitarily stressed as being more significant.

Even more blatant bias can be found in the reporting of results of PDM research. For example, borderline results may be exaggerated; particular findings from a given study which would change the overall interpretation may be omitted; and negative studies may not be mentioned at all.

Thus it is not surprising to encounter such incredible statements as:

There is hardly a study in the entire literature which fails to demonstrate the satisfaction in work is enhanced or that other generally acknowledged beneficial consequences accrue from a genuine increase in workers' decision-making power (Blumber, 1968, p. 123, italics omitted)

For a similarly sanguine view, see Kahn, 1974. The design, interpretation, and reporting of PDM research is not without its critics, however, as we shall see later in this essay (e.g., Leavitt, 1965; Lowin, 1968; Mansbridge, 1973; Strauss, 1963).

The pro-PDM bias in other countries, especially Europe, is even more extreme that that in the United States (Dachler and Wilpert, 1978; Strauss & Rosenstein, 1970). The ideological preference for PDM is more explicit, more universal, and more openly political in tone. Consequently, there has been less interest in small-scale research projects involving direct participation on the shop floor and more emphasis on establishing PDM, often called "industrial democracy," by force of law or government decree (McInnes, 1976).

These laws often require the formation of "works (or worker) councils" which assure representation of the rank and file on committees or boards which make or are consulted on decisions regarding the firm as a whole (Sturmthal, 1964). Germany is the most well-known advocate of "codetermination" through works councils (Raskin, 1976). Similar or related mechanisms exist in France, Israel, Sweden, Yugoslavia, and many other countries (Derber, 1970; Lammers, 1967; Schregle, 1970; Strauss and Rosenstein, 1970; Tannenbaum, 1974; Teague, 1971; Van de Vall and King, 1973). Radical legislative proposals along the same lines are now being considered in England (McInnes, 1977).

Since it is asserted (in differing degrees) by both American and European advocates of PDM that it is a moral issue, it is worth discussing the arguments on which this assertion is based. Again we must distinguish between the United States and Europe, since the arguments are somewhat different in each case, although at root (as we shall see) they stem from a common premise.

As noted above, the arguments offered in favor of PDM in the United States are less openly ideological than in Europe. The moral argument, when used, is often implicit and goes approximately as follows: The United States is a democracy; democracy is good; PDM is democratic; therefore PDM in business organizations is good (or good for the United States).

This argument has two major fallacies. First, the United States is not a democracy. It is a constitutional republic. A pure democracy, in the original Greek sense, is a system of government based on unlimited majority rule.[3] In contrast, our form of government protects the rights of the minority (i.e., of the individual) through constitutional guarantees.

Second, it does not follow logically that the form of organization appropriate to the government of a country is equally appropriate to the management of institutions within that country. For example, a government has a monopoly over the use of physical force, but no one would suggest that business organizations be given the power to use force, since the result would be anarchy and the destruction of all rights. Similarly, republics give their citizens the right to elect government representatives as a means of insuring that their rights will continue to be protected. This is necessary precisely because a government has the power to use physical force and therefore the potential to violate rights. In contrast, a legally mandated majority vote of the employees of a business organization as a means of electing directors and making decisions would not protect rights. Rather, it would abrogate the rights of the owners of the business to use their property as they see fit.

The rights of employees do not go unprotected in a free economy. Employees who are defrauded by owners can seek redress in the courts. Those who disagree with their employer's policies are also free to attempt to change them through persuasion, to organize unions, to quit the company, to seek employment elsewhere, or to start their own companies. Employees are also protected by competition among companies for reputation. Those which become known as "good companies to work for" are able to attract the best employees and gain a competitive advantage, thereby forcing other companies to improve their policies.

The degree of respect for property rights is precisely the issue that most fundamentally separates the United States and Europe. In the United States there is still a degree of respect for such rights, and therefore PDM has not been legislated within business organizations to the degree that it has in Europe.

Advocates of compulsory PDM in Europe are often socialists (Strauss and Rosenstein, 1970), who see PDM as a useful method of helping to abolish private property. (Some leftists, however, oppose participation schemes, fearing that they will reduce the revolutionary fervor of the masses or will prevent workers from opposing management).

Pursuing the issue a step further, let us consider the grounds on which the socialists advocate socialism. The pragmatic answer would be that socialism promotes economic prosperity and the general welfare. However, this cannot be accepted as the real motive, since there is overwhelming evidence that socialism cannot achieve prosperity in theory (Von

Mises, 1951/1962) and does not achieve it in practice (Kaiser, 1976; Smith, 1976).

Furthermore, under socialism the individual citizen does not even achieve influence. Since all property is owned by the state, if a consumer or employee is dissatisfied with products, wages, or working conditions, there is nothing he can do about it except to become head of state. Contrast this with capitalism, where businesses must adjust their products, prices, and working conditions to what the market demands (i.e., the individual monetary vote of the citizens) or go out of business.

If socialism cannot and does not achieve material prosperity, then why is it so widely advocated in preference to capitalism?

Rand (1964, 1966) has identified the basic conflict between capitalism and socialism as a *moral* one. Capitalism rests on the morality of egoism, the doctrine that man is an end in himself and not a means to the end of others. Socialism is advocated in the name of altruism, the doctrine that man is a means to the ends of others, an object of sacrifice, e.g., to the State, the Party, the Race, God, society, etc. Socialists, in short, consider socialism to be morally superior to capitalism whether it works (economically) or not.

However, the issue is not consciously expressed in this form in the PDM or industrial democracy literature, but rather in a form which derives from this basic moral conflict. Egoism implies individualism and rights (including property rights), i.e., the freedom to act on and reap the consequences of one's individual choices. Altruism implies the sacrifice of individuality and freedom. One way to sacrifice individuality is to make everyone equal.

Socialism, (pure) democracy, and participation are all ultimately advocated in the name of *egalitarianism* as a moral ideal. One article (written in the United States!) actually likened the perfect participative organization to an *amoeba* in which there are no separate parts or elements and all actions are totally integrated (Preston and Post, 1974).

Thus, contrary to a commonly held view (e.g., Strauss and Rosenstein, 1970), PDM in the last analysis is not advocated from two entirely different viewpoints; i.e., human relations theory in the United States and socialist ideology abroad, but in the name of the *same* moral ideal of equality in each case. The only difference is that the implications of egalitarianism have not been carried as far in the United States due to the strength of the conflicting ideal of individualism.

Let us now consider briefly the issue of equality. (We are indebted to Rand, 1971a, for the arguments to follow.) As the Founding Fathers viewed it, the principle that "all men are created equal" referred to equality before the law. It meant that all men, by virtue of being human, possessed the same "natural rights" (to life, liberty, and the pursuit of happiness. Rand, 1964, has defined a right as "a moral principle defining and sanctioning a man's freedom of action in a social context. . . . It means freedom from physical compulsion, coercion or interference by other men" pp. 93–94). Equality in this context has a strictly *political* meaning. It was *not* meant to imply that all men had the same degree of intelligence, knowledge, character, or ambition, since they obviously do not.

Nor would *voluntary* participation in decision making among men of differing ability make them equal in their performance, since, given that all share the same rational goal, the more able men will exert the most influence on the decision (e.g., see Mulder and Wilke, 1970). In a free economy the more competent and ambitious men will ordinarily reach positions of higher responsibility and receive higher wages than those with less ability and motivation.

One could only insure equality among men with respect to actions and results by *force,* i.e., by *abrogating,* through political means, the freedom of men to act (e.g., such as throttling the potential entrepreneur under socialism) or the freedom to keep or use the fruits of their actions (e.g., through taxation, wage price controls, and the progressive abrogation of property rights in a welfare state).

Note that this form of forced equality of actions or results is diametrically *opposed to* and incompatible with the view of political equality held by the Founding Fathers. Forced equality makes men either equal in having *no* rights, or *unequal* before the law since the less competent receive special, unearned favors at the expense of the more competent.[4]

What conclusions, then, can be drawn about the status of PDM as a moral issue?

While it is not the purpose of this essay to prove any particular theory of morality or of rights (but see Rand, 1964), *if* one accepts the premise that man has rights (including property rights, since these are inseparable from the right to life), then equality achieved through forced participation (including pure democracy) must be rejected as *immoral* on the grounds that it negates rights.

Alternatively, if one holds that PDM should be established voluntarily on the grounds that it promotes the U.S. form of government, this must be rejected as false, since PDM is not analogous to republican forms of government and since it does not follow logically that all methods appropriate to governing countries are appropriate to running private organizations.

Finally, even if equality of actions and results as such were held to be a morally desirable goal, voluntary participation does not attain it, since, with respect to knowledge and motivation, men are not equal.

Thus, we must conclude that, unless one denies the existence of rights, the arguments offered in support of the view that participation is desirable on moral grounds alone are not logically defensible.

This leaves only one other basis on which to advocate participation. It could be promoted on practical grounds, i.e., on the premise that the use of PDM will make organizations function more efficiently. As noted earlier, this argument has been widely promulgated by social scientists, especially those from the United States; however, since there has been considerable bias in the analysis and reporting of the results, it is time for one more look. The remainder of this essay will be devoted to exploring the evidence offered in support of the view that PDM fosters greater organizational effectiveness than other methods of managing and decision making.

Before we can discuss the research findings, however, we must clearly identify what the concept of participation means in an organizational context and the reasons it is alleged to be effective.

II. DEFINING PARTICIPATION

Despite the intellectuals' ideological attachment to PDM, there is surprisingly little consensus as to its exact meaning. As one writer observed, "Workers' participation has become a magic word in many countries. Yet almost everyone who employs the term thinks of something different" (Schregle, 1970, p. 117). One writer, for example, identifies PDM as active (ego) involvement (Allport, 1945); another argues that it entails a feeling of obligation to work for the best interests of the group (Schultz, 1951). American social scientists define PDM as a specific managerial style while European writers see it as a legally mandated mechanism for employees to influence organizational decisions (Strauss and Rosenstein, 1970). Some writers see PDM as including delegation (Lowin, 1968; Sashkin, 1976; Sorcher, 1971; Strauss, 1963; Tannenbaum, 1962), while others view them as separate phenomena (Davis, 1963). Several approaches identify PDM with group involvement or group decision making (Davis, 1957, 1963, 1967). A common element in many definitions is the concept of equalization of influence or power sharing (Heller and Yukl, 1969; Lammers, 1967; French, Israel, and As, 1960; Leavitt, 1965; Tannenbaum, 1974).

To resolve this confusion, let us begin by consulting the dictionary. The 1961 *Oxford English Dictionary* defines participation as:

1. The action or fact of partaking, having or forming a part of; the partaking of the substance, quality, or nature of some thing or person. 2. The fact or condition of

sharing in common (with others or with each other); association as partners, partnership, fellowship; profit-sharing. b. A taking part, association, or sharing (with others) in some action or matter.

The first definition would not be particularly useful in an organizational context, since it would make participation virtually synonymous with organizational membership as such. The second definition is more pertinent. It implies, first, that there must be at least two persons involved, and second, that something must be shared in common among these persons. Thus PDM may be defined in essence as "joint decision making," in agreement with Tannenbaum and Massarick (1950) and Vroom (1960).

Note that this definition does not logically necessitate decision making by groups of subordinates; it could involve just one supervisor and one subordinate. Furthermore, this definition does not necessitate that the sharing be equal, but only that there be *some* degree of sharing. Finally, the definition does not specify the content of what is shared. The example given in the dictionary is profit sharing, which refers to the *results* of the common effort. The concept of PDM itself refers specifically to participation in the process of reaching decisions.

PDM vs. Delegation

In the opinion of the present writers, the above definition of participation *excludes* delegation. The process of delegation, in an organizational context, involves *assigning* specific responsibilities to subordinates. The result is not a "sharing in common" with others, but rather an explicit division of labor which is determined hierarchically. The subordinate does not participate in the decision to delegate, nor does the supervisor participate in the decisions which are delegated. PDM would be involved (in addition to delegation) if the supervisor and subordinate decided together how much responsibility the subordinate would be given. While delegation could, then, be achieved participatively, it need not be. Thus participation does not logically imply delegation, nor vice versa.

The failure to separate these two concepts, in our opinion, has led to serious confusion in the PDM literature. The merging of PDM with delegation has led to a merging of the human relations school with the cognitive growth school (for example, see Susman, 1976). While there is nothing wrong with combining elements of these two schools of thought in practice, they are conceptually distinct. The human relations school stresses the importance of developing good supervisor-subordinate relationships (through PDM) and cohesive work groups in order to satisfy man's social needs. The cognifive growth school advocates job enrichment through delegating individual responsibility in order to satisfy man's need to grow in his knowledge, efficacy, and individuality (Herzberg, 1966).

Furthermore, PDM as a technique implies no specific content; it is simply a *method* of reaching decisions. Job enrichment, in contrast, explicitly involves the restructuring of the work task so as to provide additional mental challenge (although in practice other elements may be involved, e.g., see Locke, 1975).

As noted above, participation could be used (and has been used, e.g., see Locke, Sirota, and Wolfson, 1976) as a method of introducing job enrichment, but it could also involve joint decision making regarding many other issues, such as pay. (When PDM is used to introduce job enrichment, the kinds of job changes made may deviate considerably from what job enrichment theory specifies, e.g., see Seeborg, 1978). Job enrichment sometimes involves the development of work groups, but it need not do so. Similarly, jobs may be enriched successfully without using participation (Ford, 1969).

Let us now consider some dimensions on which PDM can vary.

Types of PDM
First, PDM can be *forced* or *voluntary*. The force, when it exists, is applied by law or government decree (e.g., codetermination); partially forced PDM would occur in cases where it results from a contract between management and labor but where management legally is compelled to bargain (e.g., American labor unions); voluntary PDM would occur where management initiates the idea of PDM and the employees agree to it (e.g., most Scanlon Plans), or vice versa.

Second, PDM can be *formal* or *informal*. Formal PDM involves the creation of officially recognized decision making or bargaining bodies, e.g., unions, committees, councils, boards, whereas informal PDM is based on the personal relationship between each supervisor or manager and his subordinates.

Third, participation may be *direct* or *indirect* (Lammers, 1967). Direct PDM is usually of the shop floor variety where each employee has the opportunity to assert his views, whereas indirect PDM involves the election of representatives who speak for the employees as members of higher-level committees and decision-making bodies.

Typically, these three dimensions form a pattern. Forced PDM tends to be formal and indirect and is most common in Europe, whereas voluntary PDM tends to be informal (except for unions) and direct and is more common in the United States (Dachler and Wilpert, 1978; Lammers, 1967; Tannenbaum, 1974; Van de Vall and King, 1973).

Participation may also involve parties outside the organization, e.g., stockholders, government bodies, and national and international unions.

Degree, Content, and Scope of PDM

Participation can vary in *degree* (Sadler, 1970; Tannenbaum and Schmidt, 1958; Vroom and Yetton, 1973). The standard continuum goes from *no participation* (supervisors tell the employees what to do, although they may or may not explain the reasons); to various degrees of *consultation* (the supervisors or managers consult the employees either before or after making a tentative decision, and then make the final decision themselves); to *full participation* (in which supervisors become group members and vote with their subordinates as equals).

PDM can also vary in *content,* according to the type of issue involved. The types of decisions which might be included in PDM schemes generally fall into four broad categories:

1. *Routine personnel functions:* hiring, training, payment method, discipline, performance evaluation.
2. *Work itself:* task assignments, work methods, job design, goal setting (including production level), speed of work.
3. *Working conditions:* rest pauses, hours of work, placement of equipment, lighting.
4. *Company policies:* layoffs, profit (or productivity)- sharing, general wage level, fringe benefits, executive hiring, capital investments, dividends, general policy making.

In many studies the specific content of the decisions made under participative management are not specified in much detail, but it is our impression that in studies involving voluntary, informal, direct PDM, the participation is typically restricted to issues in the first three categories (especially the second). In contrast, where forced, formal, indirect PDM is involved, the decisions may also include issues in the fourth category. Higher-level committees focus mainly on organizational policies while the lower-level committees are more concerned with the work itself and working conditions.

One writer has estimated that codetermination, as practiced in Germany, entails the greatest degree and scope of participation, followed by workers' management in Yugoslavia,[5] and joint consultation as practiced in Britain (Van de Vall and King, 1973).

Another aspect of scope, which has been given little attention in the literature, is the *stage* of problem solving at which PDM occurs, e.g., the discovery of problems, the generation of proposed solutions, the evaluation of proposed solutions, and the choice of solution. Vroom (1969) and Wood (1973) have observed that PDM may be more effective at some stages than at others.

III. THE PSYCHOLOGICAL BASIS OF PARTICIPATION EFFECTS

The benefits alleged to result from PDM by workers fall into two major categories. The first includes increased *morale and job satisfaction* and their frequent concomitants, reduced turnover, absenteeism, and conflict. The second category includes outcomes pertaining directly to *productive efficiency*, e.g., higher production, better decision quality, better production quality, and reduced conflict (again) and costs (Argyris, 1955; Davies, 1967; Davis, 1957; French, Israel, and As, 1960; Lammers, 1967; Likert, 1961, 1967; Lowin, 1968; Maier, 1973; Rosenfeld and Smith, 1967; Sadler, 1970; Schultz, 1951; Strauss, 1963; Strauss and Rosenstein, 1970; Tannenbaum, 1966, 1974; Tannenbaum and Massarik, 1950; Vroom, 1969).

Less frequently discussed are the *mechanisms* by which PDM will bring about these alleged benefits. With respect to morale, the simplest explanation is that allowing participation will increase the likelihood that the employee will get what he wants (Mitchell, 1973), or satisfy his motives (French, Israel, and As, 1960; Lowin, 1968; Strauss, 1963), i.e., that he will attain his values—value attainment being the direct cause of job satisfaction (Locke, 1976a).

If value attainment is, in fact, the mechanism involved, it becomes problematic as to the conditions under which PDM will achieve it. It depends, first, upon what it is that the employee wants, and, second, upon whether PDM actually brings it about.

If the employee wants simply to *express* his views (Argyris, 1955), then PDM will, if practiced, virtually always bring it about. If the employee wants respect or dignity (Davis, 1957), PDM may attain it, providing the participative method used corresponds to his concept of dignity. If actual influence in the decision process is what the employee desires, then satisfaction will depend upon *how much* influence is actually exerted (or perceived as being exerted) in relation to the amount desired. In many cases an employee may participate in discussions but have little influence over the final decision (Hoffman, Burke, and Maier, 1965; Mulder, 1959).

If full equality with supervisors or managers is what the employee wants, then it is unlikely that he will be satisfied by PDM. We have not found a single case (even in the Israeli kibbutzim; Tannenbaum, 1974) in which a managerial hierarchy was totally dissolved in practice through PDM. In many cases the employee may want tangible benefits from PDM such as more pay. This desire is recognized explicitly in the Scanlon Plan (to be discussed in the next section), which combines PDM with monthly bonuses for increased efficiency.

With respect to the mechanisms causing increases in productive efficiency, the probable factors are more varied. They can be divided into two subcategories—*cognitive* and *motivational*.

A major cognitive factor is the increase in information, knowledge, and creativity that will allegedly be brought to bear on organizational problems as the result of PDM (Davis, 1963; Lammers, 1967; Lowin, 1968; McGregor, 1944; Miles, 1965; Rosenfeld and Smith, 1967; Schultz, 1951; Strauss, 1963). This, in turn, has been attributed to improved upward communication (Nicol, 1948; Rosenfeld and Smith, 1967; Sashkin, 1976; Schultz, 1951; Strauss, 1963). This argument assumes, of course, that the subordinates actually have knowledge which the supervisor lacks and which is relevant to the problem. (We shall return to this very important issue later in this essay.) It is also assumed that the process of group decision making, when it is involved, will lead to better utilization and integration of knowledge than individual or consultative decision making, but again the evidence on this issue is equivocal (Vroom, 1969).

A second cognitive factor involves the greater understanding on the part of the employees who are to execute the decisions which will allegedly result from PDM (Lawler and Hackman, 1969; Maier, 1967; Rosenfeld and Smith, 1967; Vroom, 1969). This may involve such factors as greater goal clarity, a fuller grasp of the methods to be used in accomplishing the work, or a more thorough understanding of the reasons for organizational changes, decisions, and policies. Again, the validity of this premise depends upon whether greater understanding *is* actually achieved by PDM and the degree to which full understanding is crucial to job performance.

The most widely discussed motivational mechanism of PDM is reduced resistance to change (Coch and French, 1948; Davis, 1957; French, Israel, and As, 1960; French, Ross, Kirby, Nelson, and Smyth, 1958; Kahn, 1974; Lawrence, 1971; Marrow and French, 1945; Rosenfeld and Smith, 1967; Schultz, 1951). This, in turn, has been attributed to greater trust on the part of employees (Davis, 1963; Lawler, 1975; Lammers, 1967), which results from being consulted about proposed changes.

Another possible cause of reduced resistance which has rarely been mentioned is the greater feeling of control (and reduced anxiety) which may result from PDM. For example, an employee who thought a particular change would not be beneficial to his welfare or would not promote efficiency would have the opportunity to protest and perhaps modify the proposed policy.

The positive side of the coin of reduced resistance is increased acceptance of and commitment to changes or decisions, including goals (Davis, 1963; Lawler and Hackman, 1969; Meyer, Kay, and French, 1965; Rosenfeld and Smith, 1967; Sashkin, 1976; Sorcher, 1971; Strauss, 1963; Vroom, 1969). This, in turn, has been attributed to a greater degree of ego

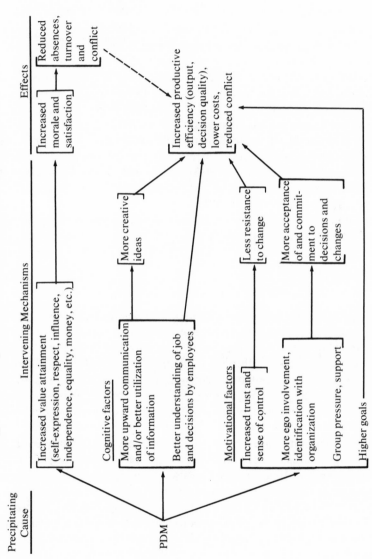

Figure 1. Proposed Effects and Mechanisms of PDM.

Precipitating
Cause

PDM

Intervening Mechanisms

Increased value attainment
(self-expression, respect, influence,
independence, equality, money, etc.)

Cognitive factors

More upward communication
and/or better utilization
of information

Better understanding of job
and decisions by employees

More creative
ideas

Motivational factors

Increased trust and
sense of control

Less resistance
to change

More ego involvement,
identification with
organization

Group pressure, support

More acceptance
of and commit-
ment to
decisions and
changes

Higher goals

Effects

Increased
morale and
satisfaction

Reduced
absences,
turnover
and
conflict

Increased productive
efficiency (output,
decision quality),
lower costs,
reduced conflict

279

involvement or identification with the organization induced by PDM (Allport, 1945; Lewin, 1952; Patchen, 1964; Tannenbaum and Massarik, 1950; Vroom, 1964) and to the effects of group pressure or "support" in cases where group decision making was involved (Lewin, 1947, 1952; Mitchell, 1973; Patchen, 1964; Rubenowitz, 1962; Sashkin, 1976; Strauss, 1963; Vroom, 1969). Again these arguments involve a number of problematic assumptions, e.g., that PDM will necessarily lead to greater ego involvement than other methods of managing (e.g., incentives, delegation); that ego involvement is necessary for effective performance; that group pressures will always be in the direction of goal acceptance; and that group members will always give in to group pressures.

A final motivational mechanism, which was not originally predicted by PDM theorists but which has been found in a number of cases, is goal level. It has been argued that PDM groups may set goals which are higher than those which would have been assigned by management (Dowling, 1975; Latham and Yukl, 1975b).

The mechanisms discussed above and their hypothesized interrelationships are shown in Figure 1.

IV. RESEARCH ON PARTICIPATION

The PDM literature is so enormous that to achieve a "complete" review is virtually impossible. We have not made any special attempt to include unpublished material in this review, nor have we included many foreign articles or any not written in English. We have also excluded studies which involve more general dimensions of supervisory style (e.g., consideration), studies involving mainly delegation, and PDM studies in educational settings which have been summarized elsewhere (Anderson, 1959). Within these limits we believe that our literature research was thorough. While we cannot claim to have included every PDM study ever done in the United States, we have not knowingly omitted any significant study. Thus we believe that our review gives an accurate indication of what is known about the effectiveness of PDM.

As noted in the previous section, two broad classes of criteria have been used to evaluate PDM effectiveness—satisfaction or morale, and productive efficiency, including decision quality. Each of these criteria are discussed separately within each of the four subsections to follow.

The four subsections are: laboratory studies, correlational field studies, multivariate experimental field studies, and univariate (controlled) experimental field studies. This classification was based on what we judged to be the two most fundamental differentiating characteristics among the studies—those pertaining to external validity and to the inference of causality. The specific strengths and/or weaknesses of each type of study will be discussed in each subsection.

Laboratory Experiments

While a laboratory environment enhances the experimental control of variables and therefore allows causal inferences to be made, the artificiality of the situation may limit generalizability to real organizational settings. For example, lab studies are short term rather than long term. The subjects usually consist of college students who may have different attitudes and values from workers (cf. Litwin and Stringer, 1968). Furthermore, volunteer subjects may be more prone than employees to do what they think the experimenter wants (although these two situations may not be as different as is typically assumed; Locke, 1978). Finally, less is "at stake" in an artificial lab situation as compared to the real world with respect to such issues as self-esteem and success in one's career.

Despite these potential drawbacks, the results of some laboratory studies have been found to generalize to field settings (e.g., Latham and Yukl, 1975a). Thus findings of such experiments may be valuable in identifying causal relationships which could apply to genuine organizations.

Productive Efficiency

Several studies by Maier and his associates have investigated the effects of training group leaders in democratic leadership techniques on group problem-solving quality and decision acceptance. With subjects role playing an assembly-line situation, Maier (1950) found that decision quality[6] and decision acceptance were greater under leaders trained to use democratic techniques than under untrained leaders. Untrained leaders obtained full agreement to their solution in 62.1 percent of their attempts as compared to 100 percent for the most highly trained leaders.

In a later study, Maier (1953) found results similar to those cited above. All groups role-played a problem involving the introduction of more efficient methods of doing assembly work. Half the groups with untrained leaders accepted the change in work methods implied by the foremen while half rejected his recommendation. Of the groups with trained leaders, 59.1 percent accepted the change and 4.5 percent refused to make any changes. The remaining 36.4 percent of the groups compromised and accepted other changes. Maier and Sashkin (1971), using the same technique, found that democratic leadership led to better decision quality.

Decision quality, according to Maier, is better in democratically led groups because this leadership style encourages group members to express and consider conflicting points of view. These differences in opinion can promote, through discussion and integration, creative problem solving. A study by Hoffman, Harburg, and Maier (1962) tested this claim. Role playing a work method change situation, it was found that conflict of opinion increased the frequency of high-quality decisions. Such decisions, however, were obtained more often in groups intent on change than in those indifferent to it.

Two other studies, however, failed to replicate the findings of Maier and his associates. McCurdy and Eber (1953) studied groups composed of authoritarian and democratic subjects. Their leaders were coached in either democratic or authoritarian patterns of behavior. Authoritarian groups were somewhat more effective than the democratic groups in speed of problem solving on a group maze, although the differences were not significant. Similar results were reported in an earlier study by McCurdy and Lambert (1952).

Torrance (1953) reported that bomber crew members who were critiqued in a performance evaluation by structured methods exhibited greater performance improvement than those crew members who were critiqued by a less authoritarian method.

An experiment by Shaw (1955) investigated the problem-solving performance by four-man groups in different communciation nets under authoritarian and democratic leadership. Problems were solved faster and with fewer errors under authoritarian leaders. This was attributed mainly to the reduction in "saturation" (input and output communication requirements) which resulted from the more directive style.

A series of three group problem-solving ("Twenty Question") experiments by Calvin, Hoffman, and Harden (1957) compared democratic and authoritarian leadership styles. There was no consistent trend in favor of either style; however, the less intelligent subjects performed better under authoritarian leaders while the more intelligent ones performed slightly better under democratic leadership.

Lanzetta and Roby (1960) found that both time and error scores for problem-solving groups were better under participative than directive styles of leadership. They noted, however, that the best-performing groups were those in which leadership and influence in decision making were congruent with ability differentials of the group's members. Thus, participation was effective when the most able members of the group exerted the most influence in decision making.

Using an air controller simulator, Kidd and Christy (1961) studied three patterns of leader behavior—laissez-faire, active monitoring (autocratic), and direct participant (participative)—on several measures of air-controller effectiveness. It was found that leadership style led to a distinct trade-off effect between performance criteria. The laissez-faire pattern allowed the controller to concentrate on maintaining a rapid flow through the system but produced high error scores. The active monitoring pattern, on the other hand, reduced error scores but inhibited the rapidity of flow. The participative style led to intermediate results.

Using a complex management game, Dill, Hoffman, Leavitt, and O'Mara (1961) studied the effects of an egalitarian distribution of influence in a laboratory simulation of organizations on the quality of problem solving and profit. Using six simulated organizations, they found that the two firms with the highest profit showed the highest dispersion of influence and were the least egalitarian. They noted that profitability was also affected by the "perceptual consonance" of the members of the work group, that is, the expectations of group members as to who should exert the greatest influence in decision making.

The effects of task structure on leadership style and group performance in problem solving were studied by Shaw and Blum (1966). Eighteen groups were assigned to either a directive or nondirective leader where they were given three tasks to solve which differed in structure. Using time scores as the measure of performance, it was found that directive leadership was more effective than nondirective leadership when the task was highly structured. Tasks that were unstructured and required varied information were more effectively solved when a nondirective leadership style was used.

The "Twenty Questions" parlor game was used by Katzell, Miller, Rotter, and Venet (1970) to examine the effects of leadership and other variables on group problem solving. They found that directive leadership was somewhat more effective than nondirective leadership in attaining solution quality. The authors attribute its effectiveness to a higher member interaction rate among the groups with directive leaders.

Finally, a study by Hannan (in press) compared the effects of participatively set and assigned goals on performance. The task was evaluating credit card applications using a formula. Hannan found that participative goal setting was marginally related to increased goal acceptance, especially for difficult goals. However, PDM was not related to performance, which was determined instead by ability and the subjects' personal goals.

The findings of the laboratory experiments discussed above are summarized in Table 1.

Satisfaction

In a study of democratic and autocratic leadership, Gibb (1951) found that subjects were more satisfied with leader behavior when it was more autocratic and when the group was more structured and unified. Satisfaction with the leader was correlated .76 with autocratic leadership and −.60 with group freedom.

Shaw (1955) found that morale was significantly higher under democratic leaders than under autocratic leaders. He noted that morale was heavily influenced by independence, which is reduced by authoritarian leadership. Democratic leadership, on the other hand, increases independence and thus morale.

Table 1. Summary of Results of Laboratory Experiments on PDM

Study	Task	Manipulation or Comparison	Results Prod.	Results Sat.
Maier (1950)	Role play of an assembly-line situation	Trained in PDM vs. untrained	+	−
Gibb (1951)	Various activities	Democratic vs. autocratic		
McCurdy & Lambert (1952)				
McCurdy & Eber (1953)	Group maze	Democratic vs. autocratic	0	
Torrance (1953)	Group maze	Democratic vs. autocratic	0	
Maier (1953)	Performance appraisal	PDM vs. autocratic	−	
Shaw (1955)	Role play in work change	Trained in PDM vs. untrained	+	
Calvin, Hoffman & Harden (1957)	Arithmetic problems	Democratic vs. autocratic	−	+
Fox (1957)	"Twenty Questions"	Democratic vs. autocratic	0	
Mulder (1959)	Decision making in a conference situation	Positive vs. negative		+
Lanzetta & Roby (1960)	15 problem-solving tasks	High vs. low power	+	+
Kidd & Christy (1961)	Mechanical task	PDM vs. directive		
Dill, Hoffman, Leavitt & O'Mara (1961)	Air controller simulator	Laissez-faire, vs. direct participant vs. active monitoring	0	
Shaw & Blum (1966)	Management simulation game	Dispersion of influence	−	−
Katzell, Miller, Rotter, & Venet (1970)	Problem solving	Directive vs. nondirective	0	
Maier & Sashkin (1971)	"Twenty Questions"	Directive vs. nondirective	−	
Wexley, Singh, & Yukl (1973)	Role play in work change	Trained in PDM vs. untrained	+	+
Hannan (in press)	Appraisal interviews	Tells vs. problem solving (PDM)	[a]	
	Goal setting	Participative vs. assigned	0	+
PDM Superior			(+) 4	5
No difference or contextual			(0) 6	0
PDM Inferior			(−) 4	2

[a]This study used an attitude rather than a performance measure of motivation; since we considered this weak evidence for a laboratory study, the performance results were not included in the analysis.

The effects of "positive" and "negative" leadership styles on member satisfaction in conference groups were explored by Fox (1957). What he called positive and negative styles can be classified as participative and autocratic, respectively. The positive style of leadership induced a more permissive and friendlier group atmosphere, greater member satisfaction with the leader, and greater member satisfaction with and acceptance of group decisions than the negative leadership style.

Mulder (1959) investigated the relationships between power-exertion, self-realization, and satisfaction of members of problem-solving groups. In this experiment two degrees of power-exertion (defined as the passing on of essential information in a group, i.e., expert power) and two degrees of self-realization (defined as the responsibility for the completion of one's own task) were compared. The power variable is related to PDM in that it pertains to the amount of influence a person has. In each condition, fifteen problem-solving tasks were completed. It was found that only the power variable had any effect on satisfaction. Those subjects in the "more power" condition showed significantly greater satisfaction than those in the "less power" condition.

Dill, Hoffman, Leavitt, and O'Mara (1961) found in their simulation that the two organizations with the highest morale had the greatest dispersion of influence and were the least egalitarian. These companies had also the highest degrees of "perceptual consonance" as to the roles of each member.

Katzell, Miller, Rotter, and Venet (1970) in a study of groups solving the "Twenty Questions" parlor game found that satisfaction with the task was significantly less for those subjects under highly directive leaders than under less directive leaders.

In an experiment using simulated appraisal interviews, Wexley, Singh, and Yukl (1973) found a positive association between degree of PDM and satisfaction with the interview.

The findings from the laboratory experiments discussed above are also summarized in Table 1. It can be seen in the lower right corner of the table that with respect to performance, there is no overall difference between PDM and more directive methods. With respect to satisfaction, the results favor PDM, although in two cases, satisfaction was higher in the groups with more directive leaders.

Correlational Field Studies
The studies in this subsection were all conducted in field settings, but this advantage was more than offset by the limitations involved in the use of the correlational method. (We define the term "correlational" broadly in this context to mean any method which correlates observed differences on one attribute or behavior with observed differences in another without

having manipulated the hypothesized cause.) Such studies may suggest causal relationships but cannot prove them.

Productive Efficiency

A classic study by Katz, Maccoby, and Morse (1950) investigated the relationship between the productivity of clerks in an insurance company and various leadership characteristics. Twelve pairs of work groups which performed the same type of work but which differed in their productivity were studied. The supervisors of the high-producing groups employed less "close" and more "general" supervision, were rated as less "production centered" and more "employee centered"; exercised better judgment; were more rational and less arbitrary; and were more democratic and less authoritarian than supervisors of low-producing sections. In addition, the high-producing supervisors reported that they were supervised less closely by their own superiors and were more satisfied with the degree of their authority and responsibility than the low-producing supervisors.

An obvious problem with this study is the failure to separate the effect of PDM from those of the other factors correlated with high work group productivity.

Katz, Maccoby, Gurin, and Floor (1951) attempted to replicate the insurance company study in three divisions of a railroad company. Thirty-six pairs of sections of maintenance workers who were similar with respect to working conditions but different in their level of productivity were compared.

As in the insurance study, high-producing foremen were described by their subordinates as more "employee centered" and less "production centered" than the low-producing foremen. In addition, high-producing foremen spent more time in supervising; planned more often and were better at it; spent more time teaching their men new techniques; and gave more constructive and less punitive comments to subordinates than supervisors of low-producing sections. However, unlike the insurance study, there was no relationship between productivity and PDM.

Two divisions of a naval research laboratory were surveyed by Weschler, Kahane, and Tannenbaum (1952). One division was headed by a permissive leader, the second by a restrictive leader. The productivity of the permissively led division was rated considerably higher than that of the restrictively led division by the division members. However, according to a more objective evaluation of performance by five superiors and two staff people who were more familiar with the *actual* performance of the two divisions, the performance of the restrictively led division was rated slightly higher. (No statistical tests were employed to test this difference.)

Adams (1952) reported that members of bomber crews performed better under medium than either high or low degrees of equalitarian leadership.

Berkowitz (1953) observed seventy-two small decision-making (conference) groups in business and government organizations. Unlike most PDM studies, the leaders of these groups were not assigned but were chosen by the group's members. It was found that PDM was negatively but not significantly correlated with performance. Berkowitz explains though that those members who chose the leader perceived his role as the *sole* leader and not as a member of a participating group. Since there was an expectation of role differentiation between the designated leader and the group members, participation was not viewed by the members as being legitimate and did not facilitate performance.

In a medical laboratory study by Pelz (1952), participative leaders who interacted daily with their subordinates were rated as being more effective than directive leaders but less effective than leaders who delegated responsibility. No substantial differences between styles emerged when there was less than daily contact, however.

In a study of assembly departments in a manufacturing firm, Argyle, Gardner, and Cioffi (1958) found democratic leadership style to be positively correlated with productivity only in those departments where direct incentive pay was not in force. In incentive departments, no relationship was found.

Mullen (1965) compared the leadership styles of three division managers in a large automobile insurance company and its relationship to each division's efficiency and effectiveness. Each manager exhibited a distinct leadership style: one was democratic, another laissez-faire, and the third was authoritarian. Using several measures of efficiency (e.g., policy issuance rate), it was found that all three divisions were operating at an equally high rate of efficiency. (No statistical tests were employed.)

The measures of effectiveness (cost and profitability) indicated that all three divisions had experienced a decline during the study, in line with the trend of the entire automobile insurance industry during this period. Although the trend was in the same direction for all the divisions, the absolute dollar figures indicated, surprisingly, that the division with the laissez-faire manager was the only one which made a profit in this interval. However, a more accurate measure of effectiveness (average closing cost of bodily injury claims) did not show any differences between divisions.

Vroom (1960), using a sample of 108 supervisory personnel, found that participation was significantly correlated with performance for the total sample, but the correlations were significantly higher for supervisors high in independence than for supervisors low in independence, and for those low in authoritarianism than high in authoritarianism. When need for independence and authoritarianism were combined, the relationship between participation and performance was significant only for those high in need for independence and low in authoritarianism.

In an attempted replication of Vroom's study with consumer finance office managers, Tosi (1970) found no relationship between perceived PDM and performance for any group.

Mahoney (1967) surveyed the judgments of eighty-four managers in thirteen organizations as to the factors that they perceived to be related to organizational effectiveness. Democratic leadership was not related to perceived effectiveness whereas supervisory control, defined as the supervisor being "on top of things," was perceived as being positively related to effectiveness. These findings, of course, were not validated by reference to any objective measures of performance for any of the organizations studied.

A similar study by Stagner (1969) found results in agreement with those of Mahoney. Using a questionnaire mailed to 500 vice presidents of 125 large firms, Stagner found that corporate profits were perceived as being associated with more formality (bureaucratic routines), centralization of decision making, and less personalized management.

Patchen (1970) administered questionnaires to 834 employees in two Engineering Divisions and three Steam Power Plants of the Tennessee Valley Authority (TVA) in order to investigate the effects of involvement in the work environment. To help create a sense of involvement, top TVA management encouraged employees to participate in decisions affecting their work. The major mechanism for such participation was direct work group participation and the TVA'a cooperative conferences and committees where representative (indirect) participation in a wide range of work problems was encouraged. At the time of this study the cooperative program had been in operation almost two decades.

Control over work methods emerged as the one factor which had sizeable associations with all attitudinal indicators of job motivation (no "hard" criteria were obtained). Influence in goal setting was also positively related to job motivation. However, the strength of this relationship was moderated by the opportunity that an employee had for achievement. Patchen comments ". . . that people will value more highly goals which they help to set for themselves . . . and . . . this is likely to be so particularly when attainment of the goal has significance as a personal achievement" (Patchen, 1970, p. 239). Clearly there is some confounding of PDM with job enrichment (delegation) in this study.

The effects of participation in a Management by Objectives program were examined by Carroll and Tosi (1973) in a questionnaire study. They found no overall correlation between degree of influence in the goal setting process and employee effectiveness (as rated by managers). However, a supplementary analysis found a marginal correlation between PDM and performance for those employees who were accustomed to PDM (a finding which replicates that of French, Kay, and Meyer, 1966).

Schuler (1977) found no overall correlation between PDM and performance among a sample of manufacturing employees.

The findings from the correlation field studies discussed above are summarized in Table 2.

Satisfaction

Weschler, Kahane, and Tannenbaum's (1952) naval research laboratory study reported that 63.3 percent of the members of the division headed by the permissive leader were "well satisfied" with their job, whereas 39.3 percent of the members of the division headed by the restrictive leader were "well satisfied" with their job. In addition, the employees' perceptions of morale in their immediate work group, the division, and the laboratory were considerably higher in the permissively run division than in the restrictively run division.

A research laboratory setting was also used by Baumgartel (1956, 1957) to study three leadership styles (participative, directive, and laissez-faire). It was found that job satisfaction was higher under participative leaders than either the directive or laissez-faire leaders.

Foa (1957) studied crews and officers of eighteen ships of an Israeli shipping company under permissive and authoritarian supervision. It was found that the group members' expectations toward discipline moderated the relationship between satisfaction and leadership style. Groups with authoritarian expectations were equally satisfied with either leadership style, whereas groups with permissive expectations were more satisfied with permissive leaders. Authoritarian leaders satisfied authoritarian crews more than permissive crews and the trend was in the same direction for permissive leaders. The term "permissive," of course, which was based on attitude toward strict discipline, is not identical in meaning to participative.

In the study of small decision-making (conference) groups by Berkowitz (1953), satisfaction with the conference was negatively and significantly correlated with PDM.

Argyle, Gardner, and Cioffi (1958) did not directly measure satisfaction in their study of assembly departments, but reported two measures usually associated with it, absenteeism and turnover. There was no relationship between turnover and leadership style, but democratic supervision was positively related to low absenteeism. This relationship was significant for the combined departments but not within the incentive departments.

Vroom (1960), in his delivery company study, found that participation

Table 2. Summary of Results of Correlational Field Studies of PDM

Study	Task	Manipulation or Comparison	Results Prod.	Results Sat.
Katz, Maccoby & Morse (1950)	Clerical work	Democratic vs. authoritarian	+	+
Katz, Maccoby, Gurin & Floor (1951)	Railroad maintenance	Democratic vs. authoritarian	0	
Weschler, Kahane & Tannenbaum (1952)	Scientific research	Permissive vs. restrictive	−	
Adams (1952)	Bomber crews	Degrees of equalitarianism	0	−
Berkowitz (1953)	Conference groups	Degrees of PDM	0	
Pelz (1956)	Scientific research	Participative, laissez-faire, & directive styles	0	0
Foa (1957)	Shipping crews	Authoritarian vs. permissive		+
Baumgartel (1956,1957)	Scientific research	PDM, directive, and laissez-faire	0	
Argyle, Gardner & Cioffi (1958)	Assembly	Democratic style	+,0	
Vroom (1960)	Delivery work	Degree of authoritarianism	0	+,0[a]
Vroom & Mann (1960)	Delivery work	Degree of authoritarianism		+,0
Mullen (1965)	Insurance	Democratic, laissez-faire, and authoritarian styles		0
Ley (1966)	Production work	Degree of authoritarianism		+
Mahoney (1967)	Managing	Democratic vs. controlling	−	+
Stagner (1969)	Managing	Degree of formality, centrality, and personalization	−	
Patchen (1970)	Engineering and maintenance work	Degree of control of job by employee	+	0
Sadler (1970)	All job levels and functions	Tells, sells, consults, joins		0

Study	Setting	Variable		
Tosi (1970)	Office management	Degree of PDM	0	+
Miles & Ritchie (1971)	Managing	Degree of consultation	0	+
Carroll & Tosi (1973)	MBO program	Degree of influence in goal setting		+,0
Ruynon (1973)	Chemical work	Degree of PDM		+
Alutto & Acito (1974)	All job functions	Degree of PDM deprivation		+
Falcione (1974)	Industrial work	Degrees of PDM		+
Lischeron & Wall (1974)	Local governments	PDM at various levels		+,0
Schuler (1977)	Manufacturing	Degree of PDM	0	+,0
		(+) PDM Superior	3	13
		(0) No difference or contextual	10	8
		(−) PDM Inferior	3	1

[a] +,0 indicates a significant main effect which only applied to one sub-group when the sample was subdivided.

was positively related to supervisors' attitudes toward their jobs. However, when subjects were grouped simultaneously, on the basis of independence and authoritarianism, there was a significant positive correlation between PDM and satisfaction only in the high need for independence, low authoritarian subgroup. Tosi's (1970) replication study, however, found an overall correlation between PDM and job satisfaction, but no moderating effects for either variable.

In another study, Vroom and Mann (1960) investigated the effects of leader authoritarianism on employee attitudes. Using two samples of employees, the authors found that employees in small work groups which were characterized by a great deal of interaction among workers and between workers and their supervisors, and by a high degree of interdependence, had more positive atittudes toward equalitarian leaders. Employees had more positive attitudes toward authoritarian leaders, however, in large work groups whose members worked independently and where interaction between them and their supervisor was infrequent.

Mullen's (1965) insurance company study used measures of job satisfaction, absenteeism and turnover. The results indicated that democratic leadership was associated with employee satisfaction, but not with turnover or absenteeism. The author speculates that although there was less satisfaction in the divisions run by the laissez-faire and authoritarian leaders, other factors in the job situation (e.g., good wages, pleasant working conditions, job security, etc.) were far more important than the leadership style in determining turnover and absenteeism.

Using a sample of production employees, Ley (1966) found that turnover was positively correlated ($r = .76$) with the related authoritarianism of foremen.

Miles and Ritchie (1971) found that managers who felt that they were more frequently consulted by their bosses were more satisfied with them than were managers who were less frequently consulted.

Using a sample of hourly employees in a chemical company, Runyon (1973) found a positive correlation between perceived degree of PDM and satisfaction with supervision. However, when the sample was subdivided on the basis of "locus of control" (Rotter's I-E scale), the correlation was positive for those high in Internal locus but negative for those high in External locus.

Degree of PDM deprivation was found to be negatively correlated with a number of dimensions of job satisfaction by Alutto and Acito (1974), although the results may have been confounded somewhat by age.

In a study of a large industrial organization, Falcione (1974) found a significant positive correlation between participation and satisfaction with the immediate supervisor. The author points out though that the participation utilized in this study was consultative and rarely did the employees have a final say in the decisions. Nor surprisingly, participation explained a very small amount of the variance in satisfaction. A second factor which explained a considerably greater variance was the perceived credibility of the supervisor. Falcione speculates that ". . . a supervisor who allows for subordinate participation but has low credibility in the eyes of his subordinates may be viewed as a 'weak' supervisor. A supervisor who allows for subordinate participation and is viewed as a high credible source may have a much more satisfied group of subordinates" (Falcione, 1974, pp. 51–52).

Lischeron and Wall (1974) examined the effect of participation at various organizational levels on several dimensions of satisfaction (organization, pay, promotion, work itself, superior, coworkers). A survey was administered to 127 blue-collar workers in four divisions of a local government department in England. Satisfaction measures were taken for perceived participation in decisions at medium and distant levels in the organization. Degree of perceived participation was significantly and positively correlated to overall job satisfaction. It was also found that those workers who were least satisfied expressed the strongest desire to participate.

In his study of the TVA, Patchen (1970) measured specific aspects of satisfaction (pay, promotion, immediate supervisor, and coworkers) as well as overall job satisfaction. Neither the overall index of satisfaction nor specific aspects of satisfaction were associated with participation for the organization-wide cooperative program. In fact, the associations tended to be negative. The effects of participation in immediate work-group decisions on satisfaction were not reported.

In a comparison of different types of employees in two British companies, Sadler (1970) found that most preferred a "consults" leadership style (i.e., moderate participation) to other styles (tells, sells, joins); however, the highest degree of satisfaction was associated with getting the style of leadership that the employee wanted, regardless of the actual style.

Schuler's (1977) results seem to indicate an overall relation between work satisfaction and PDM, but after subgrouping, the relationship held only where high levels of role conflict and/or role ambiguity existed. Possibly PDM was experienced as a means of coping with these sources of frustration.

The findings from the corrrelational field studies discussed above are summarized in Table 2. The results compiled at the bottom of the table show that with respect to performance there is no overall difference between PDM and non-PDM leadership styles. With respect to satisfaction, the results are more favorable to PDM, even though over 40 percent of the studies did not show any positive effect.

Multivariate Experimental Field Studies

Many of the studies to be discussed in this subsection are considered to be "classics" in the PDM literature. All were conducted in field settings and together encompass a time span of more than 50 years. Since they are all experimental studies, cause-and-effect inferences can be made. However, since all of the studies in this group involved the manipulation of at least one major variable in addition to PDM, conclusions about the efficacy of PDM as such are impossible to draw.

Except where logical continuity dictates otherwise, the studies will be examined in approximate chronological sequence; in cases where groups of studies are involved, the date of the earliest study in the group will be used to determine the sequence. Since most of these studies claim to have achieved improvements in both productivity and morale, these criteria will not be specifically separated as in the other subsections.

Mayo's "first inquiry"   While Elton Mayo is most well known as the director of the Hawthorne studies, his first field experiment was conducted in the mule spinning department of a textile mill near Philadelphia in 1924 (Mayo, 1924, 1945/1970). The department suffered from a 250 percent annual turnover rate, low morale, and productivity which was too low to allow the workers to benefit from an incentive bonus which was contingent upon attaining a minimum level of efficiency. The work week consisted of five 10-hour days with a 45-minute break each day for lunch. Interviews with the employees of the mule spinning department revealed that they suffered from severe fatigue.

As a result of these consultations with the workmen, Mayo suggested allowing four 10-minute rest periods spaced throughout the working day. A proposal to try out this idea on one-third of the workmen in the department was accepted by management. As a result, turnover dropped and morale improved. When the rest periods were given to the remaining department members (with cots being provided for them to lie on) overall production increased and the workers made the bonus for the first time. Later the rest periods were taken away, whereupon morale and production dropped. Subsequently they were reinstated with the added benefit that groups of employees could decide among themselves who would rest and when, providing they all received their four breaks. Again morale and efficiency increased. Turnover eventually dropped to between 5 and 6 percent.

Mayo's original (1924) report of this study attributed the results to the rest pauses, which appeared to reduce fatigue and "pessimistic revery." However, in his later (1945/1970) post-Hawthorne interpretation, he stressed the role of social factors such as participation, but admitted that too many factors were changed to make any definitive conclusion possible. While consultative participation was involved in the study, this only entailed asking the employees how they felt. Later the employees were delegated the authority to schedule the rest pauses themselves, group formation being encouraged for this purpose. There were also some technological improvements made in the department. But the central factor seems clearly to have been the introduction of the rest pauses as such. If the workmen had been consulted and the rest pauses *not* introduced, it seems very doubtful whether the increase in morale and productivity would have occurred.

The Hawthorne studies   These studies conducted in the late 1920s and early 1930s under Mayo's direction (Roethlisberger and Dickson, 1939/ 1956) are probably the most famous experiments ever conducted in the social sciences. They have been widely interpreted as demonstrating the value of PDM in industrial settings (Blumberg, 1968; Davis, 1967; French, 1950; Kahn, 1975; White and Lippitt, 1960), although contrary interpretations have been offered (Carey, 1967; Lawler, 1975; Parsons, 1974). The former interpretation was based primarily on the results of the Relay Assembly Test Room and Mica Splitting Test Room experiments. Let us examine what actually occurred in these experiments, following the incisive critique of Carey (1967).

Stage I of the Relay Assembly Test Room study included the introduction of a simpler, less-varied task, a new incentive system (more closely geared to individual effort), improved feedback, looser, more considerate supervision (which later became more harsh when productivity did not rise as expected), rest pauses, reduced hours of work, and the replacement of two of the five girls in the original group. No substantial increases in productivity occurred in this group until the two lowest producers were removed and replaced by two other, highly motivated women (whereupon supervision again became considerate). Thus the most plausible conclusion to be drawn from this study is that selection (or transfer) is an effective method of raising productivity! Carey implies that about 25 percent of the overall 30 percent increase in productivity obtained in Stage I was caused by this factor.

In Stage II of the Relay Assembly study, the only change made from the original conditions of work was the introduction of an incentive system. This led to an immediate increase of nearly 13 percent in productivity and a decrease of 16 percent when it was later withdrawn.

In the Mica Splitting Test Room study, there was, of course, a different task, and the hours of work were reduced. There was also more considerate, participative supervision. The Hawthorne investigators claimed that a 15 percent increase in productivity resulted from the introduction of the new supervisory style. Carey (1967), however, demonstrates that this result is purely fictional, since the investigators had changed their implicit definition of the term "productivity" between this and the previous studies. In the Mica Splitting Test Room there was an increase in *rate* of production, but no increase in *total* production due to the decrease in hours of work. Even the finding of an increase in rate is suspect, because the investigators included different time intervals in calculating the production of each worker, a fundamental breach of proper experimental design.

The most legitimate conclusion to be drawn from the Hawthorne studies is that there is some evidence for the beneficial effects of selection and for the effects of incentives (Lawler, 1975) on productivity. Even these findings are not conclusive because of the absence of control groups.

McCormick's multiple management (McCormick, 1938, 1949) In 1932, Charles P. McCormick inherited his uncle's position as president of a well-known spice company. Since the company was in some financial trouble, he immediately initiated a number of changes, including an increase in pay for the workmen of $2.00 a week, a reduced work week, and possibly (this is not clear in his writings) a profit-sharing plan, and a promise of no layoffs. Later new equipment was introduced and tea was served during rest breaks.

However, McCormick is known mainly for his idea of the introduction of a Junior Board of Directors (later supplemented by a Sales Board and a Factory Board). These boards consisted of junior executives who were elected by other members of the board. Their function was to develop ideas to improve the management of the company and to propose these changes to upper management (e.g., the senior Board of Directors). The development of these boards, according to McCormick, led not only to better training for the junior executives, but to a more objective merit system, increased competition, and a clearer promotion ladder.

McCormick claimed that the long-term benefits of these changes included the development of higher employee motivation and morale (including labor peace), the deradicalization of young executives with Marxist ideas, and the development of many good ideas which, when put into practice, produced greater profitability for the company.

While McCormick presents a convincing case for the value of the Junior Board (and other boards) which were a form of consultative management, the effects of the boards as such cannot be isolated from the effects of the other changes which he introduced. Certainly the changes in pay, hours of work, and job security (made in the midst of the Depression) must have had a major effect on the motivation and morale of the rank-and-file employees.

The Scanlon plan   Devised by steelworker and union officer Joseph Scanlon in the late 1930s, this plan is used by many companies today (Frost, Wakeley, and Ruh, 1974). It consists of two main elements, a participation system which involves the formation of committees composed of workers and managers to develop and evaluate ideas to improve efficiency and lower costs, and an equity system which provides the employees with monthly bonuses based on the degree of labor efficiency attained in relation to a previously-agreed-upon standard.

The results of a number of studies indicate that this plan has often been effective in increasing both employee income and company productivity. Some of the failures have been attributed to higher-level managers not allowing the employees sufficient opportunities for PDM (Frost, Wakeley, and Ruh, 1974). However, since the restrictions on PDM are often associated with a decrease in or lack of bonus payments (Geare, 1976), it is impossible to say which is cause and which is effect, or whether they are mutually reinforcing.

It should be mentioned in this context that another company-wide incentive plan, Fein's Improshare (Fein, 1976a, 1976b, 1977), has claimed results which certainly match those obtained by the Scanlon Plan, and yet Improshare makes no *formal* provision for PDM. Fein argues that his incentive plan results in greater employee effort and cooperation.

One well-known application of the Scanlon Plan at Donnelly Mirrors (Donnelly, 1977) involved even more complexities than does a normal Scanlon Plan. One modification was to include savings in the materials and operating supplies in the same pool as the savings in labor. Another was a guarantee against layoffs which resulted from changes in technology or work organization. Third, interlocking work teams, based on Likert's ideas, were introduced to encourage communication and cooperation. Fourth, all employees were represented at higher levels through a company-wide committee. And fifth, the machinery was improved. As with many Scanlon Plans, the results have been impressive, but the changes made were so numerous that the effect of PDM, per se, cannot be determined.

Lewin's group-decision experiments   These well-known experiments conducted by Lewin and his colleagues in the mid-1940s involved comparisons of the effectiveness of different methods of changing food habits (Lewin, 1947, 1952). The experiments compared the lecture method with that of group discussion and decision (a form of PDM); both methods were used to attempt to get housewives to serve more of such foods as beef hearts, kidneys, milk, cod liver oil, orange juice, etc. It was concluded from the experiments that the group discussion method was considerably more effective than the lecture method. Levine and Butler (1952) replicated these results in a factory situation where the problem involved reducing rating bias.

Numerous flaws in the design of Lewin's studies are evident. For example, no independent assessment of behavioral changes was obtained. Apparently the housewives' reports were taken at face value despite the existence of obvious pressures toward false reporting. Bennett (1955) observed that the discussion groups differed in at least six respects from the lecture groups: (1) group discussion vs. lecture, (2) decision made vs. none required, (3) 100 percent consensus vs. unknown consensus, (4) public vs. private decision, (5) specification of the time period during which action was to be taken vs. no specification, and (6) announcement that a follow-up would be made regarding any behavior taken vs. no announcement.

Bennett designed an experiment to separately determine the effects of the first four factors while holding the last two constant, and using an objective action criterion (volunteering to participate in a psychological experiment). Only one major effect was found—that of coming to a decision, a procedure which could occur either in the presence or in the absence of group discussion. Perceived degree of group consensus (but not objective degree of consensus) showed a weak relationship to action.

Sociotechnical systems   This approach, developed by Trist, Bamforth, and Rice, argues that production systems cannot be efficient unless the employees' social needs (e.g., the needs to belong, to participate) are met and thus aims to integrate the technical and social components of organizations (Emery and Trist, 1960; Miller, 1975; Rice, 1953; Trist and Bamforth, 1951). A more philosophical description is given by Susman (1976), who argues that the technical world is grasped through reason while the phenomenal world, which may be unrelated to the technical world, is organized around fantasies, beliefs, and emotions. He argues that the sociotechnical systems approach tries to develop a best match between the two. (Just how one overcomes this fundamental schism between reason and reality on the one hand and whim and fantasy on the other is never made clear.)

Let us consider what sociotechnical systems analysts actually do. The most famous of these studies is that of Trist and Bamforth (1951), who argued that the introduction of the longwall method of coal mining in England failed to raise productivity as predicted because it disrupted the accustomed social relationships among and frustrated the social needs of the miners (see also Albrook, 1967). Before the changeover, the miners worked in small groups, did a variety of different jobs depending upon what was needed at the time, and were able to adjust their pace to the particular working conditions encountered.

After the changeover to the longwall method, the miners worked in crews of forty to fifty men, had more specialized tasks, had different rates of pay, and were assigned a fixed pace. The result was an inflexible, rigid, and poorly integrated work arrangement. The problem was largely solved[7] by dividing the men into work groups with each member getting the same pay, and letting them work on several different tasks as needed rather than being confined to only one specialty (Bucklow, 1966; Trist, Higgins, Murray, and Pollock, 1963). It should be noted that the friendships that developed under this system were confined to the technical requirements of job accomplishment (Emery and Trist, 1960).

The only conclusion one can draw about this study is that flexibility is superior to rigidity when the working conditions are not totally predictable. One means to achieve this, often entailed in job-enrichment studies (e.g., see Locke, Sirota, and Wolfson, 1976), is job enlargement combined with a modular working arrangement. Thus a group of employees, cross-trained in each specialty, work together on one meaningful part of a task, each person working where they are needed (and, within limits, where they prefer). There is no evidence that social factors as such have any role in determining the success of such arrangements.

The Trist and Bamforth monograph was followed two years later by Rice's (1953) studies of Indian Weaving Sheds. Before the changes which he introduced, there was extensive individual specialization, confused lines of authority, different employees working different numbers of looms ("twelve tasks performed by twenty-nine workers were performed in a total of nineteen overlapping loom groups of five kinds" Rice, 1953, p. 304). In addition, there was low mobility among pay grades, and some employees were on piece rates while others were not.

The "sociotechnical" reorganization involved cross-training the workers on several jobs (and decreasing the number of looms), clarifying authority relationships (and increasing the autonomy of the group leader), increasing the pay of some workers, putting all group members on piece rate, and increasing the running time of the looms. The eventual result was reduced damage and increased efficiency of production. Participation was used only to allow the employees to determine the composition of the work groups.

Again, so many changes were made that it is impossible to separate the effects of each. There is very little evidence that social factors as such (including PDM) played any role in the improvements obtained. Dubin

(1965) argues that the improved organization or structuring of the work was the main factor responsible for the improved performance.

A follow-up of Rice's study through 1970 (Miller, 1975) found that some of the groups had retained their efficiency during this period while others had deteriorated somewhat. A number of factors seemed to have caused the deterioration, including upper management pressure for higher production, which led to a partial breakdown of group autonomy, resentment of group members against other members who were loafing, some unfairness in the application of the pay system, decreased group cooperation (due probably to the above factors), and inadequate training.

A final example of a sociotechnical approach to job design is a more recent coal mining study conducted in the United States by Susman (1976).[8] Unlike the English study, no major problem with technological changes precipitated the experiment. The main concern in this case was with improving safety.

The changes introduced in the experimental group (three shifts of nine men each) included training each miner in several different jobs, followed by job rotation and flexible work assignments, uniform pay for all group members with the lower-paid workers being raised to the level of the higher-paid ones, increased autonomy (delegation), a new work measurement system that made it easier to keep track of performance, confining the foreman's role mainly to safety matters, and minor improvements in technical efficiency (through reducing transport delays).

The result of these changes was a noticeable decrease in safety violations and accidents for the experimental groups as compared with two control groups. This is most easily attributed to the more explicit role played by the foreman in the safety realm. The results for absenteeism and costs were less clear, the experimental groups showed improvements on both measures but so did some of the control groups.

A more recent series of sociotechnical systems experiments conducted in Norway (which will not be described in detail here, but see Emery and Thorsrud, 1976; Thorsrud, Sorensen, and Gustavsen, 1976) were just as complex as the studies described above. Key elements in most of the Norwegian experiments were job enrichment (delegation and modules), group work, and new incentive schemes in addition to PDM. Nevertheless, both of the above references classified all of the studies under the general heading of "industrial democracy'" experiments. (A few experiments reported by Thorsrud, Sorensen, and Gustavsen, 1976, which did involve only PDM—attained through employee representation on the governing boards of companies—were found not to be successful).

The Harwood-Weldon studies   This classic in the field of organizational development (Marrow, Bowers, and Seashore, 1967) describes what occurred after the Harwood Manufacturing Company purchased the Weldon Company in 1962. Weldon, founded by two dynamic entrepreneurs, had prospered until their highly centralized decision-making style began to break down in the face of growth.

Between 1962 and 1964, an enormous number of changes were initiated at Weldon by the Harwood management. In the technical realm, the modifications included changing to a system of unit production with each semi-autonomous department responsible for a single product line, a reorganization of the shipping department, replacement of equipment, and the introduction of work aids for the operators.

The selection process was also changed, with new employees being chosen by the use of previously validated tests. A more effective (vestibule) training program was introduced for new operators. Trained operators whose production was substandard were put through an earnings development program which involved individual coaching. Operators who could not meet assigned production standards after training and coaching were fired, as were workers who showed excessive absenteeism.

Work standards and incentive pay were introduced in certain departments. The minimum pay level in the plant was raised to conform with the federal minimum wage standards. Improved feedback systems were developed so that supervisors could keep better track of the progress of the work. Authority was delegated downwards at all levels. Finally, supervisors and managers were trained in the procedures of PDM. This included sensitivity training, group discussion, and the use of joint problem-solving techniques.

The result of the sum total of these changes was that by the end of 1964, the Weldon Company showed a positive return on capital investment, increased productive efficiency, higher employee earnings, and lower turnover and absenteeism. Before-after attitude measures showed only modest improvements.

While the bottom-line results are impressive, the problem is to determine precisely what caused them. After an extensive discussion, the authors attributed 11 percent of the 30 percent gain in productivity to the earnings development program, 5 percent to the dismissal of low producers, 8 percent to the improvement in interpersonal relations and use of group problem-solving methods, and 6 percent to miscellaneous factors.

The authors must be commended for trying to sort out the contribution of the various factors, but their estimates cannot be taken at face value, since their method of determining the various effect ignores such possibilities as delayed effects and interactions among the changes.

While it seems almost certain that the sum total of the organizational changes made were responsible for the effects observed, isolating the precise degree of effect produced by each change is virtually impossible from the methodology employed in this study. Thus the title used for the book which summarized this project, *Management by Participation*, must be viewed as somewhat misleading at best.

An earlier study at Harwood reported by French, Ross, Kirby, Nelson, and Smyth (1958) was also guilty of confounding variables, although not to the same degree as in the Weldon study. In the 1958 study, extensive technological changes were introduced in three plants using consultative participation. A year later production was 10 percent higher on one of the items involved in the change, but there was no improvement for the other item. Since there was no control group in which the same changes were introduced without the use of PDM, no conclusions can be drawn about its effectiveness.

Goal setting and participation studies   Bavelas conducted a separate study at the Harwood plant in which a group of employees participated in a discussion of production goals (reported in Viteles, 1953, p. 167). It was "suggested" that the group might wish to set a team goal for higher production. While the output of this group did improve as compared to a control group, the design confounded the effects of goal setting (i.e., goal level) with those of PDM, thus making it impossible to determine which factor caused the change. A similar confounding of PDM and goal setting occurred in two other successful field studies conducted by Lawrence and Smith (1955)[9] and Sorcher (1967). A larger-scale project involving the same confounding of variables was conducted at a division of North America Rockwell (Chaney and Teel, 1972). Group goal setting (with feedback) under participative leadership was introduced in forty different groups, twenty-seven of which subsequently showed performance gains. The authors acknowledge, however, that in some cases high performance was achieved when the supervisor set the new production goal unilaterally (accompanied by an explanation of his reasons for choosing it) and allowed PDM only with respect to methods and procedures.

More recent field studies using improved experimental designs which separated the effects of goal setting from those of participation have resulted in more negative conclusions regarding the efficacy of PDM than the three studies described above. The newer studies will be discussed in the next subsection. (A relevant laboratory study by Hannan, in press, was discussed earlier.)

Likert's System 4   The essence of Likert's philosophy of management is participation and group decision making (Likert, 1961, 1967). However, it is difficult to get enough information from Likert's books to determine what actually happens when System 4 is introduced. Fortunately, a report by Dowling (1975) provides considerable detail on the application of Likert's system to the Lakewood Plant at General Motors.

In addition to introducing PDM and group decision making, Likert and his colleagues introduced increased training at all levels, increased feedback especially for hourly employees, the provision of assistants for the foremen, group goal setting, the use of cross-functional business teams, and increased delegation of authority.

Dowling (1975) emphasized the importance of feedback in causing the increases in efficiency which resulted from the intervention. On the other hand, the dramatic decrease in grievances which occurred were probably due to the greater use of PDM. As with previous studies, however, so many factors were changed that the particular causes of each effect cannot be isolated.

Conclusion

The types of changes made in each of the major multivariate studies or groups of studies (omitting the Lewin group-decision experiments, which were not done in an organizational setting) are summarized in Table 3. The total number of changes made in each study is shown in the far right column. These numbers must be considered a minimum, since some changes, which seemed trivial, were not included in the table (e.g., the serving of tea by McCormick—although since tea may serve as a stimulant, it may not be so trivial!)

The frequency with which each type of change was made is shown in the bottom row of the table. While PDM was involved, of course, in all of the studies, changes in the pay system were involved in eight cases, while technological changes, delegation, and the use of work teams were introduced about half the time. On the average, there were about five changes per study or group of studies.

While one cannot deny that most of these studies demonstrated beneficial results of the interventions, it is equally undeniable that the complex nature of the changes made precludes any clear attribution of the results to PDM as such.

In this context, it is worth discussing briefly the studies of PDM conducted in Israel and Yugoslavia, where government-sponsored participation plans have been in effect for more than two decades.

Israel The Israeli plans have been sponsored by the Histadrut, the General Federation of Labor and owner/manager of many industrial enterprises. Its leaders have been ideologically motivated socialists, although this influence is alleged to be declining (Derber, 1970). The various participation schemes, organized mainly at the plant-wide level, have evolved through a series of stages (e.g., workers' committees, joint pro-

Table 3. Summary of Types of Changes Introduced in Multivariate Field Experiments on PDM

Study	Type of Change														Total Number of Categories Changed
	Selection and/or Firing and/or Added Help	Training, Coaching	Job Security (Increased)	Pay Changes	Work Simplification	Job Enlargement/ Rotation	Work Standards, Goal Setting	Feedback (Improved)	Rest Pauses	Reduced Hours	Work Teams	Delegation	PDM	Technology	
Mayo's First Inquiry (Mayo, 1924, 1945/1970)	X								X				X	X	4
Hawthorne Studies (Roethlisberger & Dickson, 1939/1956) Relay Assembly I				X	X			X	X	X			X	X	7
Multiple Management (McCormick, 1938, 1949)			X	X						X			X	X	5
Scanlon Plans															
1. General (Frost, Wakeley & Ruh, 1974)				X									X		2
2. Donnelly Mirrors (Donnelly, 1977)			X	X							X		X	X	5
Socio-Technical Systems Studies															
1. Trist & Bamforth (1951)				X		X					X	X	X		5
2. Rice (1953)		X		X		X					X	X	X	X	7
3. Susman (1976)		X		X		X		X			X	X	X	X	8
Harwood-Weldon Studies															
1. Marrow, Bowers & Seashore (1967)	X	X		X	X		X	X				X	X	X	9
2. French, et al. (1958)												X	X		2
Goal Setting & PDM Studies (Bavelas, in Viteles, 1953; Lawrence & Smith, 1955; Sorcher, 1967; Chaney & Teel, 1972)							X						X		2
Likert's System 4 (Dowling, 1975)	X	X					X	X			X	X	X		7
TOTAL	3	4	2	8	2	3	3	4	2	2	5	6	12	7	63

duction committees, plant councils, etc.), each new stage typically following the failure of the previous stage (Rosenstein, 1970).

The goals of the PDM schemes were to increase worker identification with the enterprise, to equalize influence between managers and workers, and to increase productivity. With respect to the first two goals, the consensus is that the plans have failed (Derber, 1963, 1970; Rosenstein, 1970). The Histadrut intellectuals were unable to integrate the goal of economic efficiency with the goal of total satisfaction of the work force (Rosenstein, 1970). Furthermore, plant managers have effectively resisted attempts to encroach upon their authority, arguing that they are professionals and therefore the most qualified to make higher-level decisions (Derber, 1963, 1970). The goal of increasing productivity has, in some cases, been fulfilled, but only when the workers have been offered monetary rewards in return for their participative efforts (Rosenstein, 1970). (In many cases Histadrut intellectuals have opposed rewarding individual effort or plant-wide profit-sharing on ideological grounds.)

Yugoslavia Studies of the effects of worker participation in workers' councils in Yugoslavia have shown similarly modest outcomes. Workers who are members of the councils are only slightly more satisfied with their jobs (and *more* alienated from their work) than workers who are not council members (Obradovic, 1970). Council members with a strong desire to participate in decision making are actually no more satisfied than those who are not on the council (Obradovic, French, and Rodgers, 1970). The reason for these findings is that the council members have relatively little influence compared to the enterprise managers.

Despite these modest findings, the workers' councils have been credited with causing a dramatic increase in Yugoslavia'a rate of economic growth since the mid-1950s (Glueck and Kavran, 1971). Even if these figures (published by a communist dictatorship!) can be believed, the degree of growth attained cannot necessarily be attributed to the workers' councils. More likely it was due to the greater degree of "economic freedom" (if this concept makes sense in a socialist country—see note 5) which accompanied the establishment of the workers' councils, especially the increased authority delegated to the managers of the enterprises.

In summary, the results of PDM in both Israel and Yugoslavia have been modest, and the benefits that have followed the introduction of PDM cannot clearly be attributed to this factor, since it was confounded with economic factors in both cases.

Controlled Experimental Field Studies

The studies in this group have largely avoided the drawbacks of those in the three previous subsections. Because they are field or at least quasi-field studies, they have a high degree of external validity. Because they are experimental studies, causal inferences can be drawn from the results. And because these experiments were fairly well controlled, the effects of PDM as such can be determined. In this section the experiments done with children or in quasi-field settings are discussed first.

Productive Efficiency

The Lewin, Lippitt, and White (1939) studies (reported in Lippitt and White, 1958; White and Lippitt, 1953, 1960) carried out in the late 1930s have been among the most widely reported experiments in the PDM literature. While these studies were conducted in a field setting, they did not involve adults in a work organization, but rather 11-year-olds participating in after-school clubs. The first study included two groups of five boys and girls. Each group met eleven times under either autocratic or democratic leadership. The second, more carefully controlled and more extensive study, involved four different groups of five boys, each of whom met six times under each of three different leaders. The groups alternated between autocratic and democratic leaders, although two of the groups had one session with laissez-faire leadership.

The autocratic leaders were instructed to stress giving orders, limiting freedom, and minimizing friendliness. Observers found that as compared to the democratic leaders, the autocratic leaders did give more orders, disrupting commands, praise, approval, and nonconstructive criticism. The democratic leaders were instructed not to neglect giving guidance but to allow the members freedom to do whatever they wanted and to be friendly. Observers rated the democratic leaders as giving more: guiding suggestions, stimulation of self-guidance, information, emphasis on group decision making, and as being more friendly than the autocratic leaders.

With respect to performance, the autocratically led clubs accomplished more work than those with democratic leadership while the leader was present, but the results were reversed when the leader was absent.

While these studies were well controlled by most standards (e.g., as compared to the multivariate studies described earlier), the differences in leadership style seemed to involve more than PDM as such. Democratic leadership was used as an umbrella term to include not only PDM, but knowledge, friendliness, suggestion-giving, etc. It would be instructive to isolate the effects of these difficult elements.

Another problem which limits the generalizability of these results to organizational settings is the substantial difference in age between the boys and adults. Furthermore, the goal of the boys' clubs was to have fun, whereas that of most organizations is to get work done.

In another nonorganizational field experiment, Veen (1972) studied the effects of participation on the performance of forty boys, 10 to 14 years old, participating in an 8-week (one session per week) field hockey training course. The boys were randomly assigned to one of ten groups. Five of the groups were in the participation condition and the other five in a nonparticipation condition. The participation condition was characterized by a 5-to-10-minute conference before each training session in which the boys conferred about the program. In the first three training sessions, these boys had limited involvement in the program while being given relevant knowledge. Thereafter they were allowed to decide on the content of the training. In the nonparticipation condition, the trainer made all the decisions as to program content. In contrast to Lippitt and White, Veen was careful to control for degree of leader friendliness when manipulating PDM. Performance was measured by having the subjects run through an obstacle course. Both time and error scores were used as the criterion.

It was found that scores on performance tests, given during the fourth and sixth week of the program, were significantly better for the participation group than for the nonparticipation group. (In a follow-up performance test one year later, there was no significant difference between the scores of the two groups; however, this is probably an unfair standard in view of the limited amount of PDM involved in the experimental program.)

Another study (Litwin and Stringer, 1968) used an experimental quasi-field study to test the effects of various leadership styles on organizational climate and organizational effectiveness. Using a simulated business environment, three "organizations" were formed. Fifteen subjects (college students) who were matched on various criteria were assigned to each of three organizations. All three organizations started the experiment with an equal amount of physical resources and the same position in the same market environment. The experiment was conducted over a 2-week period comprising 8 actual days of organizational life.

The leadership manipulations resulted in the formation of three climates that were significantly different. The climate of organization A (British) was autocratic. Management exercised firm control and made all decisions. Organization B (Balance), on the other hand, exhibited a democratic climate where there was a heavy emphasis on participative decision making throughout the entire organization. The third organization C (Blazer) exhibited a high degree of achievement orientation; goal setting and delegation prevailed throughout the organization.

The results of this study indicate that the leadership styles had different effects on productive efficiency. The achievement-oriented company (Blazer) significantly outperformed both the democratic (Balance) and autocratic (British) organizations. It introduced the most new products, had the highest profits, and reduced material costs the most. The autocratic organization outperformed the democratic organization with respect to profit and had the best product quality. While the democratic organization had the poorest profit showing, it had intermediate product quality.

The major limitation of this study was that it did not take place in a bona fide organizational setting. Thus, generalizability is difficult. The authors note that the simulation might have failed to include certain boundary variables which prevented practices that are common to real businesses from occurring (e.g., building-up of inventories, concrete monetary feedback, etc.).

In another quasi-field study, Seeborg (1978) measured the effects of participation in job enrichment. A 2½ day simulation of an organization was utilized with a three-level hierarchy (plant level manager, first-level supervisors, and workers). The "organization's" goal was to manufacture decision boxes, a small electronic device that could be made by either one person or manufactured on a small assembly line.

Subjects with varying work experience were recruited from business organizations. All twenty-five subjects, with the exception of five who had experience in the technology involved in the job were randomly assigned to five work groups. The remaining five were selected for supervisory roles. During the first day of the experiment, all groups worked on an assembly line. On the second day, jobs were changed for each group in one of three ways: supervisory condition—the supervisor discussed job changes with a consultant and redesigned the jobs for his work group; participative condition—the work group was given the information that the supervisor was given in the supervisory condition and the group redesigned the job; and plant manager condition—the plant manager instructed the supervisor to implement certain changes based on the changes developed by one of the participation groups.

It was found that the content of the job changes differed between the supervisory and participative conditions. The focus of change in the supervisory condition was to load the job vertically. Workers in the participative condition "never mentioned vertical loading" (e.g., increased responsibility; Seeborg, 1978, p. 91) and were concerned mainly with the social impact of the changes.

Quality of performance was the only measure of productive efficiency used. It was concluded that the supervisory condition had better quality than the participative group (no statistical tests were employed). However, it was apparently not the method of implementation per se that led to increased quality but the nature of the job changes made as a result. In the supervisory condition, vertical loading was emphasized, resulting in increased task identity and concern for the item being produced.

There have been several controlled experimental studies of PDM in bona fide work organizations. A classic study, and still one of the best designed field experiments on PDM, was reported by Coch and French

(1948). It was conducted in a pajama manufacturing plant (Harwood) experiencing high turnover and absenteeism, resistance to methods and job changes, a high grievance rate, low efficiency, restriction of output, and aggression against management. In the experiment, small groups of employees participated in evaluating and redesigning their jobs. Three experimental groups and one control group were used to implement these changes. The first experimental group was allowed indirect participation through representation, whereas the second and third experimental groups were allowed total participation by all members of the group. Following the job changes, the PDM groups learned the job changes faster than the control group and quickly surpassed prechange productivity levels. In addition, the total PDM group learned the changes faster than the representative group. The authors noted that the rate of recovery was directly proportional to the amount of participation. In a second experiment, the control group from the previous experiment used the direct participation condition in transferring to a new job. The employees' productivity recovered rapidly and to a final level of production well above the prechange level before the transfer.

The results of this study show impressive evidence for the effectiveness of PDM in decreasing resistance to job and methods changes.

Fleishman (1965) attempted to replicate the Coch and French study in a dress factory. Using one experimental and two control groups, he found results similar to that of Coch and French with one important difference—*both* the control group and the experimental PDM groups increased their productivity. However, the research methodology was problematic and might explain the results. The experimental and the control groups were not separated from one another but were intermingled back on the job. Thus it was difficult to isolate the effect of the participation manipulation because of a possible transfer effect to the control group.

In a Norwegian replication of the Coch and French study, five experimental groups with varying degrees of participation and one control group were compared (French, Israel, and As, 1960). There were no significant differences among the groups in productivity. This may be traced in part to the unilateral exclusion by management of the central issues of rate setting and production level from PDM. Thus, only matters peripheral to productivity were discussed by the employees in the PDM conditions.

Morse and Reimer (1956) conducted a study in one department of a nonunionized industrial organization which had four parallel divisions engaged in relatively routine clerical work. This study was impressive in that PDM was manipulated not in small subgroups as in other studies but by changing the structure of four divisions of a company. In two of the divisions, PDM was increased by increasing the involvement of the rank and file in decision making. In the other two divisions, unilateral decision making was increased by centralizing all decision making (HIER).

The results of the study show that productive efficiency in all four divisions increased. However, the productivity of the HIER divisions improved significantly more than that of the PDM divisions.

It is important that the measures of productivity be considered when interpreting the results. Productive efficiency was measured by total clerical costs as compared to a prechange baseline period. The increase in productive efficiency is a result of the elimination of inefficiences prior to the change. It appears that the HIER divisions had reduced clerical costs by cutting the number of employees. The PDM divisions, however, were reluctant to eliminate their own members.

An experimental study at General Electric by French, Kay, and Meyer (1966) varied the degree of participation in the goal-setting process during performance appraisal sessions on lower-level managers. The major factor affecting subsequent performance was whether performance goals were set during the interviews. There was increased acceptance of goals in both the high and low PDM conditions, although the increase was slightly higher in the former case. The effect of PDM on performance depended on several contextual factors. When the manager was used to participation based on prior experience, the high PDM manipulation led to improved performance. But when the manager was not accustomed to PDM, the high PDM manipulation led to poorer performance than low PDM for those managers who had found the preceding appraisal (given two weeks before the experiment started) threatening. Degree of experimental PDM had no differential effect on those not used to PDM who had not been threatened by the first interview.

The authors concluded that PDM works best in a context of supportiveness caused by high usual PDM. Some support was found for Vroom's (1960) finding that those high in need for independence responded more favorably to PDM.

Six skilled construction and electrical crews were given varying degrees of participation in work scheduling in a field study conducted by Powell and Schlacter (1971). They found that PDM had no effect on productivity.

Latham and Yukl (1975b) compared three methods of goal setting among independent logging crews with different levels of education. Loggers in the participative and assigned goal-setting conditions outperformed those in the "do your best" condition. However, for the uneducated sample, workers in the participative goal-setting condition also showed higher output than those in the assigned condition. In the participative condition, goal difficulty was higher, but not significantly so. The authors concluded that the superiority of participation for this sample was due to greater goal acceptance.

For the educated sample, however, there was no significant difference in performance between the participative and assigned goal-setting methods. Nor was there a significant difference in either goal difficulty or goal acceptance between the two conditions. The authors suggest that the difference in results of the two studies might be caused by education, confounding of demographic variables, or by lack of support by local management in implementing the research.

In another study by the same authors (Latham and Yukl, 1976), female typists in a large corporation were randomly assigned to participative and assigned goal-setting conditions. They found no significant differences in goal difficulty or performance between the two treatments. The authors concluded that "subordinates with the assigned goals eventually accepted them as much as subordinates who participated in goal setting" (Latham and Yukl, 1976, p. 170).

Using a sample of sales personnel, Ivancevich (1976) compared participative, assigned, and "do your best" goal-setting groups. Four quantitative measures of performance (called frequency index, order/call ratio, direct-selling cost, and market potential) were taken for each group at four periods of time (baseline, 6 months, 9 months, and 12 months). It was found that over the 12-month period, significant positive changes in the four performance measures occurred for both the participative and assigned groups. No significant changes, however, were found for the "do your best" group. In comparing participative and assigned groups at the four periods of time, only three significant differences occurred. The performance of the participation group on the call frequency variable was significantly better at 6 months. However, at 12 months, there were no significant differences between the two groups, indicating that over time both methods were equally successful in increasing performance.

In a second study, Ivancevich (1977) investigated the effects of participative goal setting on the performance of skilled technicians and supervisors in three plants of a medium-size equipment and parts manufacturing organization. The employees in each plant were assigned to a different goal-setting condition (participative, assigned, and "do your best"). Four quantitative measures of performance (unexcused absenteeism, service complaints, cost of performance, and safety) were taken for each plant at four periods of time (baseline, 6 months, 9 months, and 12 months). An analysis of the data at 6 months revealed that both assigned and participative goal-setting conditions were significantly better on three of the performance criteria than the "do your best" condition.

The assigned goal employees showed significantly few complaints and had lower performance costs than the participative employees. On the other hand, the participative condition showed a significantly better safety record. There was no significant difference between the three groups with respect to unexcused absenteeism. (Performance scores for every criterion decreased between the ninth and twelfth month for both groups.)

Latham, Mitchell, and Dossett (1978), using a sample of engineers and scientists, studied the effects of participative versus assigned goal setting on both goal difficulty and goal acceptance. Specific goals were either participatively set or assigned by a supervisor. To measure goal difficulty, both an objective and a subjective measure were taken. The results indicate that some participative subjects set objectively higher goals than those with assigned goals. However, subjective goal difficulty was equal in the participative and assigned conditions. Rated goal acceptance was also the same.

Six months after the first study, performance measures were obtained on these subjects plus two control groups. An analysis of the data indicated no significant difference in performance between the participative and assigned goal-setting conditions, although both combined outperformed the control groups.

The findings from the experimental field studies discussed above are summarized in Table 4.

Satisfaction

In the Lewin, Lippitt, and White studies described earlier (White and Lippitt, 1960), the satisfaction of the democratically led groups appeared to be higher than that of the autocratically led groups. However, there is a potential problem with this conclusion. As noted earlier in this section, other traits (e.g., arbitrariness, tactlessness, coldness, etc.) besides participation were included in the manipulations. If this is the case, the subjects could have been satisfied with a "polite, considerate, or rational" person rather than a democratic leader. On the other hand, other subjects could have been dissatisfied with a "cold, insulting, tactless, or arbitrary person" rather than an autocratic leader.

Veen (1972), in his study of boys participating in field hockey training, obtained two questionnaire measures of satisfaction: general satisfaction and satisfaction with the training program. It was found using measures at 4 and 8 weeks that there was no significant difference between the participative and nonparticipative conditions as to general satisfaction. The participatively led boys, however, were significantly more satisfied with the training program than the nonparticipatively led boys.

Litwin and Stringer (1968) obtained pre- and post-manipulation measures of satisfaction of the employees of their simulated companies. Employees in the achievement oriented and participative companies were more satisfied than those in the authoritarian company but did not differ from each other.

Seeborg (1978) used the Job Diagnostic Survey (JDS) to assess various

Study	Task	Manipulation or Comparison	Results Prod.	Results Sat.
Lewin, Lippitt & White (1939), etc.	Children working on arts and crafts	Autocratic vs. democratic	0	+
Coch and French (1948)	Pajama manufacturing	Indirect PDM, Direct PDM, control	+	+
Morse and Reimer (1956)	Clerical work	PDM vs. HIER control	−	+
French, Israel and As (1960)	Manufacturing	PDM vs. control	0	+
Fleishman (1965)	Dress manufacturing	PDM vs. control	0	
French, Kay & Meyer (1966)	Engineering and manufacturing	PDM vs. assigned	0	
Litwin and Stringer (1968)	Simulated business	Democratic, autocratic, achievement	−	+
Lawler and Hackman (1969)	Janitorial work	PDM vs. assigned		+
Powell and Schlacter (1971)	Construction and electrical	Degree of PDM		
Veen (1972)	Children training in field hockey	PDM vs. control	+	+
Latham & Yukl (1975b)	Goal setting (logging)	PDM vs. assigned	$+,0^{a}$	0
Lischeron and Wall (1975)	Organizational PDM	PDM vs. control	0	0
Latham and Yukl (1976)	Goal setting (typing)	PDM vs. assigned	0	0
Ivancevich (1976)	Goal setting (sales)	PDM vs. assigned	0	0
Ivancevich (1977)	Goal setting (technicians)	PDM vs. assigned	0	−
Latham, Mitchell, and Dossett (1978)	Goal setting (engineering)	PDM vs. assigned	0	0
Seeborg (1978)	Job enrichment simulation (assembly)	PDM vs. assigned	−	+
PDM Superior			(+) 3	8
No difference or Contextual			(0) 10	5
PDM Inferior			(−) 3	1

[a] See Table 2, note a.

aspects of job satisfaction and motivation in his simulation of a job enrichment program. The subjects in the groups enriched by participatory methods were more satisfied than those whose jobs were enriched unilaterally by the supervisor or plant manager.

In the Coch and French (1948) study, turnover and grievance rates, which are concomitants of job satisfaction, were reported to be lower for employees in the PDM conditions than for those in the control group. This conclusion was based on limited turnover data and the investigators' impression of the number of positive work attitudes exhibited by the subjects. The authors argued that PDM is beneficial because it reduces frustration.

In the Norwegian replication by French, Israel, and As (1960), attitudes were measured with a fourteen-item postexperimental questionnaire. The experimental groups had more positive attitudes than the control group on ten of the fourteen items, but only three of the differences were significant. Two of these three items referred specifically to training. Two of the items which failed to reveal any difference were those which most closely represented the concept of job satisfaction—"How do you in general like to work here?" and "Do you feel that you have found the kind of job you would like to keep in the future?" One reason for these marginal results may have been the limited degree of participation experienced by the employees in the PDM conditions.

Morse and Reimer (1956) used pre- and postchange measures of satisfaction, employees' attitudes toward their supervisor and with the company and their jobs in both the PDM and HIER conditions. All fourteen measures were taken prior to the change program and then again 1 year after the program had been operating successfully.

Changes in satisfaction toward the company increased significantly in most cases in the PDM divisions whereas they decreased significantly in the HIER divisions. Overall, in the PDM divisions, nine of the fourteen prechange and postchange measures showed statistically significant differences indicating that attitudes were more positive after participation was introduced. In the HIER divisions, ten of the fourteen measures changed significantly in the opposite direction.

With respect to the measures of overall job satisfaction, no changes were observed in the scores of PDM divisions, whereas one of the two HIER divisions demonstrated a significant decrease in job satisfaction.

A field experiment by Lawler and Hackman (1969) and a follow-up by Scheflen, Lawler, and Hackman (1971) studied the effects of participation by janitorial workers in the development of a pay incentive plan to reduce absenteeism. Two experimental conditions and two control groups were used. In the first experimental condition, three autonomous work groups developed their own incentive plan viz. PDM. These plans were then imposed by the company upon the two work groups that made up the second experimental condition.

The results show, based on pre- and postmanipulation measures, that there was a significant increase in attendance only in the participative condition 16 weeks after the plan started, but there was a significant increase in the attendance of the imposed groups after a year. The authors cite three possible reasons for the claimed success of the participation condition: (1) participation caused greater commitment by the subjects to the plan; (2) subjects who participated in the plan were more knowledgeable about it; and (3) participation increased the employees' trust of the good intentions of management with respect to the plan. However, severe criticisms of the methodology of this study by Cook and Campbell (1976) make any firm conclusions suspect.

In the Powell and Schlacter (1971) study, the two groups which had the highest degree of PDM showed improved morale, but this was neutralized by a surprising increase in absenteeism.

Lischeron and Wall (1975) darried out a field experiment in Great Britain to examine the effects of employee decision making at higher levels of an organization. An experimental group of 150 subjects was formed from a prechange control group of 350 subjects on the basis of responses to a questionnaire measuring attitudes toward participation. The remaining 200 subjects were assigned to the control group. The Worker Opinion Survey, a version of the Job Description Index, modified for use with British blue-collar populations, was used to measure job satisfaction. No significant effect of participation on satisfaction was found. The authors conclude that participation may improve relationships between the worker and management, but will not necessarily increase job satisfaction.

Latham and Yukl (1976), in their study of typists, found no significant difference in job satisfaction between those subjects in participative and assigned goal-setting conditions. As compared to prechange measures, employees in both conditions experienced a significant decline in job satisfaction after the manipulations.

Ivancevich (1976), in his study of sales personnel, found opposite results. Using the Job Description Index, he found that both the assigned and participative goal-setting treatments led to increased work satisfaction over a 9-month period, although this washed out after 12 months. With respect to satisfaction with the supervisor, the assigned group showed a greater increase than the PDM group after 9 months but the difference again washed out after 12 months.

In his later study of skilled technicians, Ivancevich (1977) found that both assigned and participative goal-setting conditions led to increased satisfaction with work and with supervision. There was no significant difference in work satisfaction between the conditions. The assigned goal-setting group, however, reported significantly *higher* satisfaction with the supervision. (In both studies the satisfaction increases diminished over time.)

Finally, Latham, Mitchell, and Dossett (1978) found no significant difference in intrinsic satisfaction with goal accomplishment between the participative and assigned goal-setting groups.

The results of the studies of satisfaction are shown in Table 4. It is evident from the summary in the bottom right corner that there is no evidence that PDM is superior to more directive methods in increasing productivity. The results are more favorable to PDM with respect to the satisfaction criterion, but again over 40 percent of the studies showed no general superiority of PDM.

Conclusions

Omitting the multivariate studies, the results of all the studies discussed in this chapter are summarized in Table 5. An examination of the totals in the right-hand columns show that: (1) with respect to the productivity criterion there is no trend in favor of participative leadership as compared to more directive styles; and (2) with respect to satisfaction, the results generally favor participative over directive methods, although nearly 40 percent of the studies did not find PDM to be superior.

A striking aspect of these results is that each of the three types of study yields the same conclusion regarding the efficacy of PDM as a motivator and a satisfier.

Another remarkable aspect of these results is that they agree perfectly with the conclusions drawn from two other extensive literature reviews of PDM.[10]

In 1959, Anderson summarized the studies comparing authoritarian and democratic approaches to teaching. He found no trend with respect to a learning criterion (eleven studies favored democratic methods, thirteen showed no difference, and eight favored directive methods), but democratic approaches were generally superior with respect to a morale criterion.

Similarly, an earlier review of the general PDM literature by Stogdill (1974), which overlapped but was not identical with the literature reviewed here, concluded that participative leadership did not necessarily lead to higher performance but usually led to higher morale than directive leadership.

The consistency with which the results of PDM studies fail to show any clear trend with respect to effect on productivity (and to a lesser extent with respect to effects on morale) leads to only one possible conclusion— there is a great deal we do not yet know about the conditions under which PDM will "work." What we do know is the subject of the next section.

Table 5. Combined Results of PDM Studies[a]

Result	Type of Study							
	Laboratory		Correlational		Field		Combined	
	Prod.	Sat.	Prod.	Sat.	Prod.	Sat.	Prod.	Sat.
PDM superior	4	5	3	13	3	8	10 (22%)	26 (60%)
No difference or contextual	6	0	10	8	10	5	26 (56%)	13 (30%)
PDM inferior	4	2	3	1	3	1	10 (22%)	4 (9%)
TOTALS	14	7	16	22	16	14	46	43

[a]Based on Tables 1, 2, and 4.

V. CONTEXTUAL FACTORS DETERMINING THE EFFECTIVENESS OF PARTICIPATION

Both logic and the results of research on PDM make it unmistakably clear that its effectiveness depends on a number of contextual factors. In this respect, we are in agreement with the position of Vroom and Yetton (1973), although we will make no attempt to develop a sequential model here or to modify theirs. We will simply summarize the contextual factors that have been found or asserted to determine the effectiveness of PDM. (For other contextual approaches, see Lowin, 1968; Strauss, 1963; and Tannenbaum and Massarik, 1950.)

Individual Factors

Knowledge As noted earlier in this chapter, people are not equal in the extent of their knowledge or intelligence. Thus we cannot accept unqualified statements to the effect that "the information in a group exceeds that of any individual in the group" (Maier, 1970, p. 595). It depends upon the individuals and the nature of the task. PDM should be most helpful in generating high-quality decisions when the participants have relevant knowledge to contribute (Davis, 1963; Derber, 1963; Strauss and Rosenstein, 1970; Vroom, 1969). In cases where one member (e.g., the leader) has significantly more knowledge than the others, PDM would be wasteful of time and effort at best, and harmful to decision quality (if those with less knowledge outvoted the most knowledgeable member) or to efficiency (caused by delays) at worst.

Lanzetta and Roby (1960) found that "where the relative contribution of individuals to decision making was congruent with their relative differences in ability, performance of the group was better" (p. 147).

Furthermore, Mulder and Wilke (1970) observed that in cases where there are substantial differences in knowledge among group members, PDM leads to a progressive *decrease* in equality of influence over time. Similarly, Mansbridge (1973) has observed that group members who have the most critical and irreplaceable skills have the most power to influence decisions. The discovery that knowledge is power is not new, of course. It was identified by Francis Bacon over three centuries ago.

F. W. Taylor's system of Scientific Management was based on the principle of authority based on knowledge. Taylor did not advocate PDM, because he believed that the average untrained workman of his time did not know as much about the best way to do his job as a trained expert. Taylor did argue, however, that: "Every encouragement . . . should be given [the workman] to suggest improvements, both in methods and implements. And whenever a workman proposes an improvement, it should be the policy of the management to make a careful analysis of the new method, and if necessary conduct a series of experiments to determine accurately the relative merit of the new suggestion and of the old standard" (Taylor, 1911/1967, p. 128). While Taylor has been criticized for decades for advocating authoritarian practices, he did not advocate obedience to authority as such, but rather obedience to facts.

The significance of individual differences in knowledge has also been recognized, albeit reluctantly, by human relations advocates. Marrow, for example, writes, "Nor does participation mean that everyone decides on everything . . . its effectiveness is tied closely to how much the participants know and contribute" (1972, p. 88). Maier (1970) admits that in some cases the leaders' expertise is so great in relation to that of his followers that PDM would not be useful. Advocates of group decision-making acknowledge that the group's judgment is often inferior to that of the best individual member (Hall, 1971). Likert, too, recognizes that supervisors must be technically skilled (1961, 1967).

Vroom has aptly observed that "in no sense can the manager of a group be considered to be an average group member. He has been selected by different criteria, exposed to different training, and has access to different information than other group members" (1969, p. 230).

More recently, Maslow, a leading exponent of "humanism," expressed grave reservations about PDM on grounds identical to Taylor's: "More stress needs to be placed on the leader's ability to perceive the truth, to be correct, to be tough and stubborn and decisive in terms of the facts." Rather than stressing, "democracy, human relations and good feelings [t]here ought to be a bowing to the authority of the facts" (Maslow, 1970, p. 36).

In implicit agreement with Maslow's position, most nonsupervisory employees show no desire to make top management decisions;[11] those who want increased participation want to be involved in decisions regarding their own job and immediate work surroundings, i.e., issues about which they are likely to have pertinent knowledge (Holter, 1965).

Human relations advocates have stressed repeatedly that the importance of knowledge has been overemphasized and have urged that more emphasis be placed on gaining commitment to the decision reached, i.e., through PDM. Their assumption is that there will necessarily be resistance to decisions which are not made by the employees themselves.

While resistance to change obviously does occur, it has never been proven that such resistance is inevitable (or even typical) in the absence of PDM. Resistance might also develop because the changes are not based on valid reasons, or because the explanations or justifications given for them are unclear or incomplete (Lawrence, 1971). Studies have demonstrated that expert power (authority based on a high degree of knowledge) can lead to both high satisfaction and high performance among subordinates (summarized in Wood, 1973). Hamblin, Miller, and Wiggins (1961) found leader competence to be the single most important factor determining group morale (see also Rosen, 1969, pp. 58ff). It has also been found that status differentiation is quite compatible with both satisfaction and efficiency as long as there is consensus regarding the status structure (Heinicke and Bales, 1953).

Motivation A number of motivational factors may affect the success of PDM. First, PDM may not satisfy or conversely, directive leadership may not dissatisfy, employees who do not want or expect PDM, who lack independence and want to be told what to do (Berkowitz, 1953; Davis, 1963; Derber, 1963; Dill, 1958; Dill et al., 1961; Foa, 1957; Heller and Porter, 1966; Marrow, 1972; McMurray, 1958; Obradovic, French, and Rodgers, 1970; Singer, 1974; Strauss, 1963; Vroom, 1960, 1969). Second, it may not be effective with those who are not used to it (Carroll and Tosi, 1973; French, Kay, and Meyer, 1966; Ivancevich, 1976; Tannenbaum and Schmidt, 1958); this implies, of course, that with repeated exposure, employees could become used to it.

Third, it has been asserted frequently, on theoretical grounds, that PDM will not be effective among generally unmotivated employees, e.g., those with low job involvement, low need for achievement, or low commitment to organizational goals (Fein, 1976c; McMurray, 1958; Singer, 1974; Tannenbaum, 1974; Tannenbaum and Schmidt, 1958; Vroom, 1969). However, it is precisely among such populations that PDM has often been associated with enhanced motivation (e.g., Carroll and Tosi, 1977; Latham and Yukl, 1975b; Lawler and Hackman, 1969; Steers, 1975). Argyle et al.'s finding (1958) that PDM was related to performance and attitudes only in departments which were *not* on direct incentive pay is congruent with these results. No conclusive explanation for this apparent paradox has been offered. Possibly employees with low motivation feel powerless and unefficacious and PDM gives them a feeling of control which is manifested in setting challenging goals, higher goal commitment, and/or greater job involvement. In contrast, for more highly motivated employees, PDM may be simply redundant.

Organizational Factors
This category defined broadly includes all factors external to the participating employee, or to factors which involve an interaction among the participating members.

Task attributes   It has been asserted that highly complex unstructured tasks require PDM, because of the increased knowledge and flexibility requirements, whereas routine tasks do not (e.g., Morse and Lorsch, 1970; Shaw and Blum, 1966). While this may be true with respect to knowledge requirements, the opposite may be the case with respect to motivation. As noted above, employees on routine jobs may become more committed as a result of PDM while those at higher job levels need no such incentive. Vroom and Mann (1960) have observed that tasks which require extensive coordination among employees are performed more effectively under directive leadership than under participative methods.

Group characteristics   Two possible dangers of group participation have been clearly identified. First, PDM may increase (or release) group conflict (Derber, 1970; Gomberg, 1966; Maier, 1967; Mansbridge, 1973; Sashkin, 1976). Such conflicts may involve personality clashes, actual differences in values and goals between members, or simply resentment over the rejection of some members' ideas (Wood, 1973). In such cases, directive leadership may be an effective method of resolving the disputes (Burke, 1966).

However, this does not necessarily imply that group homogeneity is always beneficial. Hoffman, Harburg, and Maier (1962) found that differences of opinion, if utilized constructively, led to higher-quality decisions than did the absence of conflicting viewpoints. Similarly, Pelz (1956) found that daily group contact helped the productivity of a scientist only when the other scientists had different views than himself.

A second danger is that of group conformity and/or groupthink (Janis, 1972). Groups can be just as autocratic as supervisors, if not more so (McMurray, 1958), and may thereby inhibit the expression of new or unpopular ideas (Jaques, 1964).

It has also been argued that if group goals differ from organizational goals, the use of PDM will lead to poor decision quality (Vroom and Yetton, 1973), although this issue has not been widely studied. PDM advocates might argue, however, that PDM would lead to the development of more responsible group goals than existed before PDM.

Leader attributes   It has been frequently observed that supervisors themselves are often threatened by the introduction of PDM and, as a result, oppose and sabotage attempts to use it (Campbell, 1953; Derber, 1963, 1970; Maslow, 1970; Scheflen, Lawler, and Hackman, 1971). A common reason for the failure of Scanlon Plans is alleged to be the unwillingness of managers to allow PDM by the rank and file (Frost, Wakeley, and Ruh, 1974). Clearly, any successful PDM program must have the support of all levels of management.

A second requirement is that of skill in the use of PDM techniques (Hall, 1971; Maier, 1973; Vroom, 1969). Even a willing leader may not have the capacity to effectively utilize these methods without careful training.

The personality and style of the leader *in relation* to that of the group members must also be considered. There is some evidence that when leader and group have similar values, the members are more satisfied (Haythorn, 1958; Sadler, 1970), but the results with respect to performance are equivocal (McCurdy and Eber, 1953; Pelz, 1956).

Degree of contact with the leader may also affect the success of PDM (Pelz, 1956) especially where information exchange is needed.

Other organizational factors   PDM requires more time to reach a decision than more directive methods (although conceivably time could be saved in the long run if the decision under PDM were better); thus if there is pressure for an immediate decision, PDM may not be feasible (Davis, 1963; Derber, 1970; Mansbridge, 1973; Sadler, 1970; Strauss and Rosenstein, 1970; Vroom, 1969).

Organization (and group) size is pertinent, since an increase in the number of members participating results in a disproportionate increase in the potential amount of interaction. Thus increasing size affects not only the time needed to reach a decision, but increases the problem of regulation and coordination (Heller and Yukl, 1969; Hemphill, 1950), although methods have been suggested for solving this problem (cf. Likert, 1961, 1967).

Finally, it has been argued that PDM is needed most urgently under conditions of rapid or constant organizational (including technological) change (Albrook, 1967; Morse and Lorsch, 1970). Again the argument is made that employees will resist changes which are not introduced participatively. However, this claim cannot be accepted without qualification. We have argued that employees may accept change readily without PDM, especially if logical reasons are given for it and if the employees are already motivated. Furthermore, directive leaders are often in a position to respond more rapidly to change than are participative leaders. However, when organizational change involves complex knowledge requirements which cannot be mastered by a single person, PDM should be superior to unilateral decision making. PDM should also facilitate change when the changes are threatening to employees if it gives them a heightened sense of control.

VI. PARTICIPATION IN RELATION TO OTHER MOTIVATIONAL TECHNIQUES

Participation is not the only method organizations can use to motivate employees. There are at least three other major motivational techniques: money, goal setting with feedback, and job enrichment. It is undeniable that each of these techniques can be successful even when divorced from PDM. For example, incentive plans, in which pay is based upon the amount of work produced, have led to productivity increases as high as several hundred percent (Lawler, 1971); increases between 40 and 70 percent are not uncommon (Fein, 1976b). However, as in the case of PDM studies, the experimental designs are sometimes confounded with variables other than pay (e.g., the introduction of work standards).

There is also impressive evidence supporting the efficacy of assigned goals combined with feedback, as a means of improving job performance (Latham and Yukl, 1975a; Litwin and Stringer, 1968). Although the typical performance increases obtained with goal setting are not as large, on the average, as those obtained with incentives, a performance improvement of close to 50 percent was obtained in one goal setting experiment (Latham and Baldes, 1975).

Numerous successful studies of job enrichment (i.e., delegation of responsibility, modules) have been reported (Ford, 1969; Maher, 1971; Suojanen, McDonald, Swallow, and Suojanen, 1975). Many were done without the use of participation (e.g., Ford, 1969; Paul, Robertson, and Herzberg, 1969). Again the experimental changes in enrichment studies have often been multifaceted, so that the effect of enrichment as such is hard to isolate (Locke, 1975). One recent study attributed the productivity increases observed in job enrichment to goal setting (Umstot, Bell, and Mitchell, 1976) although other factors may also be involved (Locke, Sirota, and Wolfson, 1976).

Since PDM is a technique, with the particular content being optional, it can, of course, be combined with other techniques, including the three described above.

The Scanlon Plan, for example, combines PDM with pay (Frost, Wakeley, and Ruh, 1974). It is not clear whether the results of Scanlon Plans are any better than those which use plant-wide incentives without PDM such as Fein's Improshare (Fein, 1976b, 1977). The effects of PDM within the Scanlon Plan may not be motivational but cognitive. PDM may facilitate the development of useful ideas for improving efficiency, while the monetary aspect provides the incentive to develop and accept the proposed changes. PDM techniques have also been used to *develop* pay incentive plans (e.g., Lawler and Hackman, 1969; Lawler, 1975) with some success. The motivational effect in these cases is probably to increase acceptance of the plans, although cognitive factors (such as better understanding) cannot be ruled out.

Several investigators have argued that PDM will not work *unless* it is combined with or involves monetary incentives (Derber, 1963; Rosenstein, 1970; Strauss, 1963); however, as indicated earlier, the research does not necessarily support this claim.

A number of studies have compared assigned and participative goal setting (e.g., Carroll and Tosi, 1973; Hannan, in press; Ivancevich, 1976, 1977; Latham and Yukl, 1975b, 1976; Latham, Mitchell and Dossett, 1978). The results have been equivocal. In some cases employees set higher goals under PDM conditions than supervisors set for comparable subordinates under assigned conditions. However, when goal level is controlled by yoking the assigned groups to the goal levels set by the PDM groups, differences in goal acceptance rarely emerge. As noted earlier, PDM may be more beneficial if employees are accustomed to it than if they are not (French, Kay, and Meyer, 1966). It seems that in many cases the demand characteristics of the job are sufficient to ensure substantial goal acceptance (Locke, 1978), although if frequent changes in task and goal level occur, resistance to assigned goals may develop (Coch and French, 1948).

PDM has been used in some job-enrichment studies (e.g., Locke, Sirota, and Wolfson, 1976). Although there are strong opinions both in favor of using it (Hackman, 1975) and against using it (Ford, 1969), there is little evidence indicating that PDM necessarily makes enrichment more effective. One simulated field study (discussed earlier) compared the results of enriching jobs with and without PDM (Seeborg, 1978). It was found that the enrichment ideas of the PDM group did not really involve vertical loading at all but rather were focused around social changes. Furthermore, the work quality of the PDM group was poorer than that of the non-PDM, vertically loaded group. When the same changes made by the PDM group were made unilaterally by the plant manager in another experimental group, however, there was a difference in satisfaction in favor the PDM group.

Another way in which PDM and job enrichment are being combined, often promoted by "sociotechnical system" advocates (e.g., Susman, 1976, Trist, Higgins, Murray, and Pollock, 1963), is through the development of autonomous work groups. This system allows PDM among the group members, while the group as a whole is given increased responsibility for decision making. It is claimed that such a procedure is superior to individual enrichment, but there is no evidence to support this claim.

VII. CONCLUSIONS

Our analysis of the issue of employee participation in decision making has led to the following conclusions: (1) the use of PDM is a practical rather than a moral issue; (2) the concept of participation refers to shared or joint decision making, and therefore excludes delegation; (3) there are numerous mechanisms both cognitive and motivational through which PDM may produce high morale and performance; (4) research findings yield equivocal support for the thesis that PDM necessarily leads to increased satisfaction and productivity, although the evidence for the former outcome is stronger than the evidence for the latter; (5) the evidence indicates that the effectiveness of PDM depends upon numerous contextual factors; and (6) PDM is not the only way to motivate employees.

If the effects of PDM depend upon the context in which it is used, it follows that PDM might be not only ineffective in some circumstances, but might be actually harmful. For example, it could lead to excessive intragroup or intergroup conflict caused by such factors as fundamental value differences or the resentment of members whose ideas are rejected. Group cohesion fostered by PDM may work against the goals of the organization instead of for them. Conformity and groupthink fostered by group pressures could lead to poor decision quality, especially if these pressures intimidate the most knowledgeable members or lead the other members to ignore their ideas. The time requirements of PDM could result in harmful delays. The ubiquitous use of PDM could retard the development and emergence of leaders, and the leaders who do emerge may be too emotionally involved in their groups to make objective decisions, especially if the decisions are "tough" or unpopular.

That our claim of possible dangers associated with PDM is not idle speculation is demonstrated by the unfortunate experience of Nonlinear Systems (Malone, 1975). Following 8 years of rapid growth and increasing profitability as a manufacturer of electrical measuring instruments, the company, after consulting with several eminent PDM advocates and years of planning, introduced radical organizational changes aimed at making the company thoroughly democratic. One key change involved removing the authority of upper and middle management by making them merely "advisors." Time clocks, recordkeeping, the accounting department, and the inspection department were all eliminated.

After 5 years, the program was largely abandoned because of increasingly serious problems with respect to sales volume, sales costs, the failure of productivity to rise, managerial frustration, and decreased profitability. Malone writes that: "The seven vice presidents formerly had been vigorously active in the midst of daily problems and were more or less expert in their individual specialties. Under participative management these men were practically immobilized as 'sideline consultants''' (Malone, 1975, p. 58). Lower-level managers were correspondingly frustrated by the lack of guidance and training from above. Only among shop workers was morale high (because of increases in pay, fringe benefits, and job enrichment) and production stable (but not higher).

The key error caused by the overzealous application of PDM was the *separation of responsibility from knowledge*. The attempt to make everyone equal regardless of ability led to the most competent employees being prevented from acting on their judgment.

The issue was put succinctly by a former Westinghouse executive, "Good management is the rule of the best minds. It is anti-democratic [which it is in the Greek sense], although private organizations flourish best in democratic [i.e., free] countries. However, the democratic rule of the majority will frustrate and defeat any management. The crew cannot run the ship" (quoted in Derber, 1963, p. 69).[12]

One reason for our lack of knowledge regarding the conditions under which PDM will be harmful is the widespread pro-PDM bias noted earlier; researchers do not look for consequences they do not think will (or should not) occur. Another reason is the generally neobehavioristic approach to PDM research, which has focused more on stimulus and response, i.e., participation and productivity (or satisfaction), than on the mechanisms which bring the effects about (e.g., the knowledge and values of the participants).

Another unfortunate consequence of the pro-PDM bias has been the establishment of a false dichotomy. It is often implied that leadership must be either authoritarian or participative (or somewhere in between), thus closing the door to the discovery of other dimensions. Authoritarian leadership demands *obedience to authority as such,* regardless of reason or context. Full participative leadership involves joint decision making with one or more subordinates regardless of reason or context. Thus the dichotomy reduces to decision making based on the whim of the leader or the whim of the group. In neither case is the issue of knowledge addressed explicitly, although as noted earlier, PDM practitioners acknowledge its relevance implicitly by typically allowing participation only on issues about which the workers have pertinent knowledge.

The recognition of individual differences in competence implies a third type: knowledge-based leadership. If a leader clearly knows the best solution, then he should properly assert his knowledge and make the decision. *Such a procedure would not be authoritarian but authoritative.* If subordinates do not agree with his solution, such a leader should try to rationally persuade them by giving them the reasons for his position, or, if necessary and feasible, demonstrate the validity of his ideas in some concrete form (e.g., through an experiment, examples, etc.). In the course of such discussions, the leader, of course, may find that he is wrong, in which case he should change his position.

If a leader does not know the best (or any) solution to the problem, then he should properly consult those who do have such knowledge, including competent members of his own subordinate staff, his superiors, and outside experts.

In some cases there may be options; for example, several different solutions may work equally well and have equivalent costs. In such a case, responsibility could be delegated or a straight majority vote could be used.

It should be stressed that managers and experts do not always know more than their subordinates (e.g., see Thorsrud, Sorensen, and Gustavsen, 1976, p. 431). The rule of knowledge does not mean the rule of formal degrees, positions, or job titles, but by those who have the best ideas whatever their position. Human relations theorists could argue, with much justification, that many managers underestimate the knowledge and capacity of their subordinates. However, the solution is not to assume that everyone knows as much as everyone else but to determine who knows what.

The above should not be taken to imply that a competent leader should ignore motivational issues, e.g., getting employees to accept decisions, gaining commitment, promoting satisfaction, etc., but it *should* be taken to imply which goal has logical priority. Knowledge must take precedence over feelings in business organizations (Gomberg, 1966).

A profit-making organization survives by the discovery and application of knowledge relevant to its product and market. In the absence of rational leadership, the effort exerted by employees will be useless. Nor do business organizations exist for the purpose of satisfying their employees since employee feelings have no market price; the goal of such organizations is to satisfy their customers and stockholders.

It must be emphasized that there is no *fundamental* conflict between the interests of employees, stockholders, and customers. If employees are not reasonably satisfied, the organization will be plagued by high turnover, absenteeism, sabotage, strikes, and apathy and thus will not produce a good product at a reasonable cost. On the other hand, if the firm does not make a profit, there will be no jobs at all for the employees—a condition which will cause dissatisfaction!

While the goal of each *individual* is properly to achieve his own happiness, he cannot expect others (e.g., employers) to arbitrarily provide him with what he wants. Nor is what a given person wants always rational. If an organization based its decisions solely on whatever desires its employees happened to have rather than on reason and logic, it would soon go bankrupt. Applied on a wide scale, the result would be mass poverty.

From the point of view of an organization, employee satisfaction must be considered a means to an end (i.e., a necessary condition for long-term profitability), not an end in itself.

In view of this, one must ask what those who advocate PDM *solely* in the name of satisfaction wish to accomplish. For example, Susman (1976), citing various authorities, argues that PDM is consistent with an "organic" world view which is coming to replace the older "mechanical" world view.

The mechanical world view "is characterized by faith in applying rational methods to the betterment of the human condition. Each successful application of rationality builds upon previous achievements, resulting in . . . historical progress. It is also characterized by belief in an objective world that exists independently of any human observer" (Susman, 1976, p. 24).

In contrast, the organic world view holds that "harmony and equilibrium, not progress are the primary end of human activity. History is to be understood as a series of events that are recurrent and eternal rather than successively leading to a progressively better future . . . knowledge is gained through contemplation and revelation to the end of better understanding the relationship of man to the eternal and divine. God or a similar concept is considered the source of all knowledge" (Susman, 1976, p. 24).

The organic view, according to Susman, is partly the result of recent changes in political ideology which include the abrogation of property rights which are to be supplanted by worker control (socialism), the substitution of satisfaction for production as the goal of work organizations, and the replacement of the profit motive with the (undefined) principle of "social utility."

Let us examine exactly what these two world views imply and stand for. The so-called mechanical world view stands for reason, objectivity, production, individual rights, and progress. The organic view advocates feelings, mysticism, collectivism, and stagnation (i.e., poverty). In this context, the term "mechanical" is a rather bizarre label for the description of ideals which, in essence, represent the key values of the Renaissance and the American Revolution. The term "organic," in contrast, is a strangely benign designation for values which represent the essence of tribalism and the Dark Ages. If this is what participation on a wide scale will ultimately lead to, then one must ask why anyone would want it.

If participation is to be used as a tool for the furtherance of man's happiness and well-being, then it must be in a context which recognizes not only individual differences in knowledge and ability, but the primacy of reason over feelings in organizational decision making.

FOOTNOTES

1. Some of the ideas and conclusions in this paper were presented by the senior author at the Academy of Management meetings, August 1977. The authors would like to thank Dena Schneier and Ron Prestwich for their extensive help in collecting, compiling, and verifying the reference material used in this article.

2. This characterization of American intellectuals is not intended entirely as a compliment. While their desire to insure that theories apply to reality is admirable, their skepticism (due mainly to the philosophy of pragmatism) has been carried to the point where most of them believe that there are no general principles of any kind, that reason is basically impotent, and that there is no objective reality.

3. One book written in the United States has asserted explicitly that because the United States is a democracy, the majority should have the right to coerce the minority in the realm of action (White and Lippitt, 1960, pp. 295ff.). This view is now widely accepted.

4. If forced egalitarianism abrogates political and economic freedom and leads to a low standard of living, it remains to explain why some men still favor it. An extremely benevolent interpretation would be that it is due to an error of knowledge, i.e., its advocates do not understand what it implies and entails. However, Rand's (1971a) motivational interpretation is a more likely explanation, especially for many intellectuals. She attributes the advocacy of forced egalitarianism to hatred of the good for being the good, i.e., to hatred of achievement, of values, of competence. This, in turn, stems from the fact that because of their own resentment against mental effort and the resulting arrested cognitive development, they find men of achievement to be a threat to their self-esteem.

5. The claim that Yugoslavian business organizations are run democratically must be taken with an enormous dose of skepticism considering the fact that the Communist Party has absolute control over every facet of life. Sturmthal (1964) notes that limited autonomy has been permitted in Yugoslavia, but only because the Party has not seen the workers' councils as a threat to its authority. A similar experiment in Poland was rapidly terminated because the Party saw the councils as anti-Communist. Sturmthal concludes that "No attempts to 'humanize' totalitarianism . . . can bridge the gap that separates self government from an unlimited dictatorship" (1964, p. 190).

6. In most of Maier's studies, judgments regarding decision quality were based on the experimenter's opinion rather than on an external criterion. The present writers believe that the solutions judged to be best in Maier's studies would not necessarily be the best under all circumstances. The danger of making quality judgments divorced from a specific real-life context was illustrated by an experience of the junior author. He presented a standard decision-making exercise to one of his classes in which the students had to decide upon the best strategies to survive in the wilderness. When presented with the "official" solutions to the exercise, which the author of the exercise had proposed, one student who had spent a considerable amount of time in the military commented that the proposed solutions were absurd and both his survival education and experience had taught him that the best solutions were totally contrary to many of the solutions proposed in the exercise.

7. The term "solved" should be placed in a wider organizational perspective. In the first 10 years after nationalization, the British Coal Industry had a cumulative deficit of $164.6 million (data provided by the Library of Congress). Thus the application of a sociotechnical systems approach could not overcome (nor was it designed to overcome) the problem of organization-wide (or industry-wide) inefficiency.

8. Additional information regarding this study was presented at a lecture given by Dr. Susman at the University of Maryland on October 7, 1976.

9. In a personal communication, Dr. Smith indicated that very little participation actually occurred in the PDM groups; this suggests that the results were probably due to goal setting rather than PDM.

10. It should be noted, however, that while our major conclusion (that PDM is contextual) agrees with that of another survey by Filley, House, and Kerr (1976), we do not agree with their classification of the results of many of the studies they reviewed (see their tables 11-2, 11-3, 11-4). For example, they reported the Vroom and Mann (1960) study as showing a positive relationship between satisfaction and PDM, whereas the actual finding was a positive correlation in one group and a negative correlation in another. We classified this study as "0", meaning the result was contextual. As a result of classification discrepancies like this, Filley, House, and Kerr did not conclude, as we did, that the results with respect to satisfaction were more consistent than those with respect to productivity. The reader will have to judge for himself whose classifications are more accurate.

11. This statement must be qualified to apply only to U.S. workers at this point in time. Dr. Frank Landy, who recently visited Sweden, observed that in that country the ideal of egalitarianism is so strong that the demand for participation by workers is not limited by lack of knowledge. For example, university janitors have demanded (successfully) to participate in the decisions of faculty committees!

12. It should be noted that the principle of the "rule of knowledge" cannot be applied out of context to the running of a government. While a government should be run by qualified people, no amount of superior knowledge would give them the right to enslave the citizens (e.g., like Plato's philosopher kings). Such a policy would negate the right of the citizens to act on their own rational judgment when it disagrees with that of the authorities (and would stifle the discovery of new knowledge). In a private organization, in contrast, when employer and employee disagree, they are free to part company (refuse to trade). Neither has the right to use physical force to make the other comply.

REFERENCES

1. Adams. S. (1952) "Effect of Equalitarian Atmospheres upon the Performance of Bomber Crews," *American Psychologist 7*, 398.
2. Albrook, R. C. (1967) "Participative Management: Time for a Second Look," *Fortune 75*, 166–170, 199–200.
3. Allport. G. W. (1945) "The Psychology of Participation," *Psychological Review 53*, 117–131.
4. Alutto, J. A., and F. Acito (1974) "Decisional Participation and Sources of Job Satisfaction: A Study of Manufacturing Personnel," *Academy of Management Journal 17*, 160–167.
5. Anderson, R. C. (1959) "Learning in Discussions: A Resumé of the Authoritarian-Democratic Studies," *Harvard Educational Review 29*, 201–215.
6. Argyle, M., G. Gardner, and F. Cioffi (1958) "Supervisory Methods Related to Productivity, Absenteeism, and Labour Turnover," *Human Relations 11*, 23–40.
7. Argyris, C. (1955) "Organizational Leadership and Participative Management," *Journal of Business 28*, 1–7,
8. Baumgartel, H. (1956) "Leadership, Motivations, and Attitudes in Research Laboratories," *Journal of Social Issues 12*, 24–31.
9. ———. (1957) "Leadership Style as a Variable in Research Administration," *Administrative Science Quarterly 2*, 344–360.
10. Bennett, E. (1955) "Discussion, Decision, Commitment, and Consensus in 'Group Decision'," *Human Relations 8*, 251–273.
11. Bennis, W. (1966) "A Reply: When Democracy Works," *Trans-action 3*, 35–36.

12. Berkowitz, L. (1953) "Sharing Leadership in Small, Decision-making Groups," *Journal of Abnormal and Social Psychology 48*, 231–238.
13. Blumberg, P. (1968) *Industrial Democracy: The Sociology of Participation*, New York: Schocken Books Inc.
14. Bucklow, M. (1966) "A New Role for the Work Group," *Administrative Science Quarterly 11*, 59–78.
15. Burke, P. J. (1966) "Authority Relations and Disruptive Behavior in Small Discussion Groups," *Sociometry 29*, 237–250.
16. Calvin, A. D., F. K. Hoffmann, and E. L. Harden (1957) "The Effect of Intelligence and Social Atmosphere on Group Problem Solving Behavior," *Journal of Social Psychology 45*, 61–74.
17. Campbell, H. (1953) "Some Effects of Joint Consultation on the Status and Role of the Supervisor," *Occupational Psychology 27*, 200–206.
18. Carey, A. (1967) "The Hawthorne Studies: A Radical Criticism," *American Sociological Review 32*, 403–416.
19. Carroll, S. J., and H. L. Tosi (1973) *Management by Objectives: Applications and Research*, New York: Macmillan Inc.
20. ———, and ———. (1977) "Relationship of Various Motivational Forces to the Effects of Participation in Goal Setting in a Management by Objectives Program," *Industrial Relations Research Association, Proceedings of the Twenty-Ninth Annual Winter Meeting* pp. 20–25.
21. Chaney, F. B., and K. S. Teel (1972) "Participative Management—A Practical Experience," *Personnel 49*, 8–19.
22. Coch, L., and J. R. P. French (1948) "Overcoming Resistance to Change," *Human Relations 1*, 512–532.
23. Cook, T. D., and D. T. Campbell (1976) "The Design and Conduct of Quasi-Experiments and True Experiments in Field Settings," in M. D. Dunnette (ed.), *Handbook of Industrial and Organizational Psychology*, Chicago: Rand McNally & Company.
24. Dachler, H. P., and B. Wilpert (1978) "Conceptual Dimensions and Boundaries of Participation in Organizations: A Critical Evaluation," *Administrative Science Quarterly 23*, 1–39.
25. Davies, B. (1967) "Some Thoughts on 'Organisational Democracy'," *Journal of Management Studies 4*, 270–281.
26. Davis, K. (1957) "Management by Participation: Its Place in Today's Business World," *Management Review 46*, 69–79.
27. ———. (1963) "The Case for Participative Management," *Business Horizons 6*, 55–60.
28. ———. (1967) *Human Relations at Work: The Dynamics of Organizational Behavior*, New York: McGraw-Hill.
29. Derber, M. (1963) "Worker Participation in Israeli Management," *Industrial Relations 3*, 51–72.
30. ———. (1970) "Crosscurrents in Workers Participation," *Industrial Relations 9*, 123–136.
31. Dill, W. R. (1958) "Environment as an Influence on Managerial Autonomy," *Administrative Science Quarterly 2*, 409–443.
32. ———, W. Hoffman, H. J. Leavitt, and T. O'Mara (1961) "Experiences with a Complex Management Game," *California Management Review 3*, 38–51.
33. Donnelly, J. F. (1977) "Participative Management at Work," *Harvard Business Review 55*, 117–127.

34. Dowling, W. F. (1975) "At General Motors: System 4 Builds Performance and Profits," *Organizational Dynamics 3* (Winter), 23–38.

35. Dubin, R. (1965) "Supervision and Productivity: Empirical Findings and Theoretical Considerations," in R. Dubin (ed.), *Leadership and Productivity*, San Francisco: Chandler Publishing Company.

36. Emery, F., and E. Thorsrud (1976) *Democracy at Work*, Leiden, The Netherlands: Martinus Nijhoff.

37. ———. and E. L. Trist (1960) "Socio-technical Systems," in C. W. Churchman and M. Verhulst (eds.), *Management Sciences, Models and Techniques*, Vol. 2, New York: Pergamon Press, Inc.

38. Falcione, R. L. (1974) "Credibility: Qualifier of Subordinate Participation," *Journal of Business Communication 11*, 43–54.

39. Fein, M. (1976a) "Improving Productivity by Improved Productivity Sharing," *Conference Board Record 13* (July), 44–49.

40. ———. (1976b) "Designing and Operating an Improshare Plan," Hillsdale, N.J. (unpublished).

41. ———. (1976c) "Motivation for Work," in R. Dubin (ed.), *Handbook of Work, Organization, and Society*, Chicago: Rand McNally & Company.

42. ———. (1977) "An Alternative to Traditional Managing," Hillsdale, N.J. (unpublished).

43. Filley, A. C., R. J. House, and S. Kerr, *Managerial Process and Organizational Behavior*, Glenview, Ill.: Scott, Foresman and Company.

44. Fleishman, E. A. (1965) "Attitude versus Skill Factors in Work Group Productivity," *Personnel Psychology 18*, 253–266.

45. Foa, U. G. (1957) "Relation of Workers Expectation to Satisfaction with Supervisor," *Personnel Psychology 10*, 161–168.

46. Ford, R. N. (1969) *Motivation Through the Work Itself*, New York: American Management Association.

47. Fox, W. M. (1957) "Group Reaction to Two Types of Conference Leadership," *Human Relations 10*, 279–289.

48. Foy, N., and H. Gadon (1976) "Worker Participation: Contrasts in Three Countries," *Harvard Business Review 54*, 71–83.

49. French, J. R. P. (1950) "Field Experiments: Changing Group Productivity," in J. G. Miller (ed.), *Experiments in Social Process: A Symposium on Social Psychology*, New York: McGraw-Hill.

50. ———, J. Israel, and D. As (1960) "An Experiment in a Norwegian Factory: Interpersonal Dimensions in Decision-making," *Human Relations 13*, 3–19.

51. ———, E. Kay, and H. H. Meyer (1966) "Participation and the Appraisal System," *Human Relations 19*, 3–20.

52. ———, I. C. Ross, S. Kirby, J. R. Nelson, and P. Smyth (1958) "Employee Participation in a Program of Industrial Change," *Personnel 35*, 16–29.

53. Frost, C. F., J. H. Wakeley, and R. A. Ruh (1974) *The Scanlon Plan for Organization Development: Identity, Participation, and Equity*, East Lansing: Mich.: Michigan State University Press.

54. Geare, A. J. (1976) "Productivity From Scanlon-type Plans," *Academy of Management Review 1*, 99–108.

55. Gibb, C. A. (1951) "An Experimental Approach to the Study of Leadership," *Occupational Psychology 25*, 233–248.

56. Gillespie, J. (1971) "Toward Freedom in Work," in C. G. Benello and D. Roussopoulos (eds.), *The Case for Participatory Democracy: Some Prospects For a Radical Society*, New York: Grossman Publishers.

57. Glueck, W., and D. Kavran (1971) "Yugoslav Management System," *Management International Review 11*, 3–17.
58. Gomberg, W. (1966) "The Trouble with Democratic Management," *Transaction 3*, 30–35.
59. Hackman, J. R. (1975) "On the Coming Demise of Job Enrichment," in E. L. Cass and F. G. Zimmer (eds.), *Man and Work in Society*, New York: Von Nostrand-Reinhold Company.
60. Hall, J. (1971) "Decisions, Decisions, Decisions," *Psychology Today 5*, 51–54, 86–88.
61. Hamblin, R. L., K. Miller, and J. A. Wiggins (1961) "Group Morale and Competence of the Leader," *Sociometry 24*, 295–311.
62. Hannan, R. E. (in press) "The Effects of Participative versus Assigned Goal Setting on Goal Acceptance and Performance," *Journal of Applied Psychology*.
63. Haythorn, W. (1958) "The Effects of Varying Combinations of Authoritarian and Equalitarian Leaders and Followers," in E. E. Maccoby, T. M. Newcomb, and E. L. Hartley (eds.), *Readings in Social Psychology*, New York: Henry Holt.
64. Heinicke, C., and R. F. Bales (1953) "Developmental Trends in the Structure of Small Groups," *Sociometry 16*, 7–38.
65. Heller, F. A., and L. W. Porter (1966) "Perceptions of Managerial Needs and Skills in Two National Samples," *Occupational Psychology 40*, 1–13.
66. ———, and G. Yukl (1969) "Participation, Managerial Decison-making, and Situational Variables," *Organizational Behavior and Human Performance 4*, 227–241.
67. Hemphill, J. K. (1950) "Relations between the Size of the Group and the Behavior of 'Superior' Leaders," *Journal of Social Psychology 32*, 11–32.
68. Herzberg, F. (1966) *Work and the Nature of Man*, Cleveland: World Publishing Co.
69. Hespe, G., and T. Wall "The Demand for Participation Among Employees." Sheffield, England: MCR Social and Applied Psychology Unit (unpublished).
70. Hessen, R. (1968) "Campus or Battleground? Columbia is a Warning to All American Universities," *Barron's*, May 20, pp. 1ff.
71. Hoffman, L. R., R. J. Burke, and N. R. F. Maier (1965) "Participation, Influence, and Satisfaction among Members of Problem-solving Groups," *Psychological Reports 16*, 661–667.
72. ———, E. Harburg, and N. R. F. Maier (1962) "Differences and Disagreement as Factors in Creative Group Problem Solving," *Journal of Abnormal and Social Psychology 64*, 206–214.
73. Holter, H. (1965) "Attitudes toward Employee Participation in Company Decision-making Processes," *Human Relations 18*, 297–321.
74. Ivancevich, J. M. (1976) "Effects of Goal Setting on Performance and Job Satisfaction," *Journal of Applied Psychology 61*, 605–612.
75. ———. (1977) "Different Goal Setting Treatments and Their Effects on Performance and Job Satisfaction," *Academy of Management Journal 20*, 406–419.
76. Janis, I. (1972) *Victims of Groupthink*, Boston: Houghton Mifflin Company.
77. Jaques, E. (1964) "Social-Analysis and the Glacier Project," *Human Relations 17*, 361–375.
78. Kahn, R. L. (1974) "Organizational Development: Some Problems and Proposals," *Journal of Applied Behavioral Science 10*, 485–502.
79. ———. (1975) "In Search of the Hawthorne Effect," in E. L. Cass and F. G. Zimmer (eds.), *Man and Work in Society*, New York: Van Nostrand-Reinhold Company.
80. Kaiser, R. G. (1976) *Russia: The People and the Power*, New York: Atheneum Publishers.
81. Katz, D., N. Maccoby, and N. Morse (1950) *Productivity, Supervision and Morale in an Office Situation*, Part I, Ann Arbor, Mich.: Survey Research Center, Institute for Social Research, University of Michigan.

82. ——, ——, G. Gurin, and L. G. Floor (1951) *Productivity, Supervision and Morale Among Railroad Workers,* Ann Arbor, Mich.: Survey Research Center, Institute for Social Research, University of Michigan.

83. Katzell, R. A., C. E. Miller, N. G. Rotter, and T. G. Venet (1970) "Effects of Leadership and Other Inputs on Group Processes and Outputs," *Journal of Social Psychology 80,* 157–169.

84. Kidd, J. S., and R. T. Christy (1961) "Supervisory Procedures and Work-team Productivity," *Journal of Applied Psychology 45,* 388–392.

85. Lammers, C. J. (1967) "Power and Participation in Decision Making in Formal Organizations," *American Journal of Sociology 73,* 201–216.

86. Lanzetta, J. T., and T. B. Roby (1960) "The Relationship between Certain Group Process Variables and Group Problem-solving Efficiency," *Journal of Social Psychology 52,* 135–148.

87. Latham, G. P., and J. J. Baldes (1975) "The 'Practical Significance' of Locke's Theory of Goal Setting," *Journal of Applied Psychology 60,* 122–124.

88. ——, and G. A. Yukl (1975a) "A Review of Research on the Application of Goal Setting in Organizations," *Academy of Management Journal 18,* 824–845.

89. ——, and ——. (1975b) "Assigned versus Participative Goal Setting with Educated and Uneducated Woods Workers," *Journal of Applied Psychology 60,* 299–302.

90. ——, and ——. (1976) "Effects of Assigned and Participative Goal Setting on Performance and Job Satisfaction," *Journal of Applied Psychology 61,* 166–171.

91. ——, J. R. Mitchell, and D. L. Dossett (1978) "Importance of Participative Goal Setting and Anticipated Rewards on Goal Difficulty and Job Performance," *Journal of Applied Psychology 63,* 163–171.

92. Lawler, E. E. (1971) *Pay and Organizational Effectiveness: A Psychological View,* New York: McGraw-Hill.

93. ——. (1975) "Pay, Participation and Organizational Change," in E. L. Cass and F. G. Zimmer (eds.), *Man and Work in Society,* New York: Van Nostrand-Reinhold Company.

94. ——. (1976) "Should the Quality of Work Life be Legislated?" *The Personnel Administrator 21* (January), 17–21.

95. ——, and J. R. Hackman (1969) "Impact of Employee Participation in the Development of Pay Incentive Plans: A Field Experiment," *Journal of Applied Psychology 53,* 467–471.

96. Lawrence, L. C., and. P. C. Smith (1955) "Group Decision and Employee Participation," *Journal of Applied Psychology 39,* 334–337.

97. Lawrence, P. R. (1971) "How to Deal with Resistance to Change," in D. A. Kolb, I. M. Rubin, and J. M. McIntyre (eds.), *Organizational Psychology,* Englewood Cliffs, N.J.: Prentice-Hall, Inc.

98. Leavitt, H. J. (1965) "Applied Organizational Change in Industry: Structural, Technological and Humanistic Approaches," in J. G. March (ed.), *Handbook of Organizations,* Chicago: Rand McNally & Company.

99. Levine, J., and J. Butler (1952) "Lecture vs. Group Decision in Changing Behavior," *Journal of Applied Psychology 36,* 29–33.

100. Lewin, K. (1947) "Frontiers in Group Dynamics," *Human Relations 1,* 5–42.

101. ——. (1952) "Group Decision and Social Change," in T. M. Newcomb and E. L. Hartley (eds.), *Readings in Social Psychology,* New York: Holt, Rinehart and Winston, Inc.

102. ——, R. Lippitt, and R. K. White (1939) "Patterns of Aggressive Behavior in Experimentally Created 'Social Climates,'" *Journal of Social Psychology 10,* 271–299.

103. Ley, R. (1966) "Labor Turnover as a Function of Worker Differences, Work Environment, and Authoritarianism of Foremen," *Journal of Applied Psychology 50*, 497–500.
104. Likert, R. (1961) *New Patterns of Management*, New York: McGraw-Hill.
105. ———. (1967) *The Human Organization*, New York: McGraw-Hill.
106. Lippitt, R., and R. K. White (1958) "An Experimental Study of Leadership and Group Life," in E. E. Maccoby, T. M. Newcomb, and E. L. Hartley (eds.), *Readings in Social Psychology*, New York: Holt, Rinehart and Winston, Inc.
107. Lischeron, J., and T. D. Wall (1974) "Attitudes towards Participation among Local Authority Employees," *Human Relations 28*, 499–517.
108. ———, and ———. (1975) "Employee Participation: An Experimental Field Study," *Human Relations 28*, 863–884.
109. Litwin, G. H., and R. A. Stringer (1968) *Motivation and Organizational Climate*, Boston: Division of Research, Graduate School of Business Administration, Harvard University.
110. Locke, E. A. (1969) "What Is Job Satisfaction?" *Organizational Behavior and Human Performance 4*, 309–336.
111. ———. (1975) "Personnel Attitudes and Motivation," *Annual Review of Psychology 26*, 457–480.
112. ———. (1976a) "The Nature and Causes of Job Satisfaction," in M. D. Dunnette (ed.), *Handbook of Industrial and Organizational Psychology*, Chicago: Rand McNally & Company.
113. ———. (1976b) "The Case against Legislating the Quality of Work Life," *The Personnel Administrator 21* (May), 19–21.
114. ———. (1978) "The Ubiquity of the Technique of Goal Setting in Theories of and Approaches to Employee Motivation," *Academy of Management Review, 3*, 594–601.
115. ———, D. Sirota, and A. D. Wolfson (1976) "An Experimental Case Study of the Successes and Failures of Job Enrichment in a Government Agency," *Journal of Applied Psychology 61*, 701–711.
116. Lowin, A. (1968) "Participative Decision Making: A Model, Literature Critique, and Prescriptions for Research," *Organizational Behavior and Human Performance 3*, 68–106.
117. Macrae, N. (1977) "'Entrepreneurial Revolution' Ahead, According to British Journalist," *World of Work Report 2* (April), 46–48.
118. Maher, J. R. (1971) *New Perspectives in Job Enrichment*, New York: Van Nostrand-Reinhold Company.
119. Mahoney, T. A. (1967) "Managerial Perceptions of Organizational Effectiveness," *Management Science 14*, 76–91.
120. Maier, N. R. F. (1950) "The Quality of Group Decision as Influenced by the Discussion Leader," *Human Relations 3*, 155–174.
121. ———. (1953) "An Experimental Test of the Effect of Training on Discussion Leadership," *Human Relations 6*, 161–173.
122. ———. (1967) "Assets and Liabilities in Group Problem-solving: The Need for an Integrative Function," *Psychological Review 74*, 239–249.
123. ———. (1970) "The Integrative Function in Group Problem Solving," in L. R. Aronson, E. Tobath, D. S. Lehrman, and J. S. Rosenblatt (eds.), *Development and Evolution of Behavior*, San Francisco: W. H. Freeman and Company Publishers.
124. ———. (1973) *Psychology in Industrial Organizations*, Boston: Houghton Mifflin Company.
125. ———. and M. Sashkin (1971) "Specific Leadership Behaviors that Promote Problem Solving," *Personnel Psychology 24*, 35–44.

126. Malone. E. L. (1975) "The Non-Linear Systems Experiment in Participative Management," *Journal of Business 48,* 52–64.
127. Mansbridge, J. J. (1973) "Time, Emotion, and Inequality: Three Problems of Participatory Groups," *Journal of Applied Behavioral Science 9,* 351–368.
128. Marrow, A. J. (ed.) (1972) *The Failure of Success,* New York: AMACOM.
129. ———, and J. R. P. French (1945) "Changing a Stereotype in Industry," *Journal of Social Issues* December: 33–37.
130. ———, D. G. Bowers, and S. E. Seashore (1967) *Management by Participation,* New York: Harper & Row, Publishers.
131. Maslow, A. H. (1970) "The Superior Person," in W. G. Bennis (ed.), *American Bureaucracy,* Chicago: Aldine.
132. Mayo, E. (1924) "Revery and Industrial Fatigue," *Journal of Personnel Research 3,* 278–281.
133. ———. (1970) "The First Inquiry," in H. F. Merrill (ed.), *Classics in Management,* New York: American Management Association (originally published in 1945).
134. McCormick, C. P. (1938) *Multiple Management,* New York: Harper.
135. ———, (1949) *The Power of People,* New York: Harper.
136. McCurdy, H. G., and H. W. Eber (1953) "Democratic versus Authoritarian: A Further Investigation of Group Problem-solving, *Journal of Personality 22,* 258–269.
137. ———, and W. E. Lambert (1952) "The Efficiency of Small Human Groups in the Solution of Problems Requiring Genuine Co-operation," *Journal of Personality 20,* 478–494.
138. McGregor, D. (1944) "Conditions of Effective Leadership in the Industrial Organization," *Journal of Consulting Psychology 8,* 55–63.
139. McInnes, N. (1976) "People's Capitalism: in Europe, Workers Gain Seats on Corporate Boards," *Barron's* July 12, pp. 9, 16–17.
140. ———. (1977) "Boardroom Revolution? In Great Britain, the Rights of Investors Are in Jeopardy," *Barron's* February 14, pp. 7, 12.
141. McMurray, R. N. (1958) "The Case for Benevolent Autocracy," *Harvard Business Review 36,* 82–90.
142. Meyer, H. H., E. Kay, and J. R. P. French (1965) "Split Roles in Performance Appraisal," *Harvard Business Review 43,* 123–129.
143. Miles, R. E. (1965) "Human Relations or Human Resources?" *Harvard Business Review 43,* 148–163.
144. ———, and Ritchie, J. B. (1971) "Participative Management: Quality vs. Quantity," *California Management Review 13,* 48–56.
145. Miller, E. J. (1975) Socio-technical Systems in Weaving, 1953–1970: A Follow-up Study," *Human Relations 28,* 349–386.
146. Mitchell, T. R. (1973) "Motivation and Participation: An Integration," *Academy of Management Journal 16,* 670–679.
147. Morse, C., and E. Reimer (1956) "The Experimental Change of a Major Organizational Variable," *Journal of Abnormal and Social Psychology 52,* 120–129.
148. Morse, J. J., and J. W. Lorsch (1970) "Beyond Theory Y," *Harvard Business Review 48* (May/June), 61–68.
149. Mulder, M. (1959) "Power and Satisfaction in Task-oriented Groups," *Acta Psychologica 16,* 178–225.
150. ———, and H. Wilke (1970) "Participation and Power Equalization," *Organizational Behavior and Human Performance 5,* 430–448.
151. Mullen, J. H. (1965) "Differential Leadership Modes and Productivity in a Large Organization," *Academy of Management Journal 8,* 107–126.

152. Nicol, E. A. (1948) "Management through Consultative Supervision," *Personnel Journal 27*, 207–217.
153. Obradovic, J. (1970) "Participation and Work Attitudes in Yugoslavia," *Industrial Relations 9*, 161–169.
154. ———, J. R. P. French, and W. Rodgers (1970) "Workers' Councils in Yugoslavia," *Human Relations 23*, 459–71.
155. Parsons, H. M. (1974) "What Happened at Hawthorne?" *Science 183*, 922–932.
156. Patchen, M. (1964) "Participation in Decision-making and Motivation: What Is the Relation?," *Personnel Administration 27*, 24–31.
157. ———. (1970) *Participation, Achievement, and Involvement on the Job*, Englewood Cliffs, N.J.: Prentice-Hall, Inc.
158. Paul, W. J., K. B. Robertson, and F. Herzberg (1969) "Job Enrichment Pays Off," *Harvard Business Review 47*, 61–77.
159. Pelz, D. C. (1956) "Some Social Factors Related to Performance in a Research Organization," *Administrative Science Quarterly 1*, 310–325.
160. Powell, R. M., and J. L. Schlacter (1971) "Participative Management: A Panacea?," *Academy of Management Journal 14*, 165–173.
161. Preston, L. E., and J. E. Post (1974) "The Third Managerial Revolution," *Academy of Management Journal 17*, 476–486.
162. Rand, A. (1964) *The Virtue of Selfishness*, New York: New American Library (Signet).
163. ———. (1966) *Capitalism: The Unknown Ideal*, New York: New American Library.
164. ———. (1971a) "The Age of Envy," *The Objectivist 10* (7), 1–13, and 10 (8), 1–11.
165. ———. (1971b) *The "New Left: The Anti-industrial Revolution,"* New York: New American Library.
166. Raskin, A. H. (1976) "The Labor Scene, The Workers' Voice in German Companies," *World of Work Report 1*, (July) 5–6.
167. Rice, A. K. (1953) "Productivity and Social Organization in an Indian Weaving Shed," *Human Relations 6*, 297–329.
168. Roethlisberger, F. J., and W. J. Dickson (1956) *Management and the Worker*, Cambridge, Mass.: Harvard University Press (originally published in 1939).
169. Rosen, N. A. (1969) *Leadership Change and Work-Group Dynamics: An Experiment*, Ithaca, N.Y.: Cornell University Press.
170. Rosenfeld, J. M., and M. J. Smith (1967) "Participative Management: An Overview," *Personnel Journal 46*, 101–104.
171. Rosenstein, E. (1970) "Histadrut's Search for a Participation Program," *Industrial Relations 9*, 70–186.
172. Rubenowitz, S. (1962) "Job-oriented and Person-oriented Leadership," *Personnel Psychology 15*, 387–396.
173. Runyon, K. E. (1973) "Some Interactions between Personality Variables and Management Styles," *Journal of Applied Psychology 57*, 288–294.
174. Sadler, P. J. (1970) "Leadership Style, Confidence in Management, and Job Satisfaction," *Journal of Applied Behavioral Science 6*, 3–19.
175. Sashkin, M. (1976) "Changing toward Participative Management Approaches: A Model and Methods," *Academy of Management Review 1*, 75–86.
176. Scheflen, K. C., E. E. Lawler, and J. R. Hackman (1971) "Long-term Impact of Employee Participation in the Development of Pay Incentive Plans: A Field Experiment Revisited," *Journal of Applied Psychology 55*, 182–186.
177. Schregle, J. (1970) "Forms of Participation in Management," *Industrial Relations 9*, 117–122.
178. Schuler, R. S. (1977) "Role Perceptions, Satisfaction and Performance Moderated by

Organization Level and Participation in Decision Making," *Academy of Management Journal 20*, 165–169.

179. Schultz, G. P. (1951) "Worker Participation on Production Problems: A Discussion of Experience with the 'Scanlon Plan,'" *Personnel 28*, 201–210.

180. Seeborg, I. S. (1978) "The Influence of Employee Participation in Job Redesign," *Journal of Applied Behavioral Science 14*, 87–98.

181. Shaw, M. E. (1955) "A Comparison of Two Types of Leadership in Various Communication Nets," *Journal of Abnormal and Social Psychology 50*, 127–134.

182. ———, and J. M. Blum (1966) "Effects of Leadership Style upon Group Performance as a Function of Task Structure," *Journal of Personality and Social Psychology 3*, 238–242.

183. Singer, J. N. (1974) "Participative Decison-making about Work: An Overdue Look at Variables which Mediate its Effects," *Sociology of Work and Occupations 1*, 347–371.

184. Smith, H. (1976) *The Russians*, New York: Ballantine Books, Inc.

185. Sorcher, M. (1967) "Motivating the Hourly Employee," General Electric Company, Behavioral Research Service (unpublished).

186. ———. (1971) "Motivation, Participation and Myth," *Personnel Administration 34* (September/October), 20–24.

187. Stagner, R. (1969) "Corporate Decision Making: An Empirical Study," *Journal of Applied Psychology 53*, 1–13.

188. Steers, R. M. (1975) "Task-goal Attributes, n achievement, and Supervisory Performance," *Organizational Behavior and Human Performance 13*, 392–403.

189. Stogdill, R. M. (1974) *Handbook of Leadership: A Survey of Theory and Research.* New York: The Free Press.

190. Strauss, G. (1963) "Some Notes on Power-equalization," in H. Leavitt (ed.), *The Social Science of Organizations: Four Perspectives*, Englewood Cliffs, N.J.: Prentice-Hall, Inc.

191. ———, and E. Rosenstein (1970) "Workers' Participation: A Critical View," *Industrial Relations 9*, 197–214.

192. Sturmthal, A. (1964) *Workers Councils: A Study of Workplace Organization on Both Sides of the Iron Curtain*, Cambridge, Mass.: Harvard University Press.

193. Suojanen, W. W., M. J. McDonald, G. L. Swallow, and W. W. Suojanen (eds.), (1975) *Perspectives on Job Enrichment and Productivity*, Atlanta, Ga.: Georgia State University.

194. Susman, G. I. (1976) *Autonomy at Work: A Sociotechnical Analysis of Participative Management*, New York: Praeger Publishers, Inc.

195. Tannenbaum, A. S. (1962) "Control in Organizations: Individual Adjustment and Organizational Performance," *Administrative Science Quarterly 7*, 236–257.

196. ———. (1966) *Social Psychology of the Work Organization*, Belmont, Cal.: Wadsworth Publishing Co. Inc.

197. ———. (1974) "Systems of Formal Participation," in G. Strauss, R. Miles, C. C. Snow, and A. S. Tannenbaum (eds.), *Organizational Behavior: Research and Issues*, Madison, Wis.: Industrial Relations Research Association.

198. Tannenbaum, R., and F. Massarik (1950) "Participation by Subordinates in the Managerial Decision-making Process," *Canadian Journal of Economics and Political Science 16*, 408–418.

199. ———, and W. Schmidt (1958) "How to Choose a Leadership Pattern," *Harvard Business Review 36*, 95–101.

200. Taylor, F. W. (1967) *The Principles of Scientific Management*, New York: W. W. Norton & Company, Inc. (originally published in 1911).

201. Teague, B. (1971) "Can Workers Participate in Management Successfully?" *Conference Board Record* 8 (July): 48–52.
202. Thorsrud, E., B. A. Sorensen, and B. Gustavsen (1976) "Sociotechnical Approach to Industrial Democracy in Norway," in R. Dubin (ed.), *Handbook of Work, Organization, and Society,* Chicago: Rand McNally & Company.
203. Torrance, E. P. (1953) "Methods of Conducting Critiques of Group Problem-solving Performance," *Journal of Applied Psychology 37,* 394–398.
204. Tosi, H. (1970) "A Reexamination of Personality as a Determinant of the Effects of Participation," *Personnel Psychology 23,* 91–99.
205. Trist, E. L., and K. W. Bamforth (1951) "Some Social and Psychological Consequences of the Longwall Method of Coal-getting," *Human Relations 4,* 3–38.
206. ———, G. W. Higgins, H. Murray, and A. B. Pollock (1963) *Organizational Choice,* London: Tavistock.
207. Umstot, D. D., C. H. Bell, and T. R. Mitchell (1976) "Effects of Job Enrichment and Task Goals on Satisfaction and Productivity," *Journal of Applied Psychology 61,* 379–394.
208. Van de Vall, M., and C. D. King (1973) "Comparing Models of Workers' Participation in Managerial Decision Making," in D. Graves (ed.), *Management Research: A Cross-Cultural Perspective,* New York: Elsevier Scientific.
209. Veen, P. (1972) "Effects of Participative Decison-making in Field Hockey Training: A Field Experiment," *Organizational Behavior and Human Performance 7,* 288–307.
210. Viteles, M. S. (1953) *Motivation and Morale in Industry,* New York: W. W. Norton & Company, Inc.
211. Von Mises, L. (1962) *Socialism,* New Haven: Yale University Press (originally published in 1951).
212. Vroom, V. (1960) *Some Personality Determinants of the Effects of Participation,* Englewood Cliffs, N.J.: Prentice-Hall, Inc.
213. ———. (1964) *Work and Motivation,* New York: John Wiley & Sons, Inc.
214. ———. (1969) "Industrial Social Psychology," in G. Lindzey and E. Aronson (eds.), *Handbook of Social Psychology,* Reading, Mass.: Addison-Wesley.
215. ———, and F. Mann (1960) "Leader Authoritarianism and Employee Attitudes," *Personnel Psychology 13,* 125–140.
216. ———, and P. Yetton (1973) *Leadership and Decision-Making,* Pittsburgh: University of Pittsburgh Press.
217. Weschler, I. R., M. Kahane, and R. Tannenbaum (1952) "Job Satisfaction, Productivity, and Morale: A Case Study," *Occupational Psychology 26,* 1–14.
218. Wexley, K. N., J. P. Singh, and G. A. Yukl, (1973) "Subordinate Personality as a Moderator of the Effects of Participation in Three Types of Appraisal Interviews," *Journal of Applied Psychology 58,* 54–59.
219. White, R., and R. Lippitt (1953) "Leader Behavior and Member Reaction in Three 'Social Climates,'" in D. Cartwright and A. Zander (eds.), *Group Dynamics: Research and Theory,* Evanston, Ill.: Row, Peterson.
220. ———, and ———. (1960) *Autocracy and Democracy,* New York: Harper & Row, Publishers.
221. Wood, M. T. (1973) "Power Relationships and Group Decision Making in Organizations," *Psychological Bulletin 79,* 280–293.

LEADERSHIP: SOME EMPIRICAL GENERALIZATIONS AND NEW RESEARCH DIRECTIONS[1]

Robert J. House, UNIVERSITY OF TORONTO

Mary L. Baetz, UNIVERSITY OF TORONTO

ABSTRACT

This paper presents a selective review of the literature concerning leadership. The major empirical generalizations that appear supportable are identified. Theories of leadership enjoying current widespread attention and empirical support are also reviewed and subjected to critical analysis. Major issues in the leadership literature are identified and reviewed. Suggestions for future research are advanced throughout the review of the empirical and theoretical literature.

Research in Organizational Behavior, Vol. 1, pp. 341–423.
Copyright © 1979 by JAI Press, Inc.
All rights of reproduction in any form reserved.
ISBN 0-89232-045-1

". . . probably more has been written and less known about leadership than any other topic in the behavioral sciences" (Bennis, 1959, p. 259).

"After 40 years of accumulation, our mountain of evidence about leadership seems to offer few clear-cut facts" (McCall, 1976).

"It is difficult to know what, if anything, has been convincingly demonstrated by replicated research. The endless accumulation of empirical data has not produced an integrated understanding of leadership" (Stogdill, 1974, p. vii).

The above quotes suggest the conclusion that despite the fact that leadership has been the subject of speculation, discussion, and debate since the time of Plato and the subject of more than 3000 empirical investigations (Stogdill, 1974), there is little known about it. We disagree with this conclusion. It is our position that there are several empirical generalizations that can be induced from the wealth of research findings concerning leadership. Further it is our position that when viewed collectively these empirical generalizations provide a basis for the development of a theory of leadership—a theory that potentially describes, explains, and predicts the causes of, processes involved in, and consequences of the leadership phenomena. While such a theory is not presently available, it is argued here that it is possible of attainment.

This paper presents a selective review and summary of the major findings resulting from empirical research conducted to date. As the title implies, one of the purposes of this paper is to suggest new directions for leadership research. Identification of research directions is a theoretical endeavor. This paper is such an endeavor. Hopefully, this theorizing will result in some new and fruitful insights about the phenomenon of leadership. These insights, if presented in an empirically testable form, constitute directions for future research.

In reviewing the literature, we were primarily guided by Stogdill's (1974) exhaustive compilation and summary of seven decades of leadership research. In addition to Stogdill's review, we considered several others (Cartwright and Zander, 1968; Filley, House, and Kerr, 1976; Pfeffer, 1977; Fiedler and Chemers, 1974; McCall, 1976; Sims, 1977; Barrow, 1976; and Schriesheim, House, and Kerr, 1976) and approximately 150 studies published since 1972.

Our objective in conducting this review was to identify issues in the leadership literature that we judged to have significant theoretical and practical implications. To identify such issues we searched for:

1. Theories of leadership for which there is empiric support or for which there is current widespread interest.

2. Significant empirical generalizations based on repeatedly replicated findings.

3. Research suggestions by others that we judged to offer significant promise.

4. Discrepancies in results that, if reconciled, we believe would contribute to knowledge about leadership and its effects.

Upon identification of such theories, empiric generalizations, research suggestions by others, or empiric discrepancies, we then consulted the more prominent works and read them in the original. By rereading these works in the original we hoped to gain additional insights about the phenomena in question. These insights constitute the product of our review of the literature and many of the new directions for leadership research that we propose.

We begin with an examination of the construct of leadership and advance a definition of that construct that we hope is both empirically testable and operationally useful. We then discuss the question of how and what kind of effects leadership has on individual and organizational outcomes. Following the discussion of leadership effects, we review research on leadership traits, leadership behavior, and determinants of leadership behavior. This review is intended to identify issues for leadership research that have not been as yet specified, or suggestions for research, although not new to the literature, that seem not to have been adequately pursued. As will be shown, many such suggestions have been advanced earlier and appear to warrant serious reconsideration.

Following the review of the literature concerned with leadership traits, leadership behavior, and determinants of leader behavior, we turn to a consideration of current leadership theories and attempt to identify the most critical research issues raised by these theories. Throughout our review we will attempt to identify the more important research issues associated with each topic and suggest new directions for leadership research where appropriate.

LEADERSHIP AS A SCIENTIFIC CONSTRUCT

Stogdill reviewed seventy-two definitions of leadership advanced by writers from 1902 to 1967. Almost all definitions imply that leadership is a form of social influence. However, as Pfeffer points out, leadership as usually defined is not distinct from other concepts of social influence and ". . . to treat leadership as a separate concept, it must be distinguished from other social influence phenomena" (Pfeffer, 1977, p. 105).

It is the opinion of the authors that because of the unique context in which leadership takes place it is necessary to define it as a specific subset of social phenomena. To define the construct of leadership we will briefly describe the operational methods by which leadership is defined and studied in the social science literature. We will then infer from this description those variables that appear to be the defining characteristics of the leadership phenomena.

Studies commonly falling under the descriptive term "leadership research" can be grossly divided into two classifications: The first classification of studies consists of those that concern the traits, behavior, and impact of individuals who are assigned formal or legal authority to direct others. These individuals are referred to in these studies as *formal leaders*.

The second class of studies falling under the descriptive term "leadership research" consists of those studies concerned with the traits, behavior, and impact of individuals who exert significant influence over others in task groups for which there is no formally allocated authority. Individuals who are observed to exert such influence are referred to in this literature as *emergent leaders*.

For the first category of studies the independent variables are usually the specific traits or behaviors of the "formal leader." The formal leader is assumed to engage in behavior intended to influence those reporting to him or her. Further, the formal leader is assumed to have a legal or jurisdictional right to influence subordinates.

For the second category of studies—those concerned with emergent leaders—the independent variables are usually the traits or behaviors of the individuals to whom social influence is attributed. This attribution is usually measured by group member responses, after some period of interaction, to such questions as "who was the real leader?" or "who exerted most influence in the group?" The dependent variable is usually the degree to which others voluntarily comply with the influence attempts of the emergent leader. Voluntary compliance is taken by the researchers as an indication that the group members perceived the influence attempts of the emergent leader as acceptable.

Note that both formal and emergent leaders are implicitly defined in this literature in terms of two dimensions. These dimensions are: the degree to which behavior is intended to influence others and the degree to which such influence attempts are viewed as acceptable to the person who is the target of the influence attempt.

It is not assumed that group members and subordinates are consciously aware of another's intention to influence them or of the acceptability of such influence attempts at the time that the attempts occur. Rather, it is assumed for purposes of leadership research that members of a group or subordinates of a formal leader are able to report the degree to which they attribute to the leader the intention to influence them and the degree to which they view the leader's influence attempts as acceptable after some period of interaction.

Thus, the construct of leadership is defined as the degree to which the behavior of a group member is perceived as an acceptable attempt to influence the perceiver regarding his or her activity as a member of a particular group or the activity of other group members.[2] To qualify as a leader behavior it is necessary that the behavior is both perceived as an influence attempt and that the perceived influence attempt is viewed as acceptable. An action by a group member becomes an act of leadership when that act is perceived by another member of the group as an acceptable attempt to influence that person or one or more other members of that group. Thus, as Calder (1977) has pointed out, leadership is an attribution that one person makes about other persons.

It is argued here that leadership is an attribution made about the intentions of others to influence members of a group and about the degree to which that influence attempt is acceptable. The term leadership will be used throughout the remainder of this paper to refer to the construct of leadership as defined above, that is as an attribution.

In the literature to be reviewed in the remainder of this paper leadership is not explicitly defined as it is here. However, as argued above, leadership researchers seem to operationalize the construct of leadership in a manner that is implicitly consistent with our definition. Therefore, for the studies we review, we will proceed on the assumption that this implicit definition holds.

JUSTIFICATION FOR DEFINING LEADERSHIP AS A SCIENTIFIC CONSTRUCT

It was asserted above that because of the unique context in which leadership occurs it is necessary to define it as a scientific subset of social influence phenomena. Leadership takes place in groups of two or more people and most frequently involves influencing group member behavior as it relates to the pursuit of group goals. The nature of the goals, the task technology involved in achieving the goals, and the culture or broader organization in which the group exists frequently have a direct effect on the attitudes and behavior of group members. These variables frequently serve to direct, constrain, or reinforce follower attitudes and behavior. Thus they frequently moderate the relationship between leader behavior and follower responses. The moderating effect of these variables is discussed in more detail later in this paper.

Considering leadership as a scientific subset of the social influence phenomena and defining leadership as a separate construct calls attention to the unique characteristics of the environment in which it occurs. Without such attention important antecedents to leadership and moderator variables are likely to be overlooked and thus result in a less complete understanding of the phenomenon that when the construct of leadership is separately defined.

Research Implications of the Proposed Construct of Leadership

The above proposed construct of leadership has significant implications for leadership research.

First, it is possible to measure the degree to which the leadership construct is associated with outcome variables traditionally studied in leadership research. Some of these outcome variables are the subordinate's compliance with influence attempts, motivation, satisfaction, and individual or group productivity. If the proposed construct is shown to be causally implicated with such outcomes, it would be of significant theoretical and practical interest to identify the functional relationship between the attribution of leadership and the specific traits and behaviors of those to whom leadership is attributed. Further, the moderating effects of situational factors and group member characteristics on this relationship would also be of interest. Identification of such associations would permit prediction of the specific person or persons who emerge as leaders within groups. Further, identification of such associations would help explain how formal leaders gain and maintain influence over subordinates.

If specific behaviors or traits of individuals are found to be associated with the leadership construct, and if the construct is shown to be associated with significant group or individual outcomes, these findings would provide information for the design of leadership selection and training efforts.

Finally, it is possible to conduct an empirical test to determine if group members or observers of groups attribute responsibility for group outcomes to those group members whose behaviors are high on the two leadership dimensions. If this indeed is the case, we would have an understanding of how credit or blame for group outcomes is assigned to group members. Such findings have significant implications for predicting performance appraisal ratings, individual advancement in organizations, and assignment of rewards and punishments to group members.

The Effects of Leadership

Pfeffer (1977) states that "literature assessing the effects of leadership seems to be equivocal" (p. 105) and that "given the resources that have been spent studying, selecting and training leaders, one might expect that the question of whether or not leaders matter would have been addressed earlier" (p. 106).

Following is a brief review of some of the literature concerned with the question of whether leadership does or does not cause variance in organizational effectiveness or other relevant outcomes. To establish that leader behavior causes variance in outcomes it is necessary to show that: *(a)* a change in leader behavior precedes a change in the outcome and *(b)* that the relationship between leader behavior and the outcome is not caused by a third variable. While most leadership studies do not meet both of these requirements, there are several studies that do. Following is a brief summary of these studies.

Specific leadership styles, or combinations of leader behaviors, have been demonstrated as causing significant amounts of variance in: *(a)* the effort level of subordinates when not under the direct surveillance of the leader (Lewin, Lippitt, and White, 1939), *(b)* adaptability to change, and performance under conditions of change (Coch and French, 1948; Fleishman, 1965; Day and Hamblin, 1964; Schachter et al., 1961; DeCharms and Bridgeman, 1961), *(c)* levels of follower's turnover (Dansereau, Graen, and Haga, 1975; Graen and Ginsburgh, 1977; Coch and French, 1948), *(d)* absences (Coch and French, 1948), *(e)* subordinate productivity (Lawrence and Smith, 1955; Tomekovic, 1962; Delbecq, 1965; Shaw and Blum, 1966; Campion, 1968; Cammalleri et al., (1972); Wexley, Singh, and Yukl, 1973; Calvin, Hoffmann, and Hardin, 1957), *(f)* degree of subordinates learning from supervisory training efforts (Fleischman, Harris, and Burtt, 1955), *(g)* the quality of subordinates' decisions and the degree to which subordinates accept these decisions (Maier, 1963, 1970), and *(h)* subordinates' motivation (Graen et al., 1973a).

Most of the above evidence comes from field longitudinal studies at lower levels in the organizations or from laboratory studies. While these studies demonstrate that changes in leader behavior preceded changes in relevant outcomes and controlled for extraneous variables either statistically or by the use of control groups, very few of them were conducted at middle or higher levels in the organization. Notable exceptions to this statement are the findings of Meyer (1975) and Lieberson and O'Connor (1972).

Meyer found that organizational structures change as a result of selection of new leaders for middle management positions in a government bureaucratic organization. To identify the impact of changes in leadership on organizational structure Meyer measured the number of organizational structural changes made in government finance offices from 1966 to 1972. He found significantly more organizational changes occurred during that period after new leaders were selected as compared to the number of changes in offices in which there were no changes in leadership. Lieberson and O'Connor (1972) found that 31 percent of the variance in net profit on sales over 20 years for 167 large companies in thirteen industries is directly attributable to changes in top leadership in these companies. However, this finding held only when measured 2 and 3 years after the appointment of new chief executives. Thus, leadership alone accounted for approximately one-third of the variance in profit on sales in the sample studied. While other factors were also shown to contribute to measures of economic performance, such an effect of leadership is indeed profound and can hardly be dismissed as insignificant.

The studies reviewed in this section, when viewed collectively, demonstrate unequivocally that leadership can potentially influence significant variables related to organizational effectiveness and individual member satisfaction. However, there have also been longitudinal and experimental studies that show that leader behavior has little or no effect on subordinates' performance (Lowin, Hrapchek, and Kavanagh, 1969), or satisfaction (French, Israel, and As, 1960). Further, there are several studies that show that leader behavior is *caused by* the performance of subordinates (Herold, 1977; Lowin and Craig, 1968; Farris and Lim, 1969; Greene, 1976).

The above findings suggest that leadership has an effect under some conditions and not under others and also that the causal relationships between leader behavior and commonly accepted criteria of organizational performance is two-way.

Thus, the current prevailing paradigm in leadership research is a contingency paradigm. That is, it is now commonly accepted that the most fruitful approach to the study of leadership is a "situational" or contingency approach. According to this view it is necessary to specify the conditions or situational parameters that moderate the relationship between leader behavior and criteria. Further, it has also been found that the traits associated with leadership have differential impact on the behavior and effectiveness of leaders, depending on various aspects of the situation. We now turn to a consideration of these leader traits.

Trait Theory Revisited

Early leadership research is primarily concerned with the identification of traits that discriminate between leaders and nonleaders, effective leaders and noneffective leaders, or leaders at high echelons in organizations as opposed to those at lower echelons. A trait is defined as any distinctive physical or psychological characteristic of the individual to which the individual's behavior can be attributed. Traits are thus inferred from observation of an individual's behavior or from self-reported data provided by the individual in interviews or pencil and paper questionnaires or psychological tests.

Stogdill (1948, 1974) reviewed 70 years of trait research. The review covers approximately 280 published and unpublished studies and review articles. Certain traits have been consistently found to correlate positively with leadership. From his review Stogdill concludes that there is a cluster of personality traits that differentiate *(a)* leaders from followers, *(b)* effective from ineffective leaders, and *(c)* higher-echelon from lower-echelon leaders.

While none of the traits reviewed by Stogdill were found in all studies to be associated with leadership, the consistency with which some traits were found to be associated with leadership and the magnitudes of these associations is impressive. For example, the traits which show the most consistently high correlations with leadership are:

1. Intelligence.
2. Dominance.
3. Self-confidence.
4. Energy, activity.
5. Task-relevant knowledge.

The correlations between leadership and these traits have generally been in the range of .25 to .35. Frequently, the correlations have been much higher. For example, self-confidence, intelligence, and task-relevant knowledge often have correlations with leadership in the range of .40 to .50.

Consider Table 1. The number of times the traits listed in this table were found to be significantly positively associated with leadership prior to 1948 was 346. The number of times their associations were found to be either nonsignificant or negative was 57. Clearly, viewed collectively, these findings are impressive. Stogdill reports a review of 163 trait studies between 1948 and 1974. The number of times these studies revealed positive associations between the traits and leadership are shown in the third column of Table 1. Several abstractors were involved in recording the results of studies between 1948 and 1970. Stogdill states that ". . . it cannot be safely assumed that all negative findings were recorded on the abstracts. For this reason, only positive findings are reported . . ." (p. 73). Since only the positive findings are reported, it is not possible to determine the total number of studies in which a given trait is measured. Consequently, while Stogdill's later survey is suggestive, the reader cannot determine the relative number of positive versus negative findings with respect to a particular trait. For example, twenty-four studies showed that an individual's activity or energy level was related to either a measure of leader effectiveness or to discriminate leaders from followers. However, since the total number of studies concerned with this trait is not reported, one cannot judge the importance of this statistic.

Several of these findings reviewed by Stogdill (1948, 1974) are given further support from studies of emergent leaders. The traits of intelligence

Table 1.   Characteristics of Leaders (Number of Findings)

	1948		1970
	Positive	Zero or Neg.	Positive Only
	1	2	3
Physical Characteristics			
Activity, energy	5		24
Age	10	8	6
Appearance, grooming	13	3	4
Height	9	4	
Weight	7	4	
Social Background			
Education	22	5	14
Social status	15	2	19
Mobility	5		6
Intelligence and Ability			
Intelligence	23	10	25

(Mann, 1959; Bass and Wurster, 1953a, 1953b; Rychlak, 1963); dominance (Dyson, Fleitas, and Scioli, 1972; Mann, 1959; Megargee, Bogart, and Anderson, 1966; Rychlak, 1963), self-esteem (Bass, 1961), task ability (Bass, 1961; Marak, 1964; Palmer, 1962), sociability (Kaess, Witryol, and Nolan, 1961) have all been found to be associated with emergent leadership.

Note that it is only for the traits of age, appearance, height, weight, education, intelligence, ascendance or dominance, and emotional balance that there are sufficient negative or nonsignificant findings to consider disregarding these traits as predictors of leadership.

Yet, several of these latter traits that were nonsignificantly or negatively related to leadership have more recently been found to be associated with leadership under certain well-defined conditions. For example, physical prowess is found to be correlated with leadership under conditions requiring physical abilities such as in boys gangs and groups. Stogdill (1974) reports correlations of .38, .62, and .40 between athletic ability and leadership in three studies of boys' groups.

IQ was reported by Stogdill (1948) to have an insignificant or negative relationship to leadership in 10 of 33 studies. However, five studies show that leaders whose IQ is higher than that of subordinates have a significant advantage, but that extreme discrepancies between the IQ of leaders and followers mitigate against the exercise of leadership. Korman (1968) found intelligence differentiates effective first-line supervisors from ineffective ones, but that at high levels in the hierarchy there are not significant differences in intelligence between effective and ineffective managers. He attributes this inability of intelligence to differentiate among these managers to a restriction in the range of intelligence scores at high levels. Thus consideration of the discrepancy in intelligence between leader and follower and of range restriction helps to reconcile the conflicting findings with respect to this trait.

Leader dominance, a trait that had positive, negative, and nonsignificant associations to leadership in the studies reviewed by Stogdill has been found in the emergent leadership literature (Rohde, 1951) and in experimental studies (Berkowitz and Haythorn, Note 1; Borgatta, 1961) to be rather consistently predictive of leadership. The mixed findings concerning dominance as a trait associated with leadership can be explained by consideration of the measures used and the leadership situations. Several of the findings reviewed by Stogdill are based on measures of the degree to which the leaders were observed as being bossy or domineering. However, when dominance is defined as the leader's predisposition to be ascendant or assertive, as measured by the Dominance scale of the California Personality Inventory, and when the situation calls for one person to assume the role of leadership, this trait is highly predictive of individuals who exhibit behavior that is perceived by other members of the group to be acceptable attempts to influence their behavior. Megargee, Bogart, and Anderson (1966) asked pairs of high- and low-dominance subjects to work together on a manual task requiring one person to verbally communicate instructions to the other. When leadership was emphasized in the experimental instructions to the subjects, the dominant subjects assumed the leadership role in 14 of 16 pairs. When the task was emphasized and leadership was deemphasized, there was no association between dominance and the assumption of the leadership role. Subsequent studies by Megargee (1969) showed that when subjects are paired with members of their own sex, high-dominance subjects are significantly more likely to assume the leadership role. When women who are high in dominance are paired with men who are low in dominance, the women are found to make the decisions as to who should assume the leadership role. However, these women who asserted leadership in the decision-making phase of the experiment requested the male partner to assume the leadership role in the communications of instructions to complete the tasks. It was inferred from these findings that the women preferred the follower role in the instruction phase because of the cultural norm that women should be more submissive than men. Whether such findings would hold today among women with more "liberated" attitudes is yet to be established. However, the findings do suggest that cultural norms are capable of moderating the effects of personality variables on behavior.

Thus, the mixed feelings concerning leadership traits reported by Stogdill (1948, 1974) can be reconciled by consideration of the populations studied, the measures used, or the results of more recent research. This interpretation lead us to conclude that the study of leadership traits should not be abandoned. Not only are the mixed findings reconcilable, but the magnitude of the correlations between leader traits and criteria of leadership are as high and often higher than correlations between leader behavior and leadership criteria. While we agree that traits or personality variables alone account for a small amount of behavioral variance, and that the interaction of personality variables and situational variables is a more promising approach to leadership, we speculate that there are certain properties of *all* leadership situations that are present to a significant degree and relatively invariant, and that there are likely to be somewhat specific traits required in most if not all leadership situations. Following are some speculations about these possible invariant characteristics of the leadership situation.

First, leadership always takes place with respect to others. Therefore, social skills are likely always to be needed if attempted influence acts are to be viewed as acceptable by followers. Such skills as speech fluency and such traits as personal integrity, cooperativeness, and sociability are thus prime candidates for the status of leadership traits.

Second, leadership requires a predisposition to be influential. Therefore, such traits as dominance or ascendance, need for influence (Uleman, 1972), and need for power (McClelland, 1961) are also hypothesized to be associated with leadership.

Third, leadership most frequently takes place with respect to specific task objectives or organizational goals. Consequently, such traits as need for achievement, initiative, tendency to assume personal responsibility for outcomes, desire to excel, energy, and task-relevant ability are also hypothesized to be associated with leadership.

Conclusions: New Directions for Trait Research

The above brief review of leadership trait literature suggests several promising avenues for trait research.

First, it would be worthwhile to classify the studies Stogdill reported in his two reviews (1948, 1974) according to the following topics: *(a)* populations studied: sex, approximate age, other reported demographic variables; *(b)* type of tasks performed: routine vs. nonroutine, intellectual, mechanical, discussion, manual labor, athletic, etc.; *(c)* method used to measure traits: test and questionnaire responses, observation of behavior, analysis of biographical and case studies; *(d)* criterion variables: leaders vs. nonleaders, high- vs. low-echelon leaders, effective vs. ineffective leaders.

While such an endeavor would be very time-consuming and is beyond the scope of the present paper, we believe the payoff for such a secondary analysis would be very high. Based on such a classification one could analyze the findings and likely find some of the following: *(a)* those traits correlated with leadership regardless of how they are measured, *(b)* those traits correlated with leadership for only subjects of a given type of population classification or with certain demographic characteristics, *(c)* those traits correlated with leadership for only a given type task, *(d)* those traits correlated with leadership when leadership is measured by only a given kind of criterion.

To illustrate the potential payoff of such a secondary analysis, consider the differences in findings resulting from studies of children as compared to adults. Almost all of the negative or conflicting findings reported by Stogdill (1948) are based on studies of children. Thus, there is very little discrepency with respect to studies of adult leaders.

A second promising avenue for leadership trait research concerns the development of standardized scales designed *specifically* to identify and predict leadership on the basis of traits. This would require factor and item analysis of a large number of items administered to a large population. Studies by Ghiselli (1971) and Goodstein and Schrader (1963) support this suggestion. Ghiselli (1971) administered a number of personality scales to 336 middle managers performing a variety of jobs in nineteen different firms. Ratings for each manager's performance were provided by one other person, generally the manager's superior, who knew him and his work record well. Managerial responses to the personality inventory were compared with responses from 111 first-level supervisors and 238 nonsupervisory employees. Ghiselli found that the traits of intelligence, supervisory ability, initiative, self-assurance, and individuality were significantly related to the manager's organizational level and ratings of their performance. These traits differentiated between middle managers on the one hand, and first-level supervisors and nonsupervisory employees on the other. Secondly, successful managers possessed the traits to a greater degree than did less successful managers. Finally, the relationship between the trait and success was higher for managers than it was for supervisors and employees. The correlation between managerial success and the traits of supervisory ability and intelligence were .42 and .27, respectively. The correlations between initiative, self-assurance, and individuality were in the .20 range. While Ghiselli did not compute a multiple correlation of the traits and supervisory success, these correlations suggest that such a multiple correlation coefficient would be in the magnitude of .5 to .6. Clearly, if this is indeed the case, such a correlation is higher than most behavioral predictors of leadership criteria. In fact, the correlation between supervisory ability and managerial success ($r = .46$) is as high as one generally finds between behavioral variables and leadership criteria.

A study by Goodstein and Schrader (1963) is also revealing. Chi-square comparisons of the responses of 603 managers and supervisors with those of 1748 "men-in-general" indicated that 206 of the 480 California Psychological Inventory (CPI) items reliably differentiated the two groups ($p < .01$). Protocols of the respondents were then scored using the twenty items as a managerial key. This key not only reliably differentiated the total managerial group from the "men-in-general" group, but also differentiated personnel at three different levels of management: top management, middle management, and first-line supervision (all p's $< .01$). This scale also significantly correlated ($r = .23$) with ratings of success within the total management group and within the top and middle management subgroups (r's $= .25$ and $.27$, respectively).

These studies by Ghiselli (1971) and Goodstein and Schrader (1963) show significant promise and suggest that leadership traits might account for a significant proportion of unique variance in leadership.

Third, it would be useful to identify the unique interactions, or combinations of traits, that are most predictive of leadership. For example, it is very likely that dimensions of leader competence (such as intelligence, speech fluency, knowledge of task) interact with measures for one's tendency to attempt leadership behavior such as dominance or need for influence. One would expect measures of a person's tendency to attempt leadership behaviors to be positively correlated with leadership when the person is competent, but that such correlations would be nonsignificant or even negative when the person is not competent.

Fourth, upon identification of a standard set of leadership traits, one could begin to determine the behaviors correlated with the scales. Stogdill's findings indicate that several of the scales used in prior research are positively correlated with the criteria of leadership. However, little is known about the behavioral correlates of these traits. A knowledge of such correlates is necessary if we are ultimately to go beyond statistical associations between traits and leadership and to understand the processes involved.

LEADERSHIP BEHAVIOR

Having reviewed the evidence and advanced some research directions relevant to leadership trait research, we now turn to a consideration of leadership behaviors. By leadership behaviors we mean those behaviors of the group member or of the formally appointed leader that are perceived by subordinates as acceptable attempts to influence their behavior. Here we review in some detail prior experimental findings concerned with task-oriented and socioemotional-oriented leadership and field study findings associated with leader Initiating Structure and Consideration. In addition, both the experimental and field study research concerned with participative decision making is reviewed. The research relevant to these leader behaviors are reviewed in some depth because they have been most frequently found to be the major kinds of behaviors in which leaders engage (Yukl, 1971).

Of interest are the following questions:

1. Is there a well-defined set of behaviors that falls within the above definition of leadership behaviors?

2. What are the effects of such behaviors on others? What are the effects of such behaviors on group performance and organizational effectiveness?

3. Under what conditions do such behaviors constitute a contribution to individual or group performance and individual well-being?

To answer the above questions and suggest further research directions, we will review three independent bodies of literature. The first, concerned with role differentiation of group member behavior, comes primarily from the sociological literature. The second, concerned with subordinates' perceptions of formally appointed leaders, comes primarily from industrial psychology. The third, concerned with emergent leadership, comes primarily from social psychology. As will be shown, the findings resulting from these independent literatures are complementary and in many cases mutually reinforcing.

Role Differentiation in Groups

This section reviews early research concerned with the emergence of leadership roles under conditions where the group has no formally appointed leader. This research was primarily concerned with the following questions:

1. What are the important problems that a group must solve to be effective, cohesive, and to have satisfied members.

2. What are the behaviors of group members who are most likely to be attributed leadership status by others?

3. To what extent are those behaviors that are required to solve group problems divided into specialized roles, or kinds of behaviors, such that different members perform different roles?

4. Under what conditions can these different roles be performed by one member and when will they be divided among two or more members?

5. Will groups be more effective when their roles are divided or integrated?

Leader Role Differentiation

Early research by Bales and his associates (Borgatta, Bales, and Couch, 1954; Bales and Slater, 1955; Bales, 1958) clearly demonstrated that there are generally two functions to be accomplished by the small experimental discussion groups they observed: *(a)* the achievement of some specific group task, and *(b)* the maintenance or strengthening of the social relations group itself. Thus, it was suggested that leadership in a group can be described in terms of an individual's contributions to the accomplishment of these functions.

Group members' behaviors, while primarily addressed to group achievement or group maintenance functions, tend to be factor analytically divided among three distinct dimensions. These dimensions were identified from a number of studies in which members of small groups were asked to rate or choose each other on a wide variety of descriptive criteria, or in which members were assessed by observers.

Carter (1954) reviewed a series of these factor-analytic studies and described the most frequently occurring three factors as follows:

Factor 1: Individual Prominence and Achievement: Behaviors of the individual related to his or her efforts to stand out from others and individually achieve various personal goals.

Factor 2: Aiding Attainment by the Group: Behaviors of the individual intended to assist the group in achieving its goals.

Factor 3: Sociability: Behaviors of the individual related to his or her efforts to establish and maintain cordial and socially satisfying relations with other group members.

Bales (1958) refers to the three factors as "activity," "task ability," and "likeability." He concludes that ratings on these dimensions should be treated as three distinct factors, since over a large population of members, meetings, and groups, they tend to be uncorrelated with each other. Further, he concludes that a member who is high on all three of the factors corresponds to the traditional conception of a good leader, or the "great man." Bales refers to individuals who are high on activity and task-ability ratings but less high on likeability ratings as the "task specialist." Task specialists contribute primarily to leadership of the task function of the group. A member who is high in likeability but less high in activity and task-ability is referred to as the "social specialist." Social specialists contribute to leadership of the group maintenance function.

Thus, it can be concluded from the above studies that the behavior of group members can be measured along three independent dimensions. Two of these dimensions, activity and task ability, combine to influence task achievement while likeability or sociability contribute to group maintenance.

An early study by Carter, Haythorn, Shriver, and Lanzetta (1951) sheds light on the conditions under which individual prominence behavior is required. Carter et al. observed behavior of group members as they performed three different kinds of tasks: a reasoning task, a mechanical assembly task, and a discussion task. For approximately half of the groups a leader was appointed by the experimenter. For the other half no leader was appointed but individuals receiving the highest leadership ratings from group observers were considered "emergent leaders." Comparisons between the emergent leaders and appointed leaders showed that the emergent leader engages more frequently in the following behaviors: "supports or gives information regarding his proposal," "defends himself (or own proposal) from attack," "provides expression of opinion," and "argues with others." Appointed leaders were significantly lower on the above behaviors. While the appointed leaders engaged in significantly less such behavior, they were equally as effective as the emergent leaders.

These findings suggest that under conditions where there is no institutional factor that legitimatizes leader behavior such as formal appointment or title, an individual will have to engage in what Carter et al. (1951) describe as individual prominence and achievement behaviors and what Bales refers to as "activity." However, when leadership is legitimatized through the appointment process, such behavior is not as likely to occur. Thus, in terms of the leadership construct advanced in this paper, it appears that a necessary but not sufficient condition for the attribution of leadership is that the individual distinguish himself or herself from the group by engaging in behaviors intended to establish individual prominence. However, under conditions of appointed leadership, such behaviors are not required and may even be dysfunctional.

Surprisingly, this interpretation has not been tested in field research where appointed leaders have been studied. However, a study by Wofford (1970) is relevant to this interpretation. Wofford factor analyzed responses to a questionnaire consisting of 219 items. Respondents were asked to describe their immediate supervisor's behavior. The analysis yielded four factors: two concerned with task orientation, one concerned with group maintenance, and one concerned with personal enhancement. All factors except the personal-enhancement factors correlated positively with indices of morale and perceived organizational effectiveness.

The personal-enhancement factor correlated negatively $-.51$ and $-.32$ with perceived effectiveness and morale, respectively, thus suggesting support for the hypothesis that personal prominence seeking behavior of appointed leaders is likely to be negatively associated with the attribution of leadership as defined in this paper. This hypothesis is suggested for future field study research.

Are both task-oriented and socioemotional leadership always required for effective group performance? If not, when are they required and when are they not?

The answer to the above questions was found to depend upon whether the group is committed to the task or finds the task to be intrinsically satisfying (Gustafson, 1968; Gustafson and Harrell, 1970; Burke, 1967). Under conditions where the tasks are intrinsically satisfying to group members or members are committed to task accomplishment, task-oriented leadership is viewed as instrumental to group success, but there is less need for socioemotional leadership. Verba (1961) has argued that where tasks are not interesting or are not important to members such leadership is necessary to provide some form of social satisfaction. Task-oriented leadership under such conditions is likely to be resented. Therefore, socioemotionally oriented leadership is required to offset this resentment. A study by Bales (1958) is especially relevant to this issue. Task-oriented leaders tended to be disliked. This finding held most strongly among groups that were not cohesive. However, for the task-oriented leader who made it possible for members to give feedback, raise objections, qualifications, questions, and counter questions, there was no relationship between task orientation and liking. For those task-oriented leaders who did not permit such feedback, the relationship between task orientation and liking was negative.

The question of whether these two roles can be performed by one member or whether they must be divided by two or more members is also of importance. Borg (1957) found that in some groups the roles were integrated and performed by a single member while in other groups the roles were divided. Teams with two leaders made significantly lower effectiveness scores than teams with one leader. When the two leaders were mutually supportive, their groups performed more effectively than when the leaders were in competition with each other, but still not as effectively as groups with leaders who integrated both roles. Thus, it appears that groups with leaders who integrate the two roles will have what Bales and Slater (1955) refer to as "great men" as leaders. They found that when members of a group designated particular individuals as "leader," ". . . the individual is perhaps found to possess those qualities that best serve to solve both the task and social-emotional problems of the group" (p. 291.). ". . . Leadership . . . [is] attributed to that member . . . who best symbolizes the weighted combination and integration of the two more specialized functions [of task orientation and group maintenance]" (p. 298). Further support for this interpretation was found by Borgatta, Couch, and Bales (1954). These authors selected eleven "great men" out of 126 who scored high on three factors: task ability (leadership ratings received on a prior task and IQ score), individual assertiveness (activity rate received on a prior task), and social acceptability (sociometric choice on a prior task). These "great men" were each assigned to four tasks. Two new coparticipants participated in each of the four tasks. Groups led by "great men" were compared with groups led by men who were not high on all three of the above dimensions. Groups led by "great men" had higher rates of giving suggestions and agreements, lower tension, higher positive social and emotional behavior. The authors concluded ". . . thus, it may be said that great men tend to make 'great groups' in the sense that both major factors of group performance—productivity and satisfaction of the members—are increased" (p. 759). There is also evidence that when formally appointed group leaders fail to perform task-oriented behaviors, an informal leader will emerge and perform the task-oriented behaviors required for group success (Crockett, 1955; Berkowitz, 1953).

From the above findings the following empirical generalizations can be drawn:

1. Task-oriented leadership is necessary for effective performance in all working groups.

2. Acceptance of task-oriented leadership requires that the task-oriented leader allows others to respond by giving feedback, making objections, and questioning the task-oriented leader.

3. Socioemotionally oriented leadership is required in addition to task-oriented leadership when groups are not engaged in satisfying or ego-involving tasks.

4. Groups requiring both kinds of leadership behavior will be more effective when these leader behaviors are performed by one person rather than divided among two or more persons.

5. When the leadership roles are differentiated, groups will be most effective if those assuming the roles are mutually supportive and least effective when they are in conflict with each other.

6. When formally appointed leaders fail to perform the leader behaviors required for group success, an informal leader will emerge and will perform the necessary leader behaviors, provided success is desired by the group members.

These findings from the sociological literature concerned with leadership in small groups help us understand some of the conflicting findings resulting from field research. We now turn to a review of these studies.

FIELD STUDIES OF LEADER ROLE DIFFERENTIATION: CONSIDERATION AND INITIATING STRUCTURE

Studies conducted by the Leadership Group at Ohio State University suggest two dimensions of leader behavior that can be interpreted as task-oriented leadership and socioemotionally oriented leadership. These are leader Initiation of Structure and leader Consideration. Leader Initiation of Structure has been measured by three different scales. These scales are intended to measure the degree to which the leader clarifies and defines his or her own role and lets followers know what is expected of them. The Consideration scale is designed to measure the degree to which the leader pays regard to the comfort, well-being, status, and satisfaction of the followers.

These scales have been widely used for purposes of leadership research. Over fifty studies have been reported that assess the relationship between the leader's score on these two scales and subordinates' satisfaction, expectations, performance, turnover, and grievances. The findings resulting from this research are very mixed. All of the scales have been shown to be positively, negatively, or nonsignificantly related to such dependent variables. Much of the confusion with respect to correlates of the initiating structure construct has been traced to the particular scales used to measure this dimension. Initiating structure has most often been measured by one of the following instruments:

1. The Supervisory Behavior Description Questionnaire (SBDQ, Fleishman, Note 1), consisting of twenty items that inquire of subordinates about their leader's actual structuring behavior. Structure as measured by the SBDQ is intended to reflect the extent to which the leader organizes and defines relationships between himself and his group, defines interactions among group members, establishes ways to get the job done, schedules, criticizes, etc. (Fleishman, Note 2, p. 1).

2. The early Leader Behavior Description Questionnaire (LSDQ, Halpin, Note 3), containing fifteen items that ask subordinates to describe the actual structuring behavior of their leader. As measured by this instrument, Structure refers to the leader's behavior in delineating relationships between himself or herself and group members and in trying to establish well-defined patterns of communications and ways to get the job done (p. 1).

3. The revised LBDQ (Stogdill, Note 4), with ten items measuring Structure. As measured by this instrument, Structure is concerned with the actions of the leader in clearly defining his or her own role and letting followers know what is expected (p. 3).

As pointed out by Schriesheim, House, and Kerr (1976), the items comprising the three scales differ in content. The LBDQ forms consist largely of items describing a leader who actively communicates with subordinates, facilitates information exchange, and designs and structures his own work, the work of group members, and relationships among group members in their performance of work. In contrast, the SBDQ consists mainly of items describing a highly production-oriented leader who is autocratic and punitive.

Schriesheim, House, and Kerr (1976) reviewed the studies concerned with Initiating Structure and considered the results obtained from each version of this scale separately. Notice was also taken of the task and environment context in which respondents worked. As a result, apparent discrepancies and findings were reduced and three empirical generalizations were derived.

The findings from the small group research reviewed above helps to explain these empirical generalizations. Schriesheim et al. (1976) concluded that, when measured by the SBDQ, leader Initiating Structure is generally positively related to performance ratings by superiors of manufacturing first-level supervisors and to ratings of the work group performance. However, it is negatively related to satisfaction of the supervisors' subordinates. This generalization also held with regard to non-commissioned infantry officers and air force officers. A similar, although much weaker, pattern of relationships was found concerning non-manufacturing supervisors of clerical workers doing routine tasks.

Since the SBDQ scale reflects task-oriented leadership, it is not surprising that leaders high on this scale obtain higher performance ratings by superiors. However, since the items on the scale suggest that leaders who are high on it are arbitrary and punitive, it is reasonable to assume such leaders do not permit subordinates opportunities to react to the leader's Initiation of Structure. Recall that Bales (1958) found that individuals who are task oriented but do not permit others to react to them tend to be disliked by others. Thus, in the light of the above findings by Bales, it is not surprising that such leaders cause subordinates to be dissatisfied.

The differential strength of this pattern of relationships between manufacturing employees and clerical employees is also consistent with the findings from small-group research. Recall that socioemotional leadership was required to offset the negative effects of task-oriented leadership in the small group experiments when subordinates found the task intrinsically dissatisfying. Since it is likely that persons engaged in manufacturing tasks find less satisfaction with these tasks, there will be more resentment of leader Initiating Structure than among persons engaged in clerical tasks, which are likely not as dissatisfying.

Schriesheim et al. (1976) also concluded that when the revised LBDQ

Initiating Structure scale is used to measure leader behavior of first-level supervisors of nonmanufacturing employees performing routine tasks, correlations with subordinate satisfaction are positive although generally so low as to be, at best, only marginally significant. The revised LBDQ Initiating Structure scale does not include the autocratic and punitive items of the SBDQ. Thus, it is less likely that supervisors high on this scale will prevent subordinates from reacting to task-oriented behaviors. Recall that for task-oriented leaders who permit such reactions there was no relationship between task orientation and liking. Thus, the consistently low correlations between Initiating Structure and Satisfaction, as measured by the revised scale, are consistent with the above findings from small group research.

Finally, Schriesheim et al. (1976) found that high occupational level employees consistently react more favorably to leader Initiating Structure regardless of the instrument used. This generalization is consistent with the finding that task-oriented leadership is viewed as acceptable when the task is satisfying to subordinates.

Thus, the findings from the field studies using the Initiating Structure scales are consistent with those of small group research concerned with task-oriented leadership.

The mixed findings with respect to correlates of leader Consideration are also readily interpretable in the light of the small group research findings. House (1971) hypothesized that leader Consideration will have its most positive effect on satisfaction of subordinates who work on stressful, frustrating, or dissatisfying tasks. This hypothesis has been tested in ten samples of employees (House, 1971; House and Dessler, 1974; Szilagyi and Sims, 1974; Stinson and Johnson, 1974; Schuler, 1973; Downey et al., 1975; Weed, Mitchell, and Smyzer, Note 5). In only one of these studies was the hypothesis disconfirmed (Szilagyi and Sims, 1974). In addition, there are experimental studies that show that the effects of considerate leadership on performance are most positive when subordinates have previously been denied some source of satisfaction. For example, in experiments by Day and Hamblin (1964) and DeCharms and Bridgeman (1961) in which leader supportiveness resulted in increased productivity, subjects were first exposed to threatening, irritating, or frustrating treatment, and then to considerate and helpful treatment. In both studies subjects responded favorably to the considerate leader behavior.

These findings are consistent with the results of small group research, which demonstrated the need for socioemotional leadership under conditions where subordinates are neither committed to the task nor find it to be intrinsically satisfying.

Further, while leader Consideration is almost always positively associated with subordinates' satisfaction on dissatisfying tasks, it is not always associated with subordinate performance. Recall that small group research demonstrated that task-oriented leadership is required in small groups. It can be speculated that the failure of leader Consideration to be associated with subordinates' performance under conditions of dissatisfying tasks may be due to the absence of concurrent leader task-oriented behavior. This speculation, if found supported in future research, would serve to reconcile the conflicting findings with respect to performance correlates of leader Consideration.

Participative Leader Behavior

Participative leadership has been the source of significant concern and controversy. Participative leadership takes two forms: participative decision making (PDM) and participative supervision (PS).

PDM refers to efforts by leaders to ensure that all parties for whom a decision is relevant have an opportunity to influence the final decision. A decision is considered relevant to an individual if the individual is ego involved in the outcome of the decision, if the individual possesses significant information that pertains to the decision, or if the individual must be relied on to implement the decision once it has been made. Whereas PDM concerns specific decisions, PS concerns the manner of interaction between superior and subordinate or leader and follower on a continuing, day-to-day basis. When PS is practiced, the leader encourages subordinates to make suggestions concerning what work should be done and how it should be carried out. Further the leader encourages subordinates to engage in independent thinking and action with respect to such factors as problem analysis, selection of means, and planning and scheduling the work process.

While the distinction between PDM and PS is conceptually clear, the studies we review in this section do not distinguish between these two forms of participative leadership. However, it appears that the laboratory studies generally operationalize participative leadership by varying the degrees of PDM as an independent variable. In contrast, field studies generally use questionnaire responses of subordinates. These responses appear to represent the degree to which their formally appointed leader is perceived to engage in both PDM and pS, and PDM and PS are treated as a global measure of participative leadership. As will be seen, the results of both studies are quite consistent, regardless of how participative leadership is operationalized.

Contrary to some conceptions of participative leadership, participative leaders do *not* abdicate the leadership role by becoming a member of the group, *except* insofar as they contribute substantive (as opposed to process) guidance. Maier's research (Maier, 1970) on effective group problem solving serves as the basic paradigm for participative decision making. His work demonstrates that effective participative leaders exert substantial control over the interaction process among subordinates during the decision-making process. The specific behaviors in which effective participative leaders engage are:

- Share information with the participants.
- Prevent dominant personalities from having disproportionate influence.
- Solicit opinions, facts, and feelings from reticent participants.
- Assist participants in communicating with one another.
- Protect deviant opinions from being rejected prior to fair evaluation.
- Minimize blame-oriented statements.
- Redirect unfocused discussion back to the problem at hand.
- Encourage the generation of alternative solutions.
- Delay evaluation of alternatives until all have been presented.
- Guide the process of screening alternatives and selecting the solution.

Maier's work demonstrates that the above skills required to be an effective participative leader are trainable (Maier, 1949, 1963).

Mitchell (1973) recently described at least four ways in which participative leadership style can have an impact upon subordinate attitudes and performance. First, a participative climate should lead to greater clarity of the paths to various goals. Second, it enables subordinates to select goals they value, thus increasing commitment to goal attainment. Third, participants can increase their control over what happens on the job. If subordinate motivation is higher (as a result of points one and two), then having greater autonomy and the ability to carry out their intentions should lead to increased effort and performance. Finally, when people participate in the decision process, the decisions are made in the presence of others. Thus these others know what is expected, causing social pressures to have a greater impact. Thus motivation to perform will stem from internal and social factors as well as from formal external ones.

Maier (1970) has argued that participation should improve decision making, because it is through the participative process that subordinates' knowledge and expertise can be brought to bear. That is, participative decision making is an effective means of obtaining relevant information or expertise from subordinates, and thus of improving the objective quality of decisions. Further, he argues that when decisions require subordinate acceptance for their implementation, participation will increase such acceptance, because subordinates will have an opportunity to influence the decision-making process, and consequently their feelings are more likely to be expressed and respected. Thus, according to these theoretical perspectives, participation is predicted to increase productivity, quality, emotional orientation of subordinates toward their work setting (the job, the decision makers, and the organization), and subordinates' acceptance of decisions.

Filley, House, and Kerr (1976) reviewed thirty-three studies concerned with the effects of participative leadership. Nineteen of twenty laboratory experiments, correlational studies, or field experiments in which satisfaction of subordinates was measured demonstrated a positive relationship between participation and satisfaction. The single study not showing such a relationship had an unclear and uninterpretable result. Seventeen of the twenty-two studies in which productivity of subordinates was measured showed a positive relationship between participation on productivity and seven showed no relationship. In addition, over thirty laboratory studies concerned with the degree with which participative leadership results in effective decisions have been conducted by Maier (1963; 1970) and his associates. Maier defines decision effectiveness as a function of the degree to which *(a)* the decision meets the objective economic and physical requirements of the problem (the quality criterion) and *(b)* the decision is acceptable to subordinates (the acceptance criterion). Maier argues that some problems demand high-quality solutions, some demand high-acceptance solutions, and some both.

Maier and his associates required subjects to role play the part of a person who was in conflict with other group members. In these roles each person possessed critical information about the problem that no other member possessed. For the problem to be solved it was necessary that the information be shared and accurately evaluated. As might be expected under these conditions, participation resulted in higher decision quality and acceptance in *all* of the studies.

Filley, House and Kerr (1976) concluded that when subordinates task demands are clear and routine to the subordinate, participative leadership is not likely to have an effect, because there is little to participate about. These authors segregated the studies on participative leadership for which the task characteristics were controlled. These studies are reported in Table 2. From this table it can be seen that there are rather dramatic differences in the effects of participation, depending largely on the nature of the task performed. Simple tasks with nonambiguous demands do not lend themselves to participative leadership, whereas tasks that are more complex and ambiguous do. House and Mitchell (1974) have hypothesized that on such tasks subordinates are more ego involved in their task and therefore have a desire to influence decisions that affect the work they do and the manner in which they are required to do it. We recommend further tests of this hypothesis in future research.

In addition to being ego involved, subordinates performing such tasks are likely to have been assigned such tasks because of their intelligence

Table 2. Participative Leadership: Studies Controlling for Task Characteristics

Investigators	Type of Study	Type of Task	Performance[a]	Attitude[b]
Argyle, Gardner, and Coiffi (1957)	Field Correlational	Machine paced, paid by piece rate-Males	No relationship	No relationship
		Man paced, salary payment-Males	Positive	Positive
Vroom and Mann (1960)	Field Correlational	Independent-non-cooperative tasks		Negative
		Interdependent-cooperative tasks		Positive
Shaw and Blum (1966)	Lab Experiment	Low structure	Positive	
		Medium structure	Positive	
		High structure	Negative	
Phillipsen (reported in Lammers, 1967)	Field Correlational	Low mechanization	Low positive	Low positive
		High mechanization	Higher positive	Higher positive
Delbecq (1965)	Lab Experiment	Routine	Negative	
		Nonroutine	Positive	

[a] Performance based on objective indices of costs or productivity.
[b] Attitudes measured by questionnaire responses.
Source: Filley, House and Kiot, (1976).

366

and/or because they have some specialized knowledge relevant to task performances. Thus, it can be hypothesized that subordinates' intelligence or task-relevant knowledge will also moderate the relationship between participative leadership and its effects.

Filley et al. (1976) reviewed the empirical evidence concerned with the moderating effect of subordinate intelligence and knowledge level. Their findings are presented in Table 3. These studies show clearly that when subordinates' knowledge level or intelligence is high, participation has a positive effect. However, where subordinates' intelligence level or knowledge is low, participation generally has an insignificant effect on subordinates' performance and sometimes has a negative effect on their satisfaction.

Viewed collectively, the results of the studies reported in the prior two tables clearly indicate that knowledgeable subordinates or intelligent subordinates working on tasks that impose ambiguous, nonroutine demands perform more effectively under conditions of participative leadership. However, it is not clear whether subordinates performing such tasks respond more positively to participative leadership because they are more ego involved in their work or because they have higher competence (task knowledge and/or intelligence) or both. We recommend further research to clarify this issue.

A number of studies have shown that the effects of participation are also moderated by the predisposition of subordinates to participate or to gain satisfaction from the participative process. Subordinates who have high needs for independence, are nonauthoritarian, and have respect for nonauthoritarian behavior have been hypothesized to be more satisfied and to be more effective under conditions of participative leadership. The evidence supporting this argument is impressive (Delbecq, 1965; Vroom, 1959; Campion, 1968; Tannenbaum and Allport, 1956; Jacobson, 1953).

However, House (Note 6) argued that the moderating influence of subordinate personality upon relationships between participative leadership, satisfaction, motivation, and performance should not be expected to be strong when tasks are ego-involving. House (Note 6) hypothesized that the subordinates' personality will become an important moderator only when the task is not ego-involving. His reasoning is that subordinates performing non-ego-involving tasks who are not predisposed toward participative leadership will not find participation to be either intrinsically satisfying or instrumental to task success, whereas subordinates predisposed toward participation will find the task more satisfying and motivating, due to the opportunities to participate. However, when the task is ego-involving, the subordinates will have a desire to influence decisions *regardless* of whether they are predisposed by personality to participate

Table 3. Studies Testing the Hypothesis That Subject's Intelligence and Level of Knowledge Moderates the Effects of Participation

Investigators	Type of Task	Method	Individual Difference Controlled	Effects Without Controlling for Subject's Characteristics	Effects When Controlling for Subject's Characteristics	Direction of Effect of Controlling for Subject's Characteristics
Calvin, Hoffman, and Hardin (1957)	Complex Decision	Lab Experiment	Subordinate Intelligence		Positive[a]	As Predicted
Mulder and Wilke (1970)	Influence over Other in Complex Problem Solving	Lab Experiment	Leader Knowledge	Not Measured	Positive[b]	As Predicted
Cammalleri et al. (Note 5)	Complex Problem Solving	Lab Experiment	Leader Knowledge	Not Measured	Positive[a]	As Predicted
Kolaja (1965)*	Participative Yugoslav Workers' Council	Field Observation	Desire to Participate and Expertise	Not Measured	Positive	As Predicted
Brockmeyer (1968)*	Participative Yugoslav Workers' Council	Field Observation	Desire to Participate and Expertise	Not Measured	Positive	As Predicted

*Cited in Mulder and Wilke, measurement of effectiveness of participation on workers' committees.
[a] Performance measured in terms of laboratory task completion or accuracy.
[b] Performance measured in terms of amount of attitude change of subjects induced by confederate.
Source: Filley, House and Kerr (1976)

or not. To date, one major investigation (Schuler, 1976) has tested this prediction. Subjects were 354 employees in an industrial manufacturing organization. Personality variables, the amount of participative leadership, task characteristics, and job satisfaction were assessed. As predicted, in nonrepetitive, ego-involving tasks, employees (regardless of their personality) were more satisfied under participative leaders than nonparticipative leaders. In repetitive tasks that were less ego-involving, however, the amount of authoritarianism of subordinates moderated the relationship between leadership style and satisfaction, as hypothesized. Specifically, low authoritarian subordinates were more satisfied under participative leaders than under nonparticipative leaders. This study, together with the theoretical hypothesis advanced by House, appears to explain why findings by Tosi (1970) and Wexley, Singh, and Yukl (1973) did not find a moderating effect of subordinate personality on the relationship between participative leadership and subordinate satisfaction and performance. In both studies the subjects were performing rather routine simple tasks that are not likely to be ego-involving.

Unfortunately, Schuler's study only dealt with subordinate satisfaction as an outcome. Future research directed at replicating Schuler's study and incorporating measures of subordinate motivation and performance is required before the hypothesis can be claimed to be strongly supported.

In summary, research on participation suggests a rather parsimonious set of empirical generalizations that explain and predict the conditions under which participation leads to increased decision effectiveness and increased satisfaction, motivation, and performance of subordinates.

The above research strongly supports the hypothesis that such leadership will be most effective under conditions where tasks are ego-involving, ambiguous, and nonroutine. Further, participative leadership is most effective when subordinates have sufficient competence to contribute to the participative process. Specifically, subordinates' level of intelligence and knowledge about the issue at hand will determine whether the process can result in improved decision making. When task demands are not ego-involving, subordinates' predisposition to participate is hypothesized to moderate the degree to which the participative process will be satisfying to subordinates. On such tasks only for subordinates with a positive disposition toward participation is participation predicted to enhance satisfaction. A similar interaction is hypothesized to occur with respect to subordinate performance, but additional research is needed to test this hypothesis.

In addition to the research on the three dimensions of leader behavior reviewed above (leader initiating structure, consideration, and participation), there is an emerging literature concerned with the degree to which the leader administers rewards and punishment, contingent on subordinate performance. This literature will be reviewed in a later section, along with the discussion on Operant Leadership Theory.

DETERMINANTS OF LEADER BEHAVIOR

There are several studies that show that leader behavior is determined by several individual characteristics in interaction with variables in the environment. Specific measures of individual predispositions to assume the leadership role have been shown to predict emergence of leader behavior, that is, influence attempts that are accepted by group members in leaderless groups. These measures are the Dominance scale of the California Personality Inventory (Megargee et al., 1966); the Need for Influence scale (Uleman, 1972); and the Guilford Zimmerman Ascendance scale (Guetzkow, 1968). In addition, the Leadership scale of the California Personality Inventory developed by Goodstein and Schrader (1963) has been shown to discriminate leaders from others, to correlate positively with level in the organization and not to suffer from range restriction at higher levels.

The essential ingredient that appears to be measured in the Dominance, Ascendance, and Need for Influence scales appears to be the individual's desire or willingness to assert control over others in pursuit of task accomplishment.

Studies by Megargee et al. (1966), Zdep (1966), and Zdep and Oakes (1967) illustrate how this predisposition on the part of individuals interacts with cues in the environment to predict leader behavior. Megargee et al. (1966) found that the dominance scale of the CPI did not predict the emergence of leader behavior under conditions where the need for leadership was deemphasized. However, when the need for leadership was emphasized, fourteen of sixteen high-dominance subjects in pairs of high- and low-dominance subjects assumed the leadership role.

Zdep (1969) found that in an experimental situation high-dominance subjects increased their rate of participation in response to reinforcement administered privately by the experimenter. As participation increased, so did group leadership ratings. Low-dominance subjects exhibited such little participation that Zdep found it impossible to administer the necessary reinforcements to increase their rate of participation.

Zdep and Oakes (1967) administered the CPI leadership scale in a repeated measures design to determine whether the questionnaire has a reactive effect, thereby influencing the behavior of the people who complete the questionnaire. While it was not reactive in the usual sense, it appeared to have a differential effect. Only the more ascendant participants increased thier rate of participation after having been exposed to the questionnaire. Thus, the questionnaire likely served as a stimulus to leadership for high-ascendance subjects in the same manner that the experimental instructions by Megargee et al. (1966) served to elicit leader behavior by high-dominance subjects only.

These studies by Megargee et al. (1966), Zdep (1969) and Zdep and Oakes (1967) suggest that a high need for dominance or ascendance in interaction with the environmental cues or reinforcements results in increased acts of leadership.

A study by Guetzkow (1968) shows that individuals who are high on the Guilford Zimmerman Ascendance scale were more likely to assume the leadership role in small group network studies. Individuals who assumed the leadership role in this study established themselves by having more adequate perceptions of the organizational situation than others did and by nominating themselves as leaders. Guetzkow (1968) found that such persons did not force their way into the leadership role even though they were in a position to do so by withholding strategic information from other members of the group.

Smelser (1961) found that when groups composed of pairs in which the dominant subject was assigned a dominant role and a submissive subject a submissive role groups were most productive. The least productive groups were those in which the role assignments were reversed.

Stogdill (1974) describes a study by Rohde (1951) that sheds further light on the behavior of persons high on dominance or ascendance. In five member groups, dominant members were chosen as leaders more often than submissive members and exhibited significantly more controlling behavior. In addition, they agreed and cooperated more often than submissive members.

These studies suggest that individuals with a high need for dominance or ascendance, or individuals who are high on the CPI leadership scale, are not only more likely to take the initiative in seeking a leadership role but are *not* more likely to engage in autocratic or domineering behavior. Rather, these studies suggest that such persons are likely to be helpful to others and to be instrumental in the pursuit of group goals.

Other individual characteristics associated with effective and emergent leadership were reviewed in the earlier section concerning leadership traits. The traits of self-confidence and strength of conviction likely free individuals of inhibitions to assert leadership. In addition, the trait of achievement drive, or desire to excel, likely operates as a motivator to assert leadership in task situations.

In addition to the above characteristics of individuals, it has also been shown that leader behavior is determined by factors in the environment. Crowe, Bochner, and Clark (1972) administered a leadership belief scale to 400 lower- and middle-level managers. Based on their responses, these managers were classified as either autocratic or democratic. The managers then participated in an experimental simulation in which they managed confederate subordinates who behaved either autocratically or democratically. The results showed that subordinates' influence were strong enough to bring about the response from the manager that is opposite to their own preference. Both types of managers behaved democratically with democratic subordinates and autocratically with autocratic subordinates.

There is also evidence that leaders model the behavior of their superiors and adapt to their superiors' expectations. Fleishman, Harris, and Burtt (1955) found that cahnges in supervisory behavior resulting from a training program occurred when the supervisors reported to a leader who engaged in the kind of behavior stressed in the program but did not occur when the supervisor reported to the leader who did not engage in such behavior. Weiss (1977) found that managers who are low in self-esteem described their behavior as similar to behavior of their superiors, whereas managers who are high in self-esteem show no such similarity.

Pfeffer and Salancik (1975) asked supervisors to indicate the performance expectations that their superiors and subordinates held for them. Subsequently, the same supervisors completed a questionnaire indicating how they allocated their time. The findings indicated that whether the supervisor attends more to the expectations of his superior or to those of his subordinate is a function of the following factors: *(a)* the demands to produce coming from the superior, *(b)* percent of time the superior actually engages in supervision rather than in routine tasks, *(c)* the number of persons supervised, *(d)* whether the sex of the supervisor is the same as that of the superior, and *(e)* whether task decisions are made primarily by the supervisor. Pfeffer and Salancik (1975) interpret these findings in terms of role theory. They argue that:

> organizations are composed of interdependent positions and interlocking behaviors. Occupants of these positions are exposed to expectations and social pressures of other organizational members with whom they are interdependent. With experience, the expectations and demands become known, resulting in a collective structure of behavior, and stabilize to predictable patterns. In any given position, the occupant's behavior is influenced and constrained by the social pressures immediately from other persons in the role set (p. 141).

Salancik and Pfeffer (1977) also analyzed the variance in budgets for thirty U.S. cities over a 17-year period to determine how much of that variance is attributable to the budget year, the characteristics of the city, or the behavior of the mayor. The amount of variance explained by the city, year, and mayor was 59.1 percent, 2.5 percent, and 19.5 percent, respectively. The mayor, while having little discretion, was found to have the greatest effect over expenditures that were not as involved with political interests and had most discretion over capital outlays, which were one-time allocation decisions. Thus these findings indicate that the impact of the leader (mayor) is significantly constrained by cultural and historical factors associated with the particular organizational (city) setting. While this study did not include measure of variance in the behavior of the mayors, it is likely that such variance is also significantly determined by these same situational factors.

Salancik, Calder, Rowland, Leblebici, and Conway (1975) provide further evidence to support this position. They found that the amount of influence of power or political leadership of peers in an organization derives from their positions in the social structure. The perceived influence or leadership of members was found to be a linear combination of their position on measurements specifically derived to reflect the social structure of the organizations. In one organization the major dimension of social structure was the professional status and activity of the people involved. In the other organization the major dimensions of social structure consisted of job prestige, job variety, and social similarity to the top managers of the organization. It was also found that the effective leader is one who is responsive to the demands and social system with whom he or she must interact and coordinate his or her behavior. Specifically, the supervisor's coordination with other supervisors was negatively related to a tendency to behave as the subordinate desires ($r = .91$). This study clearly indicates support for the position of Pfeffer and Salancik (1975) that a major determinant of leader behavior is the pattern of expectation and social pressures of other organizational members with whom leaders are interdependent.

Another situational factor that has been found to determine leadership is the amount of stress and ambiguity experienced by the leader and the group. Korten (1968) has argued that:

> The greater the stress and the less the clarity and general agreement on goals and paths, the greater the compulsion among the group members to give power to a central person who in essence promises to remove the ambiguity and reduce the stress (p. 357).

There are several findings that support Korten's position. For example, a leader Initiating Structure scale that included several items describing autocratic behavior of leaders (Hemphill, Note 7) has been found to be positively related to procedural clarity, but not related to performance or satisfaction of subordinates (Halpin, 1954, 1957; Halpin and Winer, 1957), performance ratings of the leader by superiors (Halpin, 1957), or team cohesiveness (Halpin, 1954, 1957; Rush, 1957) under noncombat conditions. In contrast, under combat conditions this scale was found to be positively related to the above variables (Halpin, 1954; Rush, 1957). Other studies also show that under conditions where there is a high degree of external threat, such as that experienced by air crews in combat, individuals have been shown to prefer strong (high structured, assertive, or autocratic) leaders (Mulder and Stemerding, 1963; Mulder, Rietsema, and de Jong, 1970; Sales, 1972; Torrance, 1954; Ziller, 1955). Thus these studies suggest that stress and ambiguity serve to stimulate a desire on the part of subordinates for highly structured leader behavior. Coupled with the finding by Crowe et al. (1972) that leaders responded to subordinates' preferred leader behavior, these studies suggest that stress and ambiguity for subordinates causes leaders to become more structure and more autocratic.

However, Fiedler and his associates (Fiedler and Chemers, 1974, cf. pp. 58–60 below) have argued that a leader's reaction to stress will be determined by the trait measured by the LPC scale. As discussed earlier, persons high on this scale tend to respond to stress by engaging in socioemotional leadership behavior, whereas persons low on this scale respond to stress by engaging in task-oriented, controlling, and somewhat authoritarian behavior.

The theoretical explanation offered by Korten (1968) for findings such as those cited above is that under ambiguous and stressful conditions structured or autocratic leader behavior reduces anxiety by clarifying goals and paths to goals. These behaviors result in a reduction of ambiguity and are essential to successful purposeful actions to reduce anxiety.

There is at present no systematic conceptualization of leadership environments. Moos (1973) reviews the current conceptualizations of human environments and argues for further research in this area. The same may be said for the environments of leaders. Clearly research is needed to develop a theoretical conceptualization of the leaders' environment and to determine those variables in the leader's environment that serve as cues, constraints, and reinforcers of leader behavior.

THEORIES OF LEADERSHIP

In this section we review several theories of leadership. These theories represent deductive frameworks consisting of sets of conceptual propositions from which several specific operational hypotheses can be derived.

These theories purport to describe, explain, and predict the effects of certain kinds of leader behavior and the conditions under which such behaviors will be effective. The theories reviewed here were selected for inclusion in this paper because they have received significant empirical support or because there is current widespread interest in them.

Each theory raises specific issues worthy of future research. These issues are described after a brief description and review of the evidence relevant to each theory.

Idiosyncrasy Credit Theory

This theory, advanced by Hollander (1969), attempts to explain the emergence of leadership and the determinants of leader effectiveness within groups. Leadership is considered to be an influence process. This process is considered to be effective where the leader is able to muster willing group support to achieve certain clearly specified group goals. According to the theory, group members continually evaluate the adequacy of the behavior of other group members. These evaluations are based on whether the group members have conformed to expectations.

There are two kinds of expectations. First are norms, which are expectations that are common to all group members for all other group members. Second, there are expectations that are specific to individuals or defined positions in the group. These are referred to as roles.

Group members' judgments about the individual will be positive to the extent that that individual conforms to expectations and contributes toward the group's goal. Member evaluations in terms of expectancies determine the individual's role and status in the group. Status is defined in terms of "idiosyncrasy credit." This represents an accumulation of positively disposed impressions residing in the perceptions of relevant others; it is defined operationally in terms of the degree to which an individual may deviate from the common expectancies of the group. For an individual to assert leadership, he or she must deviate from the expectations that members have for other group members. This deviant behavior is characterized as being unique or innovative and as contributing to attainment of group goals.

Early in interaction, conformity to group norms serves to maintain or increase credit, particularly as it is seen to be combined with manifest contributions to the group. At a later phase, the credits thus generated permit greater latitude for idiosyncratic behavior. Thus individuals who conform to group norms early in their exposure to the group and also show characteristics of competence will accrue credit. If one continues to amass credits, he or she attains a threshold permitting deviations from norms. However, such high-status persons are constrained by newly differentiated expectancies. These newly differentiated expectancies are those that are associated with the leadership role—expectancies that the high-status member will bé innovative in helping the group attain important goals. Thus leaders conform to group norms and yet may act to alter them by an exercise of influence through this sequential process.

Hollander states that while an individual may have established sufficient credits to display idiosyncratic behavior, he or she may not choose to do so and therefore not necessarily become a leader. Thus the theory implies that the individual emergence of leadership depends partly on the individual's predisposition to lead.

Leadership status demands conformity to the group's expectancies regarding the role, but still leaves the leader with sway in the sphere of common expectancies associated with members at large. The leader may deviate from these or bring about the reconstruction of the perceptions of his or her prior activities after generating an appropriately high level of credit. Thus Hollander implies that with continued interaction individuals in a group change their attributions about the intentions of members' behavior.

One of the implications of the theory is that a person who breeches a group norm and in the process succeeds in helping the group to achieve its goal will be judged differently from one who fails to do so. Though the nonconforming behavior of a group with high idiosyncracy credit may be perceived more readily, it is likely to be interpreted in terms of certain positive outcomes, given the development of a history of past deviations that have proven to be fruitful and innovative.

Hollander notes that high-status people are likely to be perceived more favorably than low-status people. Their motives are likely to be viewed as more benevolent and more in the interests of the group. One implication of this hypothesis is that once a strong attribution of leadership has taken place the behavior of the person to whom leadership is attributed is likely to reinforce the attribution. Consequently, high-status people are likely to have significantly more opportunities to earn idiosyncracy credits than low-status people because of this attribution process. Thus the theory implies that, other factors remaining equal, status begets status.

The theory stresses the attribute of competence in those tasks that are of importance to the achievement of the group's goals. Unlike other theories, Hollander's is the only theory that explicitly places such a stress on competence. Various kinds of competence are relevant. They may be specific technical skills or social skills, depending on what is required if the group is to operate effectively. Therefore, the individual's functional value for the group is determined by a wide variety of situational demands for varying kinds of attributes. Also, redefinitions of competence may occur periodically. What may be important in securing a goal at one time may no longer be as important after the goal is achieved.

Another attribute required of the leader is that he or she be seen by potential followers as having an identification with the group, in the sense of a close involvement in the group's activities. It is important that the leader have those attributes that suggest such identification and also that his or her behavior manifest a loyalty to the needs and aspirations of the group members.

In addition to the two attributes of competence and identification with the group, there are three behavioral processes that are hypothesized to be important in determining the effectiveness of leadership: first, providing the group with structure and goal setting; second, maintaining flexibility and adaptability in handling changing requirements as new situations develop; and third, establishing productive social relationships that manifest themselves in emotional stability, dependability, and fairness in distribution of rewards.

Specific behaviors that are hypothesized to be associated with effective leadership are: *(a)* fostering communication within the group by providing mechanisms for participation and for informing members in advance if decisions or actions will affect them, *(b)* restraint in the use of power and impulsiveness, *(c)* rewarding actions that are in the interest of the group and judiciously avoiding the rewarding of behaviors that are contrary to the group's best interests, and *(d)* communicating to other groups and to higher authority the particular desires and needs of the group.

Hollander (1964) reports three studies conducted to test specific predictions of the Idiosyncrasy Credit theory. The first study demonstrated that high-competence people who violate group norms will be tolerated more than low-competence people, and that if one violates the norms *after* having demonstrated competence that the person's influence will be sufficiently higher than if the norm violation occurred earlier.

In a second study a brief description of a person was given to 151 subjects. They were to imagine that this person belonged to any group to which the subjects belonged. Competence and length of time in the group were the major attributes varied in the description. Subjects were asked to indicate their willingness to have that person in a position of authority in the group. A rising mean score of acceptance was found for increasing degrees of competence, and the mean for "new to the group" was uniformly lower than for "in group for some while" at each degree. Subjects were also provided a description of eight possible ways the hypothetical person may behave in the group. According to prediction, two behaviors reflecting innovative action were found to be disapproved significantly less the higher the status attributed to the innovator.

A third study was conducted in which subjects worked with coworkers who characteristically either conformed to the subject's judgments in the task, anti-conformed to their judgments, or behaved independently, that is, were evidently unaffected by the subject's judgment. The effect of these three modes of coworker behavior were studied in combination with each of three competence conditions in which the subject believed: *(a)* he alone was competent on the task, *(b)* that the coworker alone was competent, or *(c)* that both were competent. The proposition that nonconforming behavior is perceived differentially by the group, depending on the amount of credit the individual has built up through previous conformity and demonstrated competence was tested. It was hypothesized that the more the credits the individual has at his or her disposal, the more positively will his or her "nonconformity" be evaluated by the rest of the group. However, credit was expected to be put to use by the facilitation and independence behavior, not for an actual negation of normative prescriptions as in anticonformity behavior. As predicted, highest perceived coworker influence occurred under the condition where the coworker behaved independently and was perceived as competent.

Idiosyncrasy Theory stresses the importance of competence as a determinant of leadership. There is abundant evidence that the leaders' ability to contribute to the achievement of group goals is a characteristic associated with leadership emergence and effectiveness (Evan and Zelditch, 1961; Hollander, 1960; Hollander, Julian, and Perry, Note 8; Julian and Hollander, Note 9) and that this ability is largely determined by the leader's intellectual, interpersonal, administrative, and technical competence.

This statement has been found to hold in a wide variety of experimental (Hamblin, Miller, and Wiggins, 1961; Julian, Hollander, and Regula, 1969) and field studies (Comery, High, and Wilson, 1955a, and 1955b; Kahn and Katz, 1960; Baumgartel, 1956; Goodacer, 1951; Greer, Gallanter, and Nordie, 1954), and it has been demonstrated in such diverse field settings as manufacturing operations, forest ranger stations, railroad operations, research endeavors, and military combat operations.

Stogdill (1974, pp. 92–93) reports a comparative analysis of fifty-two factor-analytic studies published between 1945 and 1974. He found the following to be the most frequently occurring factors associated with leadership. These are presented in order of frequency of occurrence:

1. Social and interpersonal skills.
2. Technical skills.
3. Administrative skills.
4. Leadership effectiveness and achievement.
5. Social nearness and friendliness.
6. Intellectual skills.

The studies reported above are consistent with the prediction of

Idiosyncracy Theory. Further, they indicate that there are many kinds of leader competence. To date, there is no systematic method whereby a particular kind of leader competence can be deduced from an analysis of the leadership situation. Clearly, one would expect the formal role of the leader, the tasks of subordinates, subordinates' ability level, and their needs and expectations to determine the particular kind of competence that would lead to effective leadership.

One promising research direction concerns the need for a better conceptualization for the kinds of competence that a leader needs and a method for deducing the specific kinds of competence required for specific situations. The task remains for research to identify the situational factors that determine the intellectual, interpersonal, administrative, and technical competence requirements associated with leadership.

Prior evidence is consistent with Hollander's hypotheses that individuals who have idiosyncratic credit are more likely to engage in leadership behaviors, i.e., acceptable influence attempts, and that attributions about these behaviors are more likely to be positive than for individuals who do not have idiosyncratic credit. Individuals with high status attempt to influence others more often (Knapp and Knapp, 1966; Bass, 1963) and are generally perceived as more attractive (Pepitone, 1964) and more able (Sherif, White, and Harvey, 1955; Gardner, 1956; Hamblin, Miller, and Wiggins, 1961). Group members have a greater tendency to accept disruptive and inconsistent behavior and to accept changes introduced by high-status individuals (Hollander, 1961; Sabath, 1964; Goldberg and Iverson, 1965).

However, Jacobs (1971) discusses findings from two studies that are not in accord with the theory. These suggest boundary conditions with respect to the hypothesis that the behavior of high-status individuals is generally viewed positively. In both studies high-status confederates who violated group norms *at the expense of the attainment of group goals* were found to have suffered status losses more than the theory would predict (Alvarez, 1968; Wiggins, Dill, and Schwartz, 1965). Thus it appears that deviant behavior on the part of high-status people will be tolerated only insofar as those deviations result in either positive or insignificant consequences for group well-being. These studies suggest that when deviations of high-status persons result in negative consequences for the group, such deviatior will not be tolerated and the high-status members will lose influence v .th the group. Research designed to test this boundary condition is thus recommended.

On balance, the theory enjoys inferential support from prior studies and support from three studies designed to test its explicit hypotheses. The theory is potentially capable of predicting the emergence of leaders and the processes in which leaders engage to maintain leadership status once they have emerged or are appointed in formal organizations. Clearly additional tests of the theory are warranted.

Contingency Theory

Fiedler (1967) and Fiedler and Chemers (1974) advance a theory of formal leadership that uses the interaction of leader personality (as measured by the Least Preferred Coworker scale (LPC) and situation favorability (leader-member relations, task characteristics, and leader position power) to predict effective and ineffective leaders.

Leaders who describe their least preferred coworker in favorable terms (high LPC leaders) are assumed to be relations oriented and those who describe their least preferred coworker in negative terms are assumed to be task oriented. The orientation refers to which of two needs is dominant in that leader's personality. The two needs are need for good leader-subordinate relations and need for task success.

The dimension of situation favorability is seen as a stress continuum that, in interaction with the leader orientation, elicits leader behavior that is consistent with the hierarchical need pattern of the leader. This interaction between the leader's orientation, is measured by LPC and situational favorability, will be referred to here as the need-hierarchy hypothesis. According to this hypothesis, the leader's orientation determines which leader behaviors will be exhibited when the eader is in a stressful situation (low situation favorability). Under such conditions it is assumed that the leader must sacrifice either control over task completion or pleasant relations with subordinates.

Fiedler hypothesizes that leaders have little control over the behaviors exhibited in stressful situations. According to the theory, the low LPC leader, who concentrates on task completion, is the more effective leader in the unfavorable situation because at least the job gets done. In the situations of medium favorability, research has shown that the high LPC leader is more effective, though it is not clear why. In very favorable situations, the low LPC leader is also most effective. In this situation the leader can concentrate on secondary goals, because his or her more basic needs are met. Consequently, in very favorable situations the low LPC leader exhibits considerate behavior while the high LPC leader exhibits task-relevant behaviors. Since task-relevant behaviors are redundant in highly favorable situations (characterized by low task ambiguity, high position power, and good leader-member relations), the low LPC leader is more effective and the high LPC leader is less effective in such situations.

To date there are only three published complete tests of the model. Graen, Orris, and Alveres (1971) failed to support this model. The findings by Chemers and Skrzypek (1972) supported the model in seven of eight octants. And the most recent, most complete study by Vecchio (1977) failed to support the model. All other supporting and disconfirming evidence has been based on partial tests of the model or inferred from studies not intended to test it. The conclusions that follow rest primarily on the three complete studies.

There seems to be no clear reconciliation of the findings. The Graen, Orris, and Alvares (1971) failure to support the model may be due to methodological errors as described by Fiedler (1971). However, both the by Chemers and Skrzypek (1972) and the Vecchio (1977) studies appear to meet Fiedler's suggested requirements for a rigorous test of the theory. Vecchio presents his nonsupporting findings as a clear disconfirmation of the Contingency Model. He suggests that the theory is either task or population bound. If he is correct, then it is not a theory of leadership, but rather a theory of behavior on specific tasks or a theory of the behavior of specific populations and thus is of little value because of its lack of generalizability.

However, Vecchio's acceptance of the null hypothesis seems rather hasty. The contradictory findings resulting from these three studies suggest the need for further studies. The large number of partial tests of the theory that support the theoretical predictions for various octants suggest that the theory has some predictive power and should not be cast aside lightly.

Two main difficulties arise in considering the Contingency Model. First, there is no unambiguous interpretation of the LPC measure. Second, there is no explanation presented to account for the demonstrated effectiveness of high-LPC leaders in situations of medium favorability. Each of these problems is considered below.

Future Research Directions: Contingency Theory

The meaning of LPC   There are two approaches to interpreting the LPC measure. The first is to relate this measure directly to other personality measures. Despite numerous studies attempting to relate LPC to other personality measures, none has been consistently correlated with it. A second approach is to relate the scale to observables such as expressed leader behavior and to infer from this behavior the underlying personality dimensions that are being tapped. The need-hierarchy explanation uses this approach.

Schmidt (Note 10) reviewed the research on LPC and offered a reinterpretation of the LPC measure. He noted that the hierarchy definition has received mixed support. The behavioral studies do indicate that as the situation changes, behaviors of low- and high-LPC leaders change in a pattern consistent with the predictions of the theory (see, e.g., Green,

Nebeker, and Boni, 1976). However, he noted that the reason for the behavioral change has not been demonstrated.

If the need-hierarchy explanation is correct, it should be possible to demonstrate changes in motivation consistent with changes in the favorability of the situation. Schmidt argued that one way to demonstrate these changes in motivation would be to measure changes in the correlation between leader job satisfaction and (a) the quality of group interpersonal relations and (b) the degree of task accomplishment. For example, for the high-LPC person in an unfavorable situation, Schmidt would expect a positive relationship between the leader job satisfaction and the quality of interpersonal relations (the primary need) while under favorable conditions the high-LPC leader's satisfaction should be positively related to the task accomplishment (the secondary need). However, his review of the evidence shows that the results are in agreement only with regard to the existence of a primary need for each LPC type. Under unfavorable conditions, leader job satisfaction (or affect) was shown to be positively related to the task accomplishment for low-LPC persons and to interpersonal relations for high-LPC persons. But under favorable conditions there was neither a positive relationship between leader job satisfaction and the task accomplishment for high-LPC persons nor a positive relationship between leader job satisfaction and interpersonal relations for low-LPC persons. Thus, these studies do not clearly demonstrate the emergence of a secondary goal under favorable conditions.

From this evidence Schmidt concluded that it is questionable whether there exists a multiple goal hierarchy that is tapped by the LPC measure and that the motivational hypothesis is supported only for the primary need. Schmidt reviews eight relevant studies. In the eight studies reviewed by Schmidt, the motivational explanation was not tested directly. Consequently, Schmidt's conclusions could be viewed as tentative at this stage.

Schmidt offers an alternative explanation of the LPC measure. He hypothesizes that LPC is an individual difference variable that measures the way an individual defines the job. That is, LPC is an index of which organizational functions and goals the leader identifies as relevant and important. Schmidt argues that an individual's behavior is determined by his or her primary orientation. Four studies are cited to support his hypothesis that low-LPC leaders are ". . . clearly oriented toward the task" (p. 25). Schmidt is more equivocal with respect to the evidence about the orientation of high-LPC leaders.

. . . Some studies have shown clear evidence for an interpersonal orientation for high LPC leaders . . . other studies have produced counter or null results . . . Schmidt and

Fiedler (unpublished) demonstrated that high LPC leaders are concerned with all of the elements of the general situation, including interpersonal relations with subordinates. Similarly Mitchell (1970) found a tendency for high LPC subjects to more evenly weigh the elements of the situation in their judgments, although the group's interpersonal atmosphere was clearly the more salient factor. However, a more complete explanation of the high LPC leader's orientation will require additional research. (pp. 25–26)

Schmidt hypothesizes that the difference between high- and low-LPC leaders concerns their choice of what means will help them meet their needs, not in the needs themselves. Thus, the evidence with respect to the need-hierarchy hypothesis is mixed and the hypothesis with respect to the leader's cognitive orientation remains untested. Schmidt's alternative hypothesis also warrants explicit testing.

The medium favorability situation   To date there has been no explanation offered to account for the findings that groups led by high-LPC leaders perform more effectively in moderately favorable situations. One possible explanation may be that high-LPC individuals experience less stress than low-LPCs in response to the same level of objective stressors. There is evidence that low-LPC leaders respond to stress differently than high-LPC leaders. Low-LPC leaders become more assertive, task oriented, directive, and controlling. These behaviors suggest that low-LPC leaders become more rigid than high-LPC leaders in response to the same objective level of stress induction. Further, low-LPC leaders exhibit significantly less variability in behavior than high-LPC leaders, given the same objective high level of stress (Graen and Nebeker, 1977). Since low variability is a common response to the experience of a high degree of subjective stress (Lazarus, 1966), it can be hypothesized that low-LPC leaders subjectively experience more stress in response to a given level of objective stressor than high-LPC leaders. In the medium favorability situation, either of two conditions prevail: *(a)* leader-member relations are good, the task is unstructured, and position power is weak or *(b)* leader-member relations are poor, the task is structured, and position power is strong. Different leader behaviors are likely to be required in these two situations. Yet high-LPC leaders have been found to be more effective in both situations (Sample and Wilson, 1965; Fiedler, O'Brien, and Ilgen, 1969; Fiedler and Chemers, 1974). These findings suggest that high-LPC leaders need to be and are more flexible in the medium favorability condition. If it can be assumed that high-LPC people perceive less stress than low-LPC people in the same situation, it is likely that high-LPC people will have greater flexibility and thus be able to choose the appropriate leader behaviors for the situation. Feeling less negative effects from the stress, the high-LPC individual can use the contemplative mode to diagnose situation demands and analyze what is needed. For that part of the Situation Favorability continuum under which tasks are structured and leader-member relations are poor, one would expect more considerate, less task-oriented behavior to be accepted more readily by subordinates. It is this kind of behavior that is predicted by high-LPC leaders under low stress conditions. In contrast, if low-LPC individuals perceive the same situation as highly stressful, they would be predicted, according to Contingency Theory, to engage in highly assertive, task-oriented, directive leader behavior. Such behavior is likely to be resented by those for whom the task is highly structured, especially if their relations with the leader are poor.

It is also possible that under the medium situational favorability condition leader behavioral flexibility is required. If low-LPC leaders perceive such situations as stressful, it is doubtful that they would be able to exhibit such flexibility. In contrast, high-LPC leaders experiencing less stress, would be expected to be more flexible and adapt as the situation demands.

Thus, it can be seen that, using the hypothesis of differential perception of and response to stress, an explanation for the superior performance of high-LPC leaders under conditions of medium situational favorability can be provided.

This explanation suggests the following hypotheses for future research:

1. Given a fixed amount of objective stressors, the amount of subjective stress experienced by an individual is negatively related to that individual's LPC score.

2. Leaders can consciously identify the appropriate behaviors for effective leadership under conditions of low perceived stress.

3. Leaders are less able to vary their behavior to meet situational demands as the amount of stress experienced increases.

It is necessary to distinguish between challenging a theory because it fails to predict and because it fails to explain its prediction. The Vecchio study demonstrates the need for further research into whether the Contingency Theory has predictive power. The Schmidt review and reinterpretation of LPC clearly shows a need to further explain LPC measures. Without such an explanation the Contingency Theory is method bound. That is, neither predictions nor prescriptions can be made without reliance on the LPC measure. This prevents multimethod verification of the theory and severely limits its usefulness under conditions that do not permit administration of the LPC scale.

Comparison of Idiosyncracy Theory and Contingency Theory   Idiosyncracy Theory offers a possible explanation as to why low-LPC leaders are effective in highly favorable conditions. First, assume that low-LPC leaders do in fact behave under these conditions as predicted by the theory. That is, low-LPC leaders are dominantly task oriented, assert strong control, and are very directive under unfavorable conditions. Further, assume that under favorable conditions, low-LPC leaders are dominantly relations oriented and considerate of subordinates.

According to the theory, under favorable conditions, leader-member relations are very good. In terms of Idiosyncracy Theory, good relationships would imply that the leader has high status as a result of having accumulated idiosyncracy credits. Thus, subordinates would respond favorably to such high-status leaders. Assuming that the leader is competent, the leader's requests would be readily followed by subordinates and thus the group would perform effectively.

However, Idiosyncracy Theory would make a prediction contrary to Contingency Theory with respect to the low-favorability situation. In such a situation, leader-member relations are poor, subordinate's tasks are unstructured, and the leader has low position power. In terms of Idiosyncracy Theory, poor leader-member relations imply that the supervisor would have low status as a result of not having accumulated idiosyncratic credits. According to Contingency Theory, it is necessary for the supervisor to assert strong control in a highly directive manner under unfavorable conditions. Contingency Theory would also predict this kind of behavior of the supervisor in unfavorable situations. However, since the supervisor has a low idiosyncracy balance, Idiosyncracy Theory would predict that such influence attempts by the supervisor would be rejected by group members. Further, since the supervisor has low power over group members in the unfavorable situation, the supervisor would be in no position to force his will on the group. Thus, such influence attempts by the supervisor would be ineffective. Therefore, the two theories make contrary predictions for the unfavorable situation.

Path-Goal Theory

The Path-Goal Theory of leadership is a situational theory that is deliberately phrased so that additional variables (such as personality variables) can be added as the effects of these new variables become known. The theory is based on theoretical work by Evans (1960) and has been formulated and extended by House (1971) and House and Mitchell (1974). A concise discussion of the theory appears in Filley, House, and Kerr (1976).

> Briefly, the theory consists of two propositions. The first proposition is that leader behavior is acceptable and satisfying to subordinates to the extent that they see it as either an immediate source of satisfaction or as instrumental to future satisfaction.
>
> The second proposition of the theory is that leader behavior will be motivational to

the extent that (1) it makes satisfaction of subordinate needs contingent on effective performance, and (2) it complements the environment of subordinates by providing the coaching, guidance, support, and rewards which are necessary for effective performance and which may otherwise be lacking in subordinates or in their environment.

These two propositions suggest that the leader's strategic functions are to enhance subordinates' motivation to perform, their satisfaction with the job, and their acceptance of the leader . . . The strategic functions of the leader consist of (1) recognizing and/or arousing subordinate's needs for outcomes over which the leader has some control; (2) increasing personal payoffs to the subordinates for goal attainment; (3) making the path to those payoffs easier to travel by coaching and direction; (4) helping subordinates clarify expectancies; (5) reducing frustrating barriers; and (6) increasing opportunities for personal satisfaction, contingent on effective performance. Two classes of situational variables are asserted to be contingency factors; these are *(a)* personal characteristics of subordinates, and *(b)* environmental pressures and demands which subordinates must deal with (p. 254).

Personal characteristics in part determine how subordinates react to leader behavior. Several personality characteristics of subordinates are hypothesized to moderate the relationship between the effects of leader behavior and the satisfaction of the subordinates. For example, Runyon (1973) and Mitchell, Smyser, and Weed (1974) show that a subordinate's score on the Locus of Control Scale (Rotter, 1966) moderates the relationship between participative leadership style and subordinate satisfaction. These studies showed that individuals who are low on the scale, i.e., individuals who believe that their rewards are contingent on their own behavior are more satisfied with a participative leadership style while individuals who are high on the Locus of Control Scale, i.e., individuals who believe their rewards are the result of luck or another's behavior are more satisfied with a directive style.

Another characteristic of subordinates hypothesized to moderate the effect of leader behavior is the subordinate's tendency to be authoritarian. Highly authoritarian subordinates are hypothesized to be less receptive to participative leadership and more receptive to directive or even authoritarian leadership.

A third subordinate characteristic hypothesized to moderate the effect of leader behavior is the subordinates' perceptions of their own abilities with respect to their assigned tasks. The higher the degree of perceived ability relative to task demands, the less the subordinates will view their leader directiveness, closeness of supervision, and coaching behavior as acceptable. Where the subordinate's perceived ability is high, such behavior is likely to have little positive effect on subordinate motivation and to be perceived as excessively close control.

The second aspect of the situation is the environment. The theory asserts that effects of leader behavior on the psychological status of subordinates will be contingent on other aspects of the subordinate's environment that are relevant to motivation. Three broad classifications of contingency factors in the environment are the subordinate's task, the formal authority system of the organization, and the primary work group.

Assessment of the environment conditions makes it possible to predict the kind and amount of influence that specific leader behaviors will have on the motivation of subordinates. Each of the environmental factors mentioned above could act upon subordinates in any of three ways. First, they may serve as stimuli that motivate and direct subordinates to perform necessary task operations. Second, they may act to constrain variability in behavior. Constraints may help subordinates by clarifying their expectations that efforts lead to rewards or by preventing subordinates from experiencing conflict and confusion. Constraints may also be counterproductive to the extent that they restrict initiative or prevent increases in effort from being positively associated with rewards. Third, environmental factors may serve to clarify and provide rewards for achieving desired performance. For example, it is possible for subordinates to receive the necessary cues to do their jobs and the needed rewards for satisfaction from sources other than the leader (e.g., coworkers in the primary work group). The amount of variance in subordinate motivation accounted for by leader behavior is thus hypothesized to be a function of how deficient the environment is with respect to motivational stimuli, constraints, or rewards.

With respect to the environment, the theory asserts that when goals and paths are apparent because of the routine nature of the task, clear group norms, or objective controls of the formal authority system, attempts by leaders to clarify paths and goals will be redundant and will be seen by subordinates as an imposition of unnecessarily close control. Although such control may increase performance by preventing malingering, it will also result in decreased satisfaction. The theory also asserts that the more dissatisfying the task, the more subordinates will resent leader behavior directed toward increasing productivity or enforcing compliance with organizational rules and procedures.

Finally, the theory states that leader behavior will be motivational to the extent that it helps subordinates cope with environmental uncertainties, threats from others, or sources of frustration. Such leader behavior is predicted to increase subordinate satisfaction with their job context, and to be motivational to the extent that it increases subordinate expectations that their efforts will lead to valued rewards.

This theory and the research it has generated has helped reconcile previous findings regarding relationships between managerial style and subordinate responses. For example, House (Note 6) has utilized Path Goal theory in an attempt to explain field study data by Tosi (1970) and experimental data by Wexley, Singh, and Yukl (1973) that failed to show that any differences in the effects of participation could be attributed to subordinate characteristics. House argued that the task may have an overriding effect on the relationship between leader participation and subordinate responses and that individual predispositions or personality characteristics of subordinates may have an effect only under some tasks. He assumed that when task demands are ambiguous, subordinates will have a need to reduce the ambiguity and that participative problem solving between subordinates and leaders will result in more effective decisions than when the task demands are clear. He further assumed that when subordinates are ego-involved in a task, they will be more likely to want to have a say in the decisions that affect them than when they are not ego-involved. House reasoned that whenever participation is instrumental in reducing ambiguity, or whenever subordinates are ego-involved in the task, participative leadership will be instrumental to both need satisfaction and productivity, regardless of the personality or predispositions of the subordinates. However, when subordinates are not ego-involved in their tasks and the demands are clear, participation will have no instrumental or intrinsic value, unless the subordinates are generally predisposed toward having a high degree of independence and toward respecting nonauthoritarian behavior.

Thus House hypothesized that only when tasks are unambiguous and not ego-involving will subordinate's personality or predispositions moderate the effect of participative leadership. On such tasks, subordinates not predisposed toward participative leadership will not find it either instrumentally or intrinsically satisfying. Subordinates who are predisposed toward participation will find it intrinsically satisfying and motivating. The study by Schuler (1976) described in the above section on participative leader behavior was designed to test this hypothesis. Schuler's findings were as hypothesized.

The second proposition of the theory states that leader behavior will be motivational to the extent that it complements the environment of subordinates by providing the coaching, guidance, support, and rewards that are necessary for effective performance that may otherwise be lacking in subordinates or in their environment. Environmental factors such as intragroup conflict, task characteristics, organizational size, and organizational structure have been shown to moderate the relationship between leader behavior and subordinate satisfaction and performance in a manner consistent with the theory.

Path-Goal theory advances the hypothesis that the effects of leader consideration on performance and satisfaction of subordinates will be most positive when subordinates are engaged in dissatisfying tasks, fatigued, frustrated, or under stress. The literature reviewed above in the section on leader role differentiation concerning the effects of socioemotional leadership and leader consideration is consistent with this hypothesis, as are the findings from more recent tests of this hypothesis reported in the above section concerned with leader behavior.

A recent study by Katz (in press) incorporates intragroup conflict into the theory. Katz distinguished affective intragroup conflict from substantive intragroup conflict and argued that as the different kinds of conflict vary in strength and importance the leader behavior required would also change. Specifically, he hypothesized that increasing affective conflict will precipitate the desire by group members for more considerate and less structuring leader behavior in order to satisfy the needs of group members. Similarly, increasing substantive conflict was hypothesized to evoke a desire by group members for more structuring and less considerate leader behavior.

Finally, Katz argued that either kind of conflict will generate tension and stress, as well as hinder subordinate's perceived path-goal relationships. Thus, he hypothesized that overall effectiveness will be more positively related to leader Initiating Structure and less positively related to leader Consideration under conditions of either high affective or substantive conflict.

Katz tested this hypothesis in a field study, using correlational methods and in two laboratory studies using confederate leaders, questionnaires, and objective measures of performance. His major findings were: (*a*) the need for or desire for structuring leader behavior is significantly positively related to the degree of substantive or affective conflict, and (*b*) the relationship between leader structuring and group performance was significantly higher under high-conflict conditions. Only slight evidence was in favor of a similar conclusion for leader consideration. Thus Katz's hypotheses with respect to Initiating Structure were strongly supported. However, he did not find support for the hypothesis that affective conflict will be positively related to the desire for increased leader consideration. To explain his failure to confirm the hypothesized relationship between affective conflict and desired leader consideration Katz speculates that preferences for leader Consideration may be invariably positive. Katz states that an alternative explanation may lie in the Path Goal theory in that Consideration is simply not as relevant or as meaningful a dimension as Initiating Structure for individuals who have jobs with considerable intrinsic satisfaction. Clearly this hypothesis warrants testing.

Reasoning from the second proposition of the theory has also resulted in the identification of organizational size as a variable that can be incorporated into the theory. Miles and Petty (1977) reviewed the evidence on the relationship between organization size and degree of bureaucratization and concluded that there is a strong positive relationship between these two variables. That is formalization, routinization, standardization, and specialization were concluded to be higher in large organizations. From this conclusion they reasoned that social service professionals in large organizations would view leader initiating structure as redundant with the high degree of formalization found in such organizations. They reasoned that leader consideration would be an alternative source of satisfaction or relief from the presumed dissatisfaction that occurs as a result of bureaucratization in large agencies. As predicted, these authors found the correlations between (a) leader Initiating Structure and (b) employee work satisfaction, satisfaction with coworkers, and motivation were significantly higher in small agencies than in large agencies. In the smaller agencies, the correlations were all positive, whereas in the larger agencies they were either nonsignificant or negative. Further, they found the correlations between consideration and employee satisfaction tended to be higher in large agencies than in small agencies. However, these differences were not statistically significant.

Schuler (Note 11) extended the Path Goal theory of leadership by incorporating the moderating effects of organizational structural variables into the predictions of the theory. He found that subordinate role conflict and role ambiguity are lower and expectations that performance will result in desired rewards are higher among employees doing complex tasks in an organic organizational environment than among employees doing simple tasks in mechanistic environments. He reasoned that under complex tasks in organic organizations leader Initiating Structure will not be needed. He also reasoned that under opposite conditions, leader Initiating Structure will be positively related to satisfaction. Schuler's findings supported these hypotheses indicating that when organizational or task variables provide the subordinate with the necessary role clarity and direction, leader Initiating Structure will have little effect.

Recently, the methodologies used to test the theory have been questioned. Sheridan, Downey, and Slocum (1975) and Dessler and Valenzi (1977) have argued that path analysis should be used to test the Path Goal theory when relying on correlation data. Dessler and Valenzi argue "that the use of path analysis procedures may help explain (the conflicting findings regarding the moderating effect of task structure on the relationship between Initiating Structure and satisfaction) by focusing directly on the underlying expectancy motivation linkage specified in the Path Goal theory" (p. 252).

Using path analysis, Sheridan et al. (1975) did not find support for the direct causal relationships hypothesized in Path Goal theory. Dessler and Valenzi (1977) did not find support for the hypothesis that occupational level moderates the relationship between initiation of structure and intrinsic job satisfaction. However, Dessler and Valenzi did find support for the hypothesized linkages between leader behavior and subordinates' expectancies. This contradiction in findings will need to be resolved through future research. Green (1975) attributes the disconfirming finding of Sheridan et al. (1975) to methodological deficiencies of their research design.

Path Goal theory also offers an explanation as to why low-LPC leaders are effective in conditions of either very high or very low situation favorability. Again, if we assume that low-LPC leaders do behave under these conditions as predicted by the theory, then we would expect such leaders to exhibit considerate-relations-oriented behavior under favorable conditions and controlling assertive directive behavior under unfavorable conditions. According to Contingency Theory, under favorable conditions the jobs of subordinates are highly structured. According to Path Goal theory, task-oriented–directive-path–clarifying-leader behavior is unnecessary under these conditions and would be viewed as redundant with the situation by subordinates. Further, if highly structured jobs are assumed to be more routine and thus less satisfying, supportive-relations-oriented leader behavior would be required to offset the boredom and frustration resulting from such jobs. Under such conditions, influence attempts by supportive leaders are more likely to be accepted by subordinates. Assuming the leaders are competent, if their influence attempts are accepted, the groups are likely to be more effective.

Under unfavorable conditions, the task of subordinates are unstructured. According to Path Goal theory, task-oriented leader behavior that clarifies task requirements and paths to goals is likely to be more effective. Thus, under this condition, task-oriented leadership would be seen as instrumental to subordinate goal achievement. Again, assuming that the leader is competent, such task-oriented leader behavior is likely to be more readily accepted and the group is thus likely to be more effective.

Future research directions: Path goal theory. Path Goal theory is rich with opportunities for refinement and extension. There are many variables that have been the subject of other leadership research that are not yet included in the Path Goal model. For example, as mentioned earlier, substantial research has shown that leader task competence is an important variable in predicting leader effectiveness. The theory in its present form assumes the leader to have the task competence when engaging in clarifying behaviors. Leader competence thus constitutes a boundary variable of the theory. It is hypothesized here that the predictions of the theory concerning leader path clarifying behaviors will not hold under conditions where the leader does not have the competence to clarify subordinates' task demands, or where the leader's intelligence is either below or too far above the intelligence of subordinates.

Other leader personality variables might be incorporated into the theory. For example, the traits of dominance and self-confidence are likely predictive of leaders who will initiate path-clarifying behaviors. The traits of sociability, interpersonal skills, and social participation were found by Stogdill (1948, 1974) to be related to criteria of leader effectiveness in a large number of studies (see Table 1). These traits may be used as surrogates for measures of supportive leader behavior.

Subordinate personality variables that might be investigated in addition to authoritarianism and locus of control include subordinate need for achievement, need for affiliation, and tolerance for ambiguity.

A recent study by Graen and Ginzbergh (1977) suggests another variable that might be incorporated into the theory. These authors found that when subordinates perceive their job as relevant to their future career objectives, leader supportiveness was unrelated to their subsequent tendency to resign. However, when the subordinates peceived their jobs as not relevant to career objectives, leader supportiveness had a strong negative relationship to a subsequent tendency to resign. This perception of the employee, which Graen and Ginzbergh (1977) refer to as Role Orientation, thus interacts with leader supportiveness as task satisfaction is hypothesized to interact with leader supportiveness according to Path Goal theory. It thus appears that subordinate role orientation may be a better moderator of the relationships between leader behavior and the dependent variables of the theory than the task characteristics originally specified in the theory. Research to test the relative importance of Role Orientation and task characteristics is called for.

Recent research and theorizing by Kerr (Note 12) suggests additional variables that are hypothesized to moderate relationships between leader behavior and subordinate motivation, satisfaction, and performance. Kerr has advanced a notion of "substitutes for leadership." According to this notion, there are factors in the environment that serve as sources of psychological structure or support for subordinates such that leader behavior is irrelevant to the satisfaction or performance of subordinates. Such "substitutes for leadership" are hypothesized to *negate* the leader's ability to either improve or impair subordinate satisfaction or performance. Kerr argues that what is needed, then, is a taxonomy of situations where we should not be studying leadership in the formal hierarchical sense at all. Kerr advanced a preliminary taxonomy of such substitutes. These are presented in Table 4. In addition, Kerr has developed a questionnaire designed to measure the existence of such substitutes in hierarchical field settings. While the questionnaire is still in its developmental stage, there is evidence that it is useful for its proposed purpose (Kerr, Note 7).

Kerr (note 7) distinguishes between "substitutes" and "neutralizers."

Table 4. Substitutes for Leadership

Characteristic	Will Tend to Neutralize	
	Relationship-Oriented, Supportive, People-Centered Leadership: Consideration, Support, and Interaction Facilitation	Task-Oriented, Instrumental, Job-Centered Leadership: Initiating Structure, Goal Emphasis, and Work Facilitation
of the subordinate		
1. Ability		X
2. Experience		X
3. Training		X
4. Knowledge		X
5. "Professional" orientation	X	X
6. Indifference toward organizational rewards	X	X
of the task		
7. Unambiguous and routine		X
8. Methodologically invariant		X
9. Provides its own feedback concerning accomplishment		X
10. Intrinsically satisfying	X	
of the organization		
11. Formalization (explicit plans, goals, and areas of responsibility)		X
12. Inflexibility (rigid, unbending rules and procedures)		X
13. Highly-specified and active advisory and staff functions		X
14. Closely-knit, cohesive work groups	X	X
15. Organizational rewards not within the leader's control	X	X
16. Spatial distance between superior and subordinates	X	X

A substitute is defined to be a "person or thing acting or used in place of another." In the context of leadership, Kerr uses the term to describe characteristics that render leader behavior unnecessary.

Kerr argues that the effect of neutralizers is therefore to create an "influence vacuum" from which a variety of dysfunctions may emerge. Kerr hypothesizes that the variables listed in Table 4 have the capacity to counteract leader influence. Consequently, all of these variables may be termed neutralizers. Additional research is needed to determine which are neutralizers and which are substitutes or the conditions under which each variable is a neutralizer, a substitute, or neither. There may well be specific conditions under which leadership is necessary but neutralized by environmental factors. Under such conditions the "influence vacuum" will likely result in dysfunctional consequences such as lower performance and satisfaction, increased turnover, conflict, and grievances. Clearly, a better understanding of how environmental factors operate as substitutes or neutralizers of leadership is required.

The Rational Decision Making Theory[4]

Vroom and Yetton (1973) have advanced a prescriptive theory intended to help managers meet the two criteria of effective decision solutions suggested by Maier (1963). That is, the theory is intended to help managers ensure a high quality of solutions to problems they must deal with and also obtain solutions that are acceptable to subordinates, if acceptability of solutions is important to effective implementation.

The theory is intended to be a diagnostic tool with which leaders can choose the appropriate decision-making methods for a given problem. The decision-making method prescribed by the theory ranges from autocratic decision making by the manager alone to various degrees of participation with subordinates, to joint decision making between the manager and his or her subordinates as a group. The model specifies seven properties of problems that Vroom and Yetton believe to have relevance to the determination of appropriate methods of decision making. Various combinations of the seven properties result in twenty-three problem types.

The model specifies seven decision rules that are intended to guide a leader in selecting the most appropriate decision method. A combination of seven decision rules, seven problem attributes, and twenty-three problem types constitute the normative (prescriptive) model.

Application of the seven rules permits the leader to determine the decision-making approaches that are feasible. Once the set of feasible approaches has been identified, the leader is instructed to select the single most suitable approach for a particular situation. Vroom and Yetton state, "When more than one method remains in the feasible set, there are a number of alternative decision rules which might dictate the choice among them'' (p. 37). Vroom and Yetton illustrate how additional rules are developed using the number of man hours required to solve the problem as the basis for choice. Other alternative rules for narrowing the choice among the feasible set are alluded to but are not made explicit.

The Vroom and Yetton theory is the first systematic integration of prior research findings concerning participative decision making. It represents an advance in conceptualizing about situtional factors that determine the degree to which various degrees of participative decision making will be effective. Consequently, the model represents an advance in the literature concerning participative decision making, and it is the opinion of the authors that further research on the model will result in a significant improvement of our understanding of the conditions under which the five decision-making approaches prescribed by the model are effective.

Vroom and his associates have conducted a substantial amount of research on the properties of the theory. Vroom and Yetton (1973) present a substantial amount of data concerning the decision-making approaches managers report having used in various situations, and the approaches they say they would use for hypothetical problems. These self-reports provide some indication of how well managers are able to diagnose situations in terms of the concepts of the model. These data show:

> On both standardized and recalled problems, leaders tend to use participative processes when they lack the necessary information to solve the problem by themselves and when their subordinates have a high probability of possessing that information. This use of participation is a means of protecting the quality of the decision by insuring that the decision making system contains the information needed to generate and evaluate the alternatives. The results from both methods also indicate that acceptance considerations enter into the choice of decision process. Managers tend to employ more participative styles when it is critical for their subordinates to accept the decision in order to get it effectively implemented and when the likelihood of selling an autocratic decision is low (p. 118).

Vroom (1976) and Vroom and Jago (Note 13) report tests of the theory using managers' reports of recalled successful and unsuccessful decisions. After describing the decisions and the decision method they used, the managers were trained in coding problem attributes according to the theory. The percentage of successful decisions based on a decision style within the feasible set was significantly greater than the percentage of unsuccessful decisions not within the set. The percentage of unsuccessful decisions outside the feasible set was significantly greater than successful decisions that were ouside the set. The results of the two studies were very similar, indicating agreement with the predictions of the theory in approximately 67 percent of the cases. This level of agreement is approximately twice that which would be expected by chance alone.

There were several other findings of interest as well. On both standardized and recalled problems, it was found that the model's normatively prescribed behavior exhibited substantially greater variety in decision processes than did the managers' recalled or hypothetical behavior. Thus managers describe themselves as being less flexible than the model prescribes. Further, the model makes more extensive use of the two extreme decision processes than do the managers, in that the model is both more autocratic and more participative than the managers described themselves as being.

While the evidence reported by Vroom (1976) and Vroom and Jago (Note 13) shows rather strong support for the model, there are several reservations that should be considered when evaluating its validity.

Self-report data have been shown in prior research to be biased; that is, self-reports tend to disagree with the observations of others (Bass, 1957; Besco and Lawshe, 1959; Campbell, 1956; and Graham and Gleno, 1970). Jago and Vroom (1975) collected descriptions of managers' decision behavior using self-report data from the managers and data collected from subordinates. The descriptions from the two sources did not correlate significantly, thus raising a serious question about the validity of the managers' self-report data used to test the model.

Filley, House and Kerr (1976) point out that one of the problems in applying the model is likely to be its lack of parsimony.

> The model requires the decision maker to choose one of five kinds of decision-making styles by analyzing a problem in terms of seven attributes and applying it requires a decision about how a particular decision is to be made. Application of the model in this way assumes that the choice of decision style can be made deliberately. If this is true it would probably be very time consuming to apply the model to any new decision. Consequently, the more varied the situations in which managers make decisions the more time will be required to decide how decisions are to be made.

> While Vroom & Yetton present evidence that training improves managers' abilities to apply the model appropriately to hypothetical case problems, it is possible that managers are not sufficiently rational or cognitively complex to apply such a complex decision-making model under pressures of time and stresses of work. Even if such a rational selection of decision-making methods is possible, it is questionable whether managers would or could take the time required to apply the model to normal day-to-day problems. Only additional research can answer whether or not managerial attempts to apply the model will result in more effective decision making (p. 252).

This criticism that the model lacks parsimony is primarily concerned with the prescriptive applications of the model. It may well be that such a complex model is necessary to *explain* the conditions under which various kinds of decision-making methods are appropriate, but it is questionable whether the prescriptive applications of the model can be made because of its complexity.

Field (Note 14) analyzed the feasible sets given in the Vroom-Yetton model and concluded that based on the decision rules, a more parsimonious model would suffice. Field noted that the decision style C2 (consultative—the subordinates and manager as a group share ideas, then the manager makes the decision) is in the feasible set for nineteen of the twenty-three problem situations. Of the four situations where C2 is not in this set, G2 (total group decision making with manager as a group member) is found in the feasible set. Thus Field concluded that a simple rule to guarantee a decision method in the Vroom-Yetton feasible set is as follows:

> If acceptance of the decision by subordinates is critical to effective implementation and it is not reasonably certain that subordinates would accept an autocratic decision, but they share organizational goals (or decision quality is not important) use G2, otherwise use C2 (page 27).

This simple model uses four situation attributes instead of the seven used by the Vroom-Yetton model and uses only two decision styles rather than the five of the Vroom-Yetton model.

For prescriptive purposes, Field's suggested rule is much easier to apply and results in the selection of decision methods consistent with the intent of the Vroom-Yetton model. That is, it results in decision methods that, according to the model, will protect the quality of decisions and insure that they are accepted by subordinates.

While it is argued here that the model is likely to be too complex for managers to systematically apply it, it also suffers from omission of decision rules relevant to characteristics of subordinates. In the initial development of the model, a decision rule was included for managers to consider whether subordinates had sufficient information to result in a high-quality decision. Subsequently, this decision was eliminated, because it was found by Vroom and Yetton (1963) that managers almost always assumed subordinates would have additional information, and if the additional information needed to be gathered, then any one of the other decision processes might result in a decision to do so (Vroom and Yetton, 1963, p. 187).

However, as discussed in the previous review of research on participative decision making, subordinates' knowledge relevant to the decision is a significant moderator of the relationship between participative decision making and decision effectiveness (see Table 3 for a summary of studies relevant to this issue). Further, the model does not consider the subordinates' predisposition toward engaging in the participative decision process. Again, there is substantial evidence that such predispositions moderate the relationship between participative leadership and decision effectiveness (cf. pp. 38–40 above for a review of this evidence).

The decision rule that instructs managers to determine whether subordinates can be trusted to base solutions on organizational considerations appears to result in a logical inconsistency. Purportedly, the model applies at all levels of the organization. Thus, when a manager is in the position of making a decision, he must judge subordinates according to this rule in order to determine the appropriate decision method to be used. However, when the manager at the next level above makes a decision, he must also apply the same rule. This assumes that in the first instance the manager can be trusted to base solutions on organizational considerations and in the second instance his superior must question whether he can be so trusted. Thus, there appears to be a logical inconsistency in the model in that the manager who is applying the decision rules is instructed to judge trustworthiness of subordinates, but he or she might not be trustworthy to base solutions on organizational considerations. Consequently, if the model were valid, it would maximize the interests of the manager applying it but would not necessarily maximize organizational effectiveness. In fact, the resulting decision could be contrary to the interests of the organization.

Another limitation of the model concerns the decision styles it makes available to the manager. Field notes that the field model does not provide for delegation in that the leader is actively involved in all decision styles specified by the model.

Field (Note 8) also notes that in some situations managers may wish to use a mixed strategy, such as having each individual generate alternatives and then using the entire group to evaluate the alternatives and reach agreement on one of the alternatives as the solution. There is evidence that such a process results in more alternatives and higher-quality alternatives being generated than when individuals generate alternatives in interaction with each other (Vroom, Grant, and Cotton, 1969).

In conclusion, while the Vroom and Yetton model represents the first systematic attempt to integrate participative decision making with situational variables, both the methods used to test it and the assumptions on which it is based remain to be validated. Additional tests of the model are recommended, using laboratory methods or field research methods that do not rely exclusively on self-report data.

Charismatic Theory of Leadership

Charisma is the term commonly used in the sociological and political science literature to describe leaders who by force of their personal abilities are capable of having profound and extraordinary effects on followers. These effects include commanding loyalty and devotion to the leader and of inspiring followers to accept and execute the will of the leader without hesitation or question or regard to one's self-interest. The term "charisma," whose initial meaning was "gift," is usually reserved for leaders who by their influence are able to cause followers to accomplish outstanding feats. Frequently, such leaders represent a break with the established order and through their leadership major social changes are accomplished.

Most writers concerned with charisma or charismatic leadership begin their discussion with Max Weber's conception of charisma. Weber describes as charismatic those leaders who ". . . reveal a transcendent mission or course of action which may be in itself appealing to the potential followers, but which is acted on because the followers believe their leader is extraordinarily gifted" (Dow, 1969, p. 307). Transcendence is attributed implicitly to both the qualities of the leader and the content of his mission, the former being variously described as "supernatural, superhuman or exceptional" (Weber, 1947, p. 358).

Several writers contend that charismatic leadership can and does exist in formal complex organizations (Dow, 1969; Oberg, 1972; Shils, 1965). House (1977) has explicitly advanced such a theory of charismatic leadership within organizations. The theory is based on a review of the political science, sociological, and social psychological literature. House found that many of the propositions advanced in the sociological and political science literature are implicitly testable. In addition he found several findings from the social psychological literature that help to understand the nature of charismatic leadership.

The dependent variables for the theory are: follower trust in the correctness of the leader's beliefs, similarity of followers' beliefs to those of the leader, unquestioning acceptance of the leader, affection for the leader, willing obedience to the leader, identification with and emulation of the leader, emotional involvement of the follower in the mission, heightened goals of the follower, and the feeling on the part of followers that they are able to accomplish or contribute to the accomplishment of the mission.

House argues that scores of followers on scales designed to measure these variables could serve as a basis for identifying leaders who have charismatic effects. Personality characteristics and behaviors of these leaders could be compared with those of other leaders who do not have such effects to identify characteristics and behaviors that differentiate the charismatic leaders from others.

The theory consists of seven propositions, which, collectively, are intended to explain the conditions under which charismatic leaders emerge and their personality characteristics and behavior. While none of the propositions have been explicitly tested, they all have some inferential support from prior empiric studies. Following are the seven propositions:

1. Characteristics that differentiate leaders who have charismatic effects on subordinates from leaders who do not have such charismatic effects are dominance and self-confidence, need for influence, and a strong conviction in the moral righteousness of their beliefs.

2. The more favorable the perceptions of the potential follower toward a leader, the more the follower will model: (a) the valences of the leader, (b) the expectations of the leader that effective performance will result in desired or undesired outcomes for the follower, (c) the emotional responses of the leader to work-related stimuli, and (d) the attitudes of the leader toward work and toward the organization.

3. Leaders who have charismatic effects are more likely to engage in behaviors designed to create the impression of competence and success than leaders who do not have such effects.

4. Leaders who have charismatic effects are more likely to articulate ideological goals than leaders who do not have such effects.

5. Leaders who simultaneously communicate high expectations of, and confidence in, followers are more likely to have followers who accept the goals of the leader and believe that they can contribute to goal accomplishment and are more likely to have followers who strive to meet specific and challenging performance standards.

6. Leaders who have charismatic effects are more likely to engage in behaviors that arouse motives relevant to the accomplishment of the mission than are leaders who do not have charismatic effects.

7. A necessary condition for a leader to have charismatic effects is that the role of followers be definable in ideological terms that appeal to the follower.

Several of the variables of the theory are operationalizable with existing instruments. For example, the personality traits of dominance, self-confidence, and need for influence have been frequently measured by use of psychological tests. Role modeling and its effects have been studied in the laboratory (Bandura, 1968) and field settings (Weiss, 1977). Valence and expectations are frequently measured by researchers of motivation. Motive arousal is frequently measured and also experimentally induced.

Throughout the paper on charisma, House offers suggestions for operationalizing the variables and testing the propositions. Tests of these propositions are recommended for future research. House stated "admittedly tests of the theory will require the development and validation of several new scales. However it is hoped that the propositions are at least presently testable in principle" (p. 24).

An Attribution Theory of Leadership

Calder (1977) has advanced a theory intended to explain the process by which the attribution of leadership occurs. According to this theory, leadership is a label that can be applied to behavior. Certain inherent qualities of the actor are taken as causing both the behavior and its intended effects. Judgments about leadership are made on the basis of observed behavior. Thus leadership is an inference based on behavior accepted as evidence of leadership. Leaders are not in fact leaders until there is some basis for distinguishing their behaviors. These evidential behaviors may differ from group to group. The behaviors accepted as evidence of leadership depend on the particular set of actors involved. At the very least, the predominant social class composition of a class of actors and the purpose of the group renders some behaviors more appropriate than others for leadership inferences. Members of a street-corner gang obviously focus on behaviors that are very different from those that are salient to the corporation board room. Thus, evidential behaviors must be *typical* of a class of behaviors that are *different* from those of most group members. In terms of Hollander's (1964) theory described above, this assertion would mean that one must meet the expectations that the observer has for group leaders and that these expectations are different from the expectations held of other group members.

Calder's theory is diagrammed in Figure 1. The first stage in the attribution of leadership is therefore the observation of behavior by another and the effects of this behavior. If an effect is not in actuality due to a person, it is a source of attributional error. Observed actions and effects may imply entirely different behaviors that have no basis at all in actual observation. These "inferred observations" supply indirect evidence to the attribution process.

In the second stage of the attribution process, actual and inferred observations are either accepted or rejected as evidence of leadership. Observations are first examined for distinctiveness. By definition, leadership cannot describe everyone in the group; its very meaning calls for distinctive behavior. Once distinctive behavior is observed, it is matched against expectations about how leaders should act. The observer has an implicit theory of leadership. This theory is used to interpret potential evidential behaviors and effects. The observed behavior must be consistent for a leader attribution to be made. If an observation, either actual or inferred, is made of behavior and effects that fit the observer's leadership theory;

Figure. 1   Flow diagram of the attribution model.

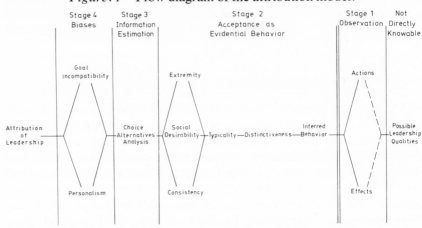

Source: Calder, B.J. An attribution theory of leadership. In Staw & Salancik (1977).

if the evidence for this holds up over time and across relevant situations; and if it is supported by the opinion of other relevant actors, it meets the requirement of consistency and thus will be attributed to the personal disposition of the observer. Coupled with this evaluation of consistency is the evaluation of extremity. First, the evidence must be judged extreme or sufficiently important to imply leadership qualities. The extremity evaluation is also important as a possible by-pass for consistency requirement. Thus, in Figure 1 extremity is depicted as operating in parallel with consistency and social desirability. Any behavior or effect that is sufficiently extreme can override alternative explanations. The requirements of distinctiveness and extremity for the attribution of leadership are consistent with Hollander's (1964) assertion that the leader must engage in idiosyncratic behavior and that this behavior must be seen as a unique or innovative contribution to group goals.

The next stage postulated in the attribution process involves what is termed "information estimation." Evidence may be acceptable and still not specifically informative about leadership qualities. An observer evaluates the evidential worth of an observed behavior and effects by comparing them to what he construes to be the *personal* alternatives to the actor. Here the observer compares the behavior performed with other things the actor might have done. The question is whether there is any evidence for "leadership" qualities causing the observed behavior versus other, perhaps, unsuspected, qualities. If the observed behavior could be explained by nonleadership (externally caused) variables as well as possible leadership (personally caused) qualities, the attribution of leadership will be weaker. Finally, if the goals of the observed person are compatible with those of the observer, there will be a stronger tendency for the attribution of leadership. Again, Hollander (1964) arrives at a similar prediction by asserting that the leader must be perceived as having a high degree of identification with the group values and goals.

The attribution theory of leadership potentially explains the psychological intraperson process of members of leaderless groups that results in the attribution of leadership to one member of the group. As described above, through the process of sociometric choice or questionnaire response, such members indicate who they believe to be the "real leader" of the group or whom they would most prefer to have as their leader.

While the theory has not been tested empirically, several of its assertions are consistent with those of Hollander's (1964) Idiosyncracy Theory of Leadership. Interestingly, the idiosyncratic propositions with which the Attribution Theory are consistent are also the propositions of Hollander's theory that have received the strongest empirical support. Further, the theory is clearly testable. If it is shown that the process described in the theory predicts the attribution of leadership, and that the attribution of leadership is associated with follower satisfaction, motivation, and performance, the theory will have made a significant contribution to the understanding of the leadership phenomena.

Operant Conditioning Theory

Scott (1977), Sims (1977), and Mawhinney and Ford (1977) have argued that behavior is predominantly determined by the contingencies of reinforcement in the environment. Thus, these authors agree that since leaders are a significant source of reward contingencies and a significant source of reward administration, leadership can best be explained in terms of the principles of operant conditioning.

According to this perspective, leadership can be regarded as a process of managing reinforcement contingencies in the work environment. Sims suggests that the supervisor can be instrumental in defining the occasion of appropriate responses such as when and how the task is to be accomplished. In the terms of reinforcement literature, the supervisor would be establishing discriminative stimuli (S^D). Sims interprets the traditional leader behavior dimension of Initiating Structure as a somewhat imprecise measure of the degree to which the leader specifies discriminative stimuli. Similarly, goal specifications by the leader specifies the "occasion for a response" of subordinates and these are therefore interpreted as discriminative stimuli. It is further argued that administration of rewards, contingent upon performance, is also necessary to specify the contingency reinforcement. Thus Initiating Structure and goal specification serve to specify "what comes before" subordinate behavior, while leader reward behavior serves to specify "what comes after" subordinate behavior.

Scott (1977) also argues for the operant perspective of leadership. Scott defines leadership as human operant behavior reinforced by its effect on the behavior of others. Scott points out that an interesting implication of his definition of leader behavior is that individuals other than those formally appointed to lead will be "leading" from time to time and that two or more individuals within a group can be simultaneously leading though possibly not the same followers, if different and incompatible follower responses are reinforcing to the several leaders. Another implication of Scott's definition of leadership is that in the typical case, the behavior of the leader, whether emergent or appointed, is under the control of certain variables in addition to the behavior of followers, whereas the behavior of followers is almost exclusively under the control of the behavior of the leader.

Mawhinney and Ford (1977) also intepret leadership within the operant paradigm. They argue that hypotheses of the Path Goal Theory of Leadership may be better explained by the principles of reinforcement and that the resulting interpretation supports the basic propositions of the Path Goal theory without supporting the validity of the motivational model on which that theory rests. These authors argue that the operant interpretation of leader effectiveness has several advantages when compared to the Path Goal theory. First, it is based on a set of empirically derived generalizations. Second, they argue that the assumption of mental activities on the part of subordinates need not be postulated to explain their behavior. Third, explanation in terms of observable behavior permits remedial action by leaders more easily. Fourth, they argue that from this paradigm conditions in formal organizations requiring leadership can be determined. Finally, they argue that using this paradigm it will be possible to predict the relative influence of leader on follower and follower on leader and thus explain reciprocal influence processes.

Stogdill (1974, Ch. 19) reviews a wide variety of evidence from the operant reinforcement literature. His review indicates that when attempts at leadership are reinforced, a higher rate of leadership acts ensue and that the leadership status of the person involved increases in the eyes of the group members. Possessing the ability to offer rewards or to reinforce the expectations of other group members constitutes an advantage toward emerging as a leader. Leadership attempts are made more frequently under high than under low probabilities of reinforcement and positive reinforcement is more effective than punishment.

In addition, Sims (1977) reviews a number of laboratory studies that demonstrate rather unequivocably that when rewards are administered contingent on performance subsequent performance is improved.

Several field studies concerning the leader's use of reward and punishment contingencies are relevant to the operant theory of leadership.

Leader reward behavior The degree to which leaders are seen by subordinates as engaging in rewarding and punitive behavior has been investigated in five studies. This research employs a questionnaire developed by Reitz (1971) entitled "The Leader Reward Behavior Questionnaire" (LRBQ). Responses to this questionnaire indicate the degree to which the leader is likely to reward effective performance and punish ineffective performance. The questionnaire is designed to identify the degree to which leaders engage in reward and punishment contingent on performance rather than noncontingent reward and punishment behavior.

Several authors have found positive relationships between leader contingent reward behavior and satisfaction of subordinates (Reitz, 1971; Sims and Szilagyi, 1975; Keller and Szilagyi, 1976). Keller and Szilagyi also found a positive relationship between leader contingent reward behavior and subordinate expectancies that their performance will lead to valued outcomes. Leader contingent punitive behavior has also been found to be positively related to subordinate's satisfaction (Reitz, 1971; Sims and Szilagyi, 1975). Sims and Szilagyi (1975) interpret these findings to mean that when role demands are ambiguous to subordinates, leader contingent punitive behavior can serve to reduce subordinates' role ambiguity, thus increasing their satisfaction. This hypothesis warrants further testing.

Contingent reward behavior has also been found to be positively related to measures of subordinate performance (Sims and Szilagyi, 1975; Hunt and Schuler, Note 15).

Two longitudinal studies have shown that positive reward behavior causes subordinate performance (Greene, Note 16; Sims, 1977). Greene also found that the relationship between punitive reward behavior and performance was reciprocal. Punitive reward behavior caused performance and a lack of performance caused the leader to use punitive reward behavior. In this study, punitive reward behavior caused dissatisfaction among low performers.

The correlations between leader contingent reward behavior and subordinate performance generally range between .40 and .50. Sims (personal communication) has found that this correlation is highest when performance measures are taken approximately 6 months after subordinates described their leaders' reward behavior. Sims found a correlation of .51 after a 6-month interval between the two measures. He found a slight decrease after 9 months and negligible correlations between leader reward behavior and performance after 18 months.

These findings based on the LRBQ, together with the laboratory experiments reviewed by Sims (1977), make an impressive case for the argument that leaders will be more effective to the extent that they administer rewards contingent on performance rather than administer rewards noncontingently, i.e., without respect to performance level of subordinates.

While the above findings concerning the leader contingent reward and punishment are consistent with the operant paradigm, they are also consistent with Path Goal theory. Leader behavior that functions as discriminative stimuli can also be considered path-clarifying behavior. As stated above, such behavior is positively correlated with subordinates' satisfaction and their expectancies of reward for performance and negatively correlated with subordinates' experience of role ambiguity. Thus the evidence clearly suggests the importance of leader contingent reward and punishment behavior. The theoretical interpretation of this evidence is a matter of substantial controversy (Scott, 1977; Evans, Note 17) and is only likely to be resolved through continued research. The major issue in this controversy concerns the use of intervening cognitive variables such as intentions, role perceptions, or expectations. The specific question is whether or not such variables, when treated as intervening variables between leader behavior and subordinate responses, account for more variance in responses of subordinates than a more parsimonious stimulus-response relationship.

Sims specifies many research questions that emerge from the operant leadership paradigm.

First, several questions are appropriate to a closed system, considering *only* the superior/subordinate dyadic relationship:
- How does Initiating Structure and goal specification by the leader (S^D) relate to the behavior of the subordinate?
- How does behavior of the subordinate relate to subsequent Initiating Structure and goal specification (S^D) by the leader?
- How does reward administration by the leader relate to subsequent behavior by the subordinate?
  —Positive versus negative reinforcers?
  —Reinforcing rewards versus nonreinforcing rewards?
  —Externally controlled versus internally controlled rewards?
  —The interaction effect of various types of reinforcers?
  —Effect of *different* schedules of reinforcement?
  —Effect of *changing* schedules of reinforcement (e.g., "Stretching the ratio")?
- What are the interaction effects between various types of discriminative stimuli and various types of reward administration?

Once we have insight into these *basic* questions, the model can be expanded to an open systems approach that includes consideration of contingencies of reinforcement stemming from nonleader sources in the work setting. The operant paradigm would still be appropriate, and would open new perspectives and new questions to be answered. For example:

- What is the result of discriminative stimuli from the leader when adequate S^P is presented from non-leader sources?
- What is the result when discriminative stimuli from the leader contradict S_Ds from other sources?
- What is the result when discriminative stimuli *must* stem from non-leader sources, but reinforcement administration is controlled by the leader?
- What is the result when rewards administered by non-leader sources reinforce behaviors that contradict those behaviors reinforced by the leader?
- What is the result when discriminative stimuli are specified by the leader, but administration of reinforcers is controlled by non-leader sources?

Even in a closed system the fundamental question yet to be answered is this: What leader behaviors serve to define effective contingencies of reinforcement at work? This is indeed a challenging question, principally because contingencies of reinforcement in the work setting are complex structures.

Once insight has been provided into this basic question, an open systems perspective would be appropriate. That is, how do leader behaviors complement, supplement, and interact with other non-leader contingencies of reinforcement? Answers to this question will go far toward building a truly comprehensive contingency theory of leadership that is based on a functional analysis of the relevant contingencies of reinforcement (Sims, pp. 82–84).

Thus, Operant Leadership theory together with the empirical evidence concerning the leader's management of reward contingencies clearly suggest another important class of leader behavior in addition to leader Consideration, Initiating Structure, and Participating. Further, the research issues raised by Sims (1977) and Scott (1977) call for a functional analysis of leader behavior in terms of the operant paradigm are suggested for future leadership research.

Finally, Mawhinney and Ford's (1977) assertions of the relative merits of the operant paradigm over Path Goal theory deserve testing in future research.

OTHER RELEVANT RESEARCH

In addition to the studies and topics reviewed above, several studies were identified that suggest additional new directions for leadership research. Since these studies do not fit together cohesively, nor do they logically fit within any of the above topics, they are reviewed here for the readers' consideration.

Graen and his colleagues (Graen, Dansereau, and Minima, 1972; Dansereau, Cashman, and Graen, 1973; Dansereau, Graen, and Haga, 1975; Graen, Dansereau, Minami, and Cashman, 1973a; Graen, Orris, and Johnston, 1973b; Graen and Cashman, 1975; Graen, 1976; and Graen and Ginsburgh, 1977) have conducted a series of studies that suggest several important directions for future research.

One series of studies concentrated on employees who had been with the organization of a period of time. Graen et al. (1972) found that the amount of structuring by the leader (as perceived by the subordinate) moderated the correlation between the amount of consideration shown by the leader (again as perceived by the subordinate) and the leader's evaluation of the subordinate's performance. It was found that under both high and low structure there was a positive correlation between perceived consideration and performance evaluation. Under moderate structure there was no correlation between these two variables.

Dansereau et al. (1973) hypothesized that the findings of the moderating effect of the structuring on the correlation between consideration and performance could be used to predict differential rates of turnover when Instrumentality and Equity theories were considered. High and low structuring leaders were assumed to be consistent in their behaviors toward particular individuals. It was also assumed that this consistency clarifies the relationship between subordinate's performance and the resulting outcomes. Finally, it was assumed that medium structure leaders do not unambiguously establish for their subordinates the contingencies between performance and outcomes. Therefore:

> . . . in the high and low structure group, if increased performance leads to increased rewards, then a lack of these rewards can be attributed in part to the individual's *own* performance level. In this situation, if the individual receives relatively unattractive outcomes, instrumentality theory predicts that he will be motivated toward leaving (perhaps moving to an organization where his same performance level will gain more attractive outcomes). In contrast, in the medium structure group a lack of rewards can not be attributed as clearly to the individual's performance level. As a result, the individual may look at the situations of his colleagues as references to help him determine if he is being treated fairly (p. 192).

Accordingly, under medium structure high performers would feel inequitably treated, since they would be contributing more than lower performers but not receiving higher rewards. Thus it was hypothesized that under medium structure high performers would leave. However, under high and low structure only low performers were hypothesized to leave. The study supported these hypotheses.

Dansereau et al., 1975, and Graen and Cashman, 1975, examined the extent to which a supervisor "is willing to consider requests from a member concerning role development" (Dansereau et al., 1975, p. 51). This willingness on the part of the leader is referred to as negotiating latitude. These studies showed that leaders very early differentiated unit members into "cadre" (with greater negotiating latitude) and "hired hands" (with less latitude). Subordinate subsequent satisfaction, turnover rates, performance, and kind and amount of superior/subordinate interaction were found to be a fucntion of the subordinate's membership in either the "cadre" or "hired hand" group.

Graen (1976) has advanced an integration of the previous studies and proposed an interpretation of the findings. Graen suggests that very early in the relationship between supervisors and subordinates the supervisor differentiates subordinates into cadre and hired hands. The basis of differentiation may be the supervisor's perception of the subordinate's incoming role orientation, i.e., the degree to which the subordinate perceives the job as contributing to a long-term career. The differentiation is necessary, because there is too much work for the supervisor to handle alone and so some of the work must be delegated to those subordinates perceived as most compatible with the leader in terms of skills, orientation, and trust. These subordinates may then become the cadre. To reward the cadre for taking on extra duties, the supervisor provides extra attention, support, and negotiating latitude to these members. Over time the hired hands come to feel rejection and respond by leaving or performing less effectively.

Support for this interpretation was presented by Graen and Ginsburgh (1977). They found that the subordinate's role orientation (the degree to which the subordinate perceives the present job relevant to a long-term career) and leader acceptance interact such that there was a decreased tendency for resignation of employees who were high on role orientation, leader acceptance, or both. Employees low on both variables had an increased tendency to resign.

These studies by Graen and his associates suggest many significant research questions.

Concerning findings on leader Initiating Structure, it is inferred that leaders who are high or low in structuring behavior more effectively communicate expectations to subordinates. Studies designed to test this inference directly would be useful. Such a study would involve determining whether subordinates of high and low structuring leaders have a clearer understanding of their leader's expectations than do subordinates of moderate structuring leaders. It is also not clear what the moderate structuring leaders do or do not do. Is it that they do not distinguish between good and bad performers? Or is it that they do not effectively communicate expectations?

The finding that subordinates are divided into cadre and hired hands has significant implications for group cohesiveness and performance.[5] When this division occurs, under what conditions does it result in a lower amount of cohesiveness among all of the people reporting to a given supervisor? Is there a higher level of cohesiveness within each of the two groups? If so, does the cohesiveness among the hired hands serve to help them in thwarting the objectives of the leader and the organization?

It is not clear what behaviors on the part of subordinates and on the part of the leaders result in the subordinates becoming members of each of these groups. Is it the behavior of the subordinates that causes the leader to be alienated from them or is it the behavior of the leader that causes the subordinates to be alienated?

It is not clear how the subordinates' role orientation is affected by the leader. Is role orientation an effect of initial leader behavior or is it a stimulus that causes leaders to treat subordinates differently?

There is not yet a clear definition of the construct of negotiating latitude. Nor is it clear whether leaders or subordinates are the best source of information for measuring this construct.

These questions raised by the research of Graen and his associates offer many fruitful areas for future research. They are especially important because they are based on one of the few systematic cumulative research efforts based on field study data. Clearly such research efforts should be extended and encouraged.

Ronan, Latham, and Kinne (1973) report a field study based on a questionnaire administered to 292 pulpwood producers. They found that goal setting is correlated with high productivity and a low number of injuries only when it is accompanied by supervision. Goal setting without immediate supervision was related to employee turnover. Supervision alone did not correlate with any performance criteria. Here supervision was measured by the number of hours the producer was on the job with his employees. This measure was in turn found to be correlated significantly with responses to questions related to giving the men instructions and explanations, providing training, using varied methods of employee payment, and having military experience; and negatively related to a response to the question "Do you have a key man in your operation?" These behaviors correlated positively with productivity and negatively with injury rate.

Supervisory behavior that was exclusively production centered, consisting of goal setting only and not related to working with the employees, correlated positively with compulsory and voluntary employee terminations. Supervisory behavior consisting of working with the men but not setting production goals and using key men did not correlate with any performance criterion. In a second study, these authors tested the hypothesis that supervision, as defined by staying on the job with men and setting a daily or weekly production goal, results in higher productivity than supervision that does not include goal setting or goal setting that is not accompanied by supervision. A questionnaire was administered to 1,000 independent producers and 892 responded. Questionnaire responses were related to the amount of cord produced per day by each producer. Based on the questionnaire responses, producers were classified according to whether they provided on-the-job supervision and engaged in goal setting. Goal setting and supervision were found to have a significant interaction effect on daily production. Thus the hypothesis was accepted.

Surprisingly, little attention has been given to supervisory goal setting in prior literature. Goal setting as a supervisory behavior has generally been included in, or implied by, questionnaire items or experimental observational categories designed to measure task-oriented leadership but not dealt with as a separate variable. The study by Ronan et al. (1973) warrants replication. If the findings continue to hold, then supervisory goal-setting behavior should be included as an explicit variable in future leadership research.

A study by Oldham (1976) also suggests several additional leader behaviors that warrant further research. Oldham developed questionnaire scales designed to measure the degree to which leaders engage in personally rewarding and personally punishing behavior contingent on performance, goal setting, designing feedback systems, placing personnel on existing jobs that challenge their skills and designing job systems that are motivational. The scales were found to have moderate independence of each other, and with the exception of the Punishment scale, were all found to be significantly related to the motivational effectiveness of middle managers as rated by store managers and assistant managers. These findings support the validity of a conceptualization of supervisory behavior consistent with present theories of motivation. Clearly the findings reported by Oldham deserve replication and extension.

A study by Stein, Geis, and Damarin (1973) suggests a methodological improvement for research concerning emergent leadership. The research by Stein et al. (1973) concerned the accuracy with which observers perceive emergent task and socioemotional leadership in small groups. One hundred and forty-nine undergraduates viewed a videotape of a group and guessed the order in which the group would rank its members on five leadership test items. Subjects were individually and collectively accurate beyond chance. Subjects' accuracy correlated .82 with the actual group members' rankings of each other. Further research using the methodology developed by Stein et al. (1973) is likely to permit a more precise identification and definition of the particular verbal and nonverbal behaviors of group members that result in leadership emergence. Earlier observation methods used in studying emergent leadership were based on a priori categories developed by the researchers. The methodology used by Stein et al. makes it possible to identify the criteria used by group members to attribute leadership to others within the group.

Finally, a methodological suggestion by Motowidlo (Note 18) deserves serious consideration. Motowidlo argues that the most commonly studied dimensions of leader behavior, mainly leader Initiating Structure, Consideration, and Participative Decision Making, may not result in increased effectiveness under certain conditions, because situational factors may cause leaders to behave in particular ways with respect to these three dimensions, thus resulting in little or no variance in behavior in these situations. Yet it is argued that there still might be wide variance in effectiveness in these situations accounted for by other dimensions of leader behavior not under the causal influence of the situational circumstance. Thus it may well be that other dimensions of leader behavior, which have not as yet been conceptualized, are related to effectiveness. Further, given that leader effectiveness is determined jointly by the leader's behavior and the situation, it is necesary to identify the leader behavior variables *in combination* with the situational variables that will account for the most variance in effectiveness. Thus Motowidlo argues that variables of leader behavior should be defined with full recognition of the situations in which they occur and variables of situations should be defined with full recognition of the leader behaviors that occurred in them. Motowidlo recommends a systematic accumulation of a large number of incidents portraying actual examples of specific leader behaviors observed in a broad range of leadership situations. Each example would include a full description of both the observed leader behavior as well as the situational context in which it occurred. The descriptions should be sufficiently complete to enable persons familiar with leadership processes in these situations to estimate reliably the degree of leader effectiveness reflected in each episode of situational leader behavior. Such situational/behavioral episodes might be gathered by "critical incident" techniques portraying highly effective and ineffective leadership. Motowidlo recommends that the episodes then be rated by another group of knowledgeable persons for level of effectiveness they reflect. The resulting descriptions of specific leader behaviors together with particular situations in which they occurred and informed esimates of the degree to which each situational leader behavior is effective could be used to conceptualize inductively combinations of behavioral and situational variables that may account for variance in estimated leader effectiveness. The conceptualizations and the specific incidents should result in speculations and hypotheses about interaction effects of behavioral situational variables, thus leading to an empirically derived and testable contingency theory of leadership. Motowidlo argues that this approach has several advantages over existing methodology.

> Since it does not assume a priori leader behavior variables, it offers a means of avoiding what threatens to be an overly narrow emphasis on the three leader behavior dimensions identified through earlier main effects research . . . While research exploring situationally contingent effects on these three behavioral dimensions may indeed yield much useful information about leadership processes, there is a need for other contingency leadership research that is not confined to these particular behavioral dimensions . . . The strategy outlined here illustrates a broad approach that appears to circumvent the difficulties and promises to enrich leadership research by introducing new variables which may yield interestingly colorful explanations of leadership effectiveness (pp. 10–11).

CONCLUSION

The purpose of the present paper is to identify important empirical generalizations that have been established through replicated research and critical research issues for future investigation. While the literature review was necessarily selective, it is believed by the authors that the material covered is representative of the broad range of leadership research. It is hoped that this paper will be helpful to others in conceptualizations about leadership and in formulations of future research endeavors concerning leadership issues.

FOOTNOTES

1. The authors are indebted to Hugh J. Arnold, Martin G. Evans, Federico Leon, Jack Ito, and Barry Staw for their helpful critiques and suggestions for revision of the first draft of this paper.

2. For purposes of this discussion a group may consist of two or more people.

3. The authors are indebted to Martin G. Evans for calling to our attention the comparisons between Contingency Theory and Idiosyncratic Credit Theory and between Attribution Theory and the Idiosyncracy Theory discussed below.

4. The model analyzed in this section is that for group problems presented in Vroom and Yetton (1973, Chapter 9) and Vroom and Jago (1976).

5. This issue of cohesiveness was suggested to the authors by Martin G. Evans.

REFERENCES

1. Alverez, R. (1968) "Informal Relations to Deviants in Simulated Work Organization: A Laboratory Experiment," *American Sociological Review 33:* 895–912.
2. Bales, R. F. (1958) "Task Roles and Social Roles in Problem-solving Groups," in Eleanor E. Maccoby, T. M. Newcomb, and E. L. Hartley, *Readings in Social Psychology,* New York: Holt, Rinehart & Winston, Inc.
3. ———, and P. E. Slater (1955) "Role Differentiation in Small Decision-making Groups," in T. Parsons et al., *Family, Socialization, and Interaction Processes,* Glencoe, Ill.: The Free Press.
4. Bandura, A. (1968) "Social Learning Theory of Identificator Process," in David A. Goslin (ed.), *Handbook of Socialization Theory and Research,* Chicago: Rand McNally & Company.
5. Barrow, J. C. (1976) "Worker Performance and Task Complexity as Causal Determinants of Leader Behavior Style and Flexibility," *Journal of Applied Psychology 61,* (4) 443–440.
6. Bass, B. M. (1957) "Leadership Opinions and Related Characteristics of Salesmen and Sales Managers," in R. M. Stogdill and A. E. Coons (eds.), *Leader Behavior: Its Description and Measurement,* Columbus: Ohio State University, Bureau of Business Research.
7. ———. (1961c) "Some Aspects of Attempted, Successful, and Effective Leadership," *Journal of Applied Psychology 45:* 120–122.
8. ———. (1963) "Amount of Participation, Coalescence, and Probability of Decision Making Discussions," *Journal of Abnormal and Social Psychology 67:* 92–94.
9. ———, and C. R. Wurster (1953a) "Effects of the Nature of the Problem on LGD Performance," *Journal of Applied Psychology 35:* 96–99.
10. ———, and ———. (1953b) "Effects of Company Rank on LGD of Oil Refinery Supervisors' Performance," *Journal of Applied Psychology 37:* 100–104.
11. Baumgartel, H. (1956) "Leadership, Motivations, and Attitudes in Research Laboratories," *Journal of Social Issues 12,2:* 24–31.
12. Bennis, W. G. (1959) "Leadership Theory and Administrative Behavior: The Problems of Authority," *Administrative Science Quarterly 4:* 259–301.
13. Berkowitz, L. (1953b) "Sharing Leadership in Small, Decision-making Groups," *Journal of Abnormal and Social Psychology 48:* 231–238.
14. ———, and W. Haythorn (1955) "The Relationship of Dominance to Leadership Choice," Crew Research Laboratory, AF Personnel & Training Reserve Center, Randolf AF Base, CRL-LN-55-8.
15. Besco, R. O., and C. H. Lawshe (1959) "Foreman Leadership as Perceived by Superiors and Subordinates," *Personnel Psychology 12:* 573–582.
16. Borg, W. R. (1957) "The Behavior of Emergent and Designated Leaders in Situational Tests," *Sociometry 20:* 95–104.
17. Borgatta, E. F. (1961) "Role-playing Specifications, Personality, and Performance," *Sociometry 24,3:* 218–233.
18. ———, S. Couch, and R. F. Bales (1954) "Some Findings Relevant to the Great Man Theory of Leadership," *American Sociological Review 19:* 755–759.
19. Burke, P. J. (1967) "The Development of Task and Social-emotional Role Differentiation," *Sociometry 30:* 379–392.
20. Calder, B. J. (1977) "An Attribution Theory of Leadership," in B. Staw and G. Salancik (eds.), *New Directions in Organizational Behavior,* Chicago: St. Clair Press.
21. Calvin, A. D., F. K. Hoffmann, and E. D. Harden (1957) "The Effect of Intelligence and

Social Atmosphere on Group Problem-solving Behavior," *Journal of Social Psychology 45:* 61–74.

22. Cammalleri, J. R., H. W. Hendrick, W. C. Pittman, Jr., H. D. Blout, and D. C. Prather (1972) "Differential Effects of Democratic and Authoritarian Leadership Styles on Group Problem Solving and Processes," Research Report 72-1, United States Air Force Academy.

23. Campbell, D. T. (1956) "Leadership and Its Effect upon the Group," Columbus: Ohio State University, Bureau of Business Research.

24. Campion, G. E., Jr. (1968) "Effects of Managerial Style on Subordinates' Attitudes and Performance in a Simulated Organizational Setting," unpublished Doctoral Dissertation, University of Minnesota.

25. Carter, L. F. (1954) "Evaluating the Performance of Individuals as Members of Small Groups," *Personnel Psychology 7:* 477–484.

26. ———, W. Haythron, B. Shriver, and J. T. Lanzetta (1954) "The Behavior of Leaders and Other Group Members," *Journal of Abnormal and Social Psychology 46:* 589–595.

27. Cartwright, D., and A. Zander (1968) *Group Dynamics: Research and Method,* third ed., New York: Harper & Row, Publishers.

28. Cashman, J., F. Dansereau, Jr., G. Graen, and W. J. Haga (1976) "Organizational Understructure and Leadership: A Longitudinal Investigation of the Managerial Role Making Process," *Organizational Behavior and Human Performance 15,* 2: 278–296.

29. Chemers, M. M., and G. J. Skrzypek (1972) "Experimental Test of the Contingency Model of Leadership Effectiveness," *Journal of Personality and Social Psychology 24:* 172–177.

30. Coch, L., and J. R. P. French (1948) "Overcoming Resistance to Change," *Human Relations 1:* 512–532.

31. Comrey, A. L., W. S. High, and R. D. Wilson (1955a) "Factors Influencing Organizational Effectiveness. VI. A Survey of Aircraft Supervisors," *Personnel Psychology 8:* 79–99.

32. ———, ———, and ———. (1955b) "Factors Influencing Organizational Effectiveness. VII. A Survey of Aircraft Supervisors," *Personnel Psychology 8:* 245–257.

33. Crockett, W. H. (1955) "Emergent Leadership in Small, Decision-making Groups," *Journal of Abnormal and Social Psychology 51:* 378–383.

34. Crowe, B. J., S. Bochner, and A. W. Clark (1972) "The Effects of Subordinates' Behavior on Managerial Style," *Human Relations 25:* 215–237.

35. Dansereau, F., Jr., J. Cashman, and G. Graen (1973) "Instrumentality Theory and Equity Theory as Complementary Approaches in Predicting the Relationship of Leadership and Turnover among Managers," *Organizational Behavior and Human Performance 10,* 1: 184–200.

36. ———, G. Graen, and W. J. Haga (1975) "A Vertical Dyad Linkage Approach to Leadership within Formal Organizations: A Longitudinal Investigation of the Role-making Process," *Organizational Behavior and Human Performance 13,* 1: 46–78.

37. Day, R. C., and R. L. Hamblin (1964) "Some Effects of Close and Punitive Styles of Supervision," *American Journal of Sociology 69,* 5: 499–510.

38. DeCharms, R., and W. Bridgeman (1961) "Leadership Compliance and Group Behavior," Technical Report, Contract N ONR 816 (11), Washington University.

39. Delbecq, A. L. (1965) "Managerial Leadership Styles in Problem Solving Conferences," *Journal of Academy of Management 8:* 32–44.

40. Dessler, G., and E. R. Valenzi (1977) "Initiation of Structure and Subordinate Satisfac-

tion: A Path Analysis Test of Path-Goal Theory," *Academy of Management Journal 20,* 2: 251–259.

41. Dow, T. E. (1969) "The Theory of Charisma," *Sociological Quarterly 10:* 306–318.
42. Downey, H. K., J. E. Sheridan, and J. W. Slocum, Jr. (1975) "Analysis of Relationships among Leader Behavior, Subordinate Job Performance, and Satisfaction: A Path-Goal Approach," *Academy of Management Journal 18:* 253–262.
43. Dyson, J. W., D. W. Fleitas, and F. P. Scioli (1972) "The Interaction of Leadership Personality and Decisional Environments," *Journal of Social Psychology 86:* 29–33.
44. Evan, W. M., and M. Zelditch (1961) "A Laboratory Experiment on Bureaucratic Authority," *American Sociological Review 26:* 883–893.
45. Evans, M. G. (1970) "The Effects of Supervisory Behavior on the Path-Goal Relationship," *Organizational Behavior and Human Performance 5:* 277–298.
46. Farris, G. F., and F. G. Lim (1969) "Effects of Performance on Leadership, Cohesiveness, Influence, Satisfaction, and Subsequent Performance," *Journal of Applied Psychology 53:* 490–99.
47. Fiedler, F. E. (1967) *A Theory of Leadership Effectiveness,* New York: McGraw-Hill.
48. ———. (1971) "Notes on the Methodology of the Graen, Orris, and Alvares Studies Testing the Contingency Model," *Journal of Applied Psychology 55:* 202–204.
49. ———, and Chemers, M. M. (1974) *Leadership and Effective Management,* Glencoe, Ill.: Scott, Foresman and Company.
50. ———, G. E. O'Brien, and D. R. Ilgen (1969) "The Effect of Leadership Style upon the Performance and Adjustment of Volunteer Teams Operating in a Stressful Environment," *Human Relations 22:* 503–514.
51. Filley, A. C., R. J. House, and S. Kerr (1976) *Managerial Process and Organizationsl Behavior,* Glenview, Ill.: Scott, Foresman and Company.
52. Fleishman, E. A. (1965) "Attitude versus Skill Factors in Work Group Productivity," *Personnel Psychology 18:* 253–266.
53. ———, E. F. Harris, and H. E. Burtt (1955) "Leadership and Supervision in Industry," Columbus: Ohio State University, Bureau of Educational Research.
54. French, J. R. P., J. Israel, and D. As (1960) "An Experiment on Participation in a Norwegian Factory," *Human Relations 13:* 3–19.
55. Gardner, G. (1956) "Functional Leadership and Popularity in Small Groups," *Human Relations 9:* 491–509.
56. Ghiselli, E. E. (1971) *Exploration in Managerial Talent,* Goodyear Publishing Co. Inc., Pacific Palisades, Calif.
57. Goldberg, H., and M. A. Iverson (1965) "Inconsistency in Attitude of High Status Persons and Loss of Influence: An Experimental Study," *Psychological Reports 16:* 673–683.
58. Goodacre, D. M. (1951) "The Use of a Sociometric Test as a Predictor of Combat Unit Effectiveness," *Sociometry 14:* 148–152.
59. Goodstein, L. D., and W. J. Shrader (1963) "An Empirically-derived Key for the California Psychology Inventory," *Journal of Applied Psychology 47:* 42–45.
60. Graen, G. (1976) "Role Making Processes within Complex Organizations," in M. D. Dunnette (ed.), *Handbook of Industrial and Organizational Psychology,* Chicago: Rand McNally & Company, Ch. 28.
61. ———, and J. F. Cashman (1975) "A Role-making Model of Leadership in Formal Organizations: A Developmental Approach," in J. G. Hunt and L. L. Larsn (eds.), *Leadership Frontiers,* Kent, Ohio: The Comparative Administration Research Institute, Graduate School of Business Administration, Kent State University.
63. ———, F. Dansereau, Jr., and T. Minami (1972) "Dysfunctional Leadership Styles," *Organizational Behavior and Human Performance 7,* 2: 216–236.

63. ———, ———, ———, and J. Cashman (1973a) "Leadership Behaviors as Cues to Performance Evaluation," *Academy of Management Journal 16*, 4: 611–623.
64. ———, J. B. Orris, and K. M. Alvares (1971) "Contingency Model of Leadership Effectiveness: Some Experimental Results," *Journal of Applied Psychology 55:* 196–201.
65. ———, ———, and T. W. Johnson (1973b) "Role Assimilation Processes in a Complex Organization," *Journal of Vocational Beahvior 3*, 4: 395–420.
66. ———, and S. Ginsburgh (1977) "Job Resignation as a Function of Role Orientation and Leader Acceptance: A Longitudinal Investigation of Organizational Assimilation," *Organizational Behavior and Human Performance 19:* 1–17.
67. Graham, W. K., and T. Gleno (1970) "Perception of Leader Behavior and Evaluation of Leaders across Organizational Levels," Experimental Publication System Ms. 144A, Issue 4.
68. Green, S. G., and D. M. Nebeker (1977) "The Effects of Situational Factors and Leadership Style on Leader Behavior," *Organizational Behavior and Human Performance 19:* 368–377.
69. ———, ———, and M. a. Boni (1976) "Personality and Situational Effects on Leader Behavior," *Academy of Management Journal 19*, 2: 184–194.
70. Greene, C. N...(1975) "Limitations of Cross-lagged Correlational Designs and an Alternative Approach," in J. G. Hunt and L. L. Larson (eds.), *Leadership Frontiers,* Kent, Ohio: Kent State University Press, pp. 121–126.
71. Greer, F. L., E. H. Galanter, and P. G. Nordlie (1954) "Interpersonal Knowledge and Individual and Group Effectiveness," *Journal of Abnormal and Social Psychology 49:* 411–414.
72. Guetzkow, H. (1968) "Differentiation of Roles in Task-oriented Groups," in D. Cartwright and A. Zander, *Group Dynamics,* Evanston, Ill.: Row, Peterson.
73. Gustzfson, D. P. (1968) "The Effect of Commitment to the Task on Role Differentiation in Small Unstructured Groups," *Academy of Management Journal 11:* 457–458.
74. ———, and T. W. Harrell (1970) "A Comparison of Role Differentiation in Several Situations," *Organizational Behavior and Human Performance 5:* 299–312.
75. Halpin, A. W. (1954) "The Leadership Behavior and Combat Performance of Airplane Commanders," *Journal of Abnormal and Social Psychology 49:* 19–22.
76. ———. (1957) "The Leader Behavior and Effectiveness of Aircraft Commanders," in R. M. Stogdill and A. E. Coons, *Leader Behavior: its Description and Measurement,* Columbus: Ohio State University, Bureau of Business Research.
77. ———, and B. J. Winer (1957) "A Factorial Study of the Leader Behavior Descriptions," in R. M. Stogdill and A. E. Coons, *Leader Behavior: Its Description and Measurement,* Columbus: Ohio State University, Bureau of Business Research.
78. Hamblin, R. L., K. Miller, and J. A. Wiggins (1961) "Group Morale and Competence of the Leader," *Sociometry 24:* 295–311.
79. Herold, D. M. (1977) "Two-way Influence Processes in Leader-Follower Dyads," *Academy of Management Journal 20*, 2: 224–237.
80. Hollander, E. P. (1960) "Competence and Cofnormity in the Acceptance of Influence," *Journal of Abnormal and Social Psychology 61:* 365–369.
81. ———. (1961) "Emergent Leadership and Social Influence," in L. Petrullo and B. M. Bass, *Leadership and Interpersonal Behavior,* New York: Holt, Rinehart and Winston, Inc.
82. ———. (1964) *Leaders, Groups, and Influence,* New York: Oxford University Press.
83. House, R. J. (1971) "A Path Goal Theory of Leader Effectiveness," *Administrative Science Quarterly 16:* 321–338.
84. ———. (1977) "A 1976 Theory of Charismatic Leadership," in J. G. Hunt and L. L.

Larson (eds.), *Leadership: The Cutting Edge*, Carbondale, Ill.: Southern Illinois University Press.

85. ——, and G. Dessler (1974) "The Path-Goal Theory of Leadership: Some Post Hoc and A Priori Tests," in J. G. Hunt and L. L. Larson (eds.), *Contingency Approaches To Leadership*, Carbondale, Ill.: Southern Illinois University Press.

86. ——, and T. R. Mitchell (1974) "Path-Goal Theory of Leadership," *Journal of Contemporary Business 5:* 81–94.

87. Hunt, J. G., and R. S. Schuler (1976) "Leader Reward and Sanctions Behavior Relations with Criteria in a Large Public Utility," Carbondale, Ill.: Southern Illinois University, Working Paper.

88. Jacobs, T. O. (1971) "Leadership and Exchange in Formal Organizations," Alexandria, Va.: Human Resources Research Organization.

89. Jacobson, J. M. (1953) "Analysis of Interpersonal Relations in a Formal Organization," unpublished Ph.D. Dissertation, University of Michigan.

90. Julian, J. W., E. P. Hollander, and C. R. Regula (1969) "Endorsement of the Group Spokesman as a Function of His Source of Authority, Competence, and Success," *Journal of Personality and Social Psychology 11:* 42–49.

91. Kaess, W. A., S. L. Witryol, and R. E. Nolan1961) "Reliability, Sex Differences, and Validity in the Leadership Discussion Group," *Journal of Applied Psychology 45:* 345–350.

92. Kahn, R., and D. Katz (190) "Leadership Practices in Relation to Productivity and Morale," in D. Cartwright and A. Zandza (eds.), *Group Dynamics Research Theory* New York: Harper and Row, Publishers.

93. Katz, R. (in press) "The Influence of Group Conflict on Leadership Effectivensss," *Organization Behavior and Human Performance*.

94. Keller, R. T., and A. D. Szilagyi (1976) "Employee Reactions to Leader Reward Behavior," *Academy of Management Journal, 19,* 4: 619–627.

95. Knapp, D. E., and D. Knapp (1966) "Effect of Position on Group Verbal Conditioning," *Journal of Social Psychology 69:* 95–99.

96. Korman, A. K. (1968) "The Prediction of Managerial Performance: A Review," *Personnel Psychology 21:* 295–322.

97. Korten, D. C. (1968) "Situational Determinants of Leadership Structure," in D. Cartwright, and A. Zander, *Group Dynamics: Research and Theory,* 3rd ed., New York: Harper and Row Publishers.

98. Lawrence, L. C., and P. C. Smith (1955) "Group Decision and Employee Participation," *Journal of Applied Psychology 39:* 334–337.

99. Lazarus, R. (1966) *Psychological Stress and the Coping Process,* New York: McGraw-Hill.

100. Lewin, K., R. Lippitt, and R. K. White (1939) "Patterns of Aggressive Behavior in Experimentally Created Social Climates," *Journal of Social Psychology 10:* 271–301.

101. Lieberson, S., and J. F. O'Connor (1972) "Leadership and Organizational Performance: A Study of Large Corporations," *American Sociological Review 37:* 117–130.

102. Lowin, A., and J. R. Craig (1968) "The Influence of Level of Performance on Managerial Style: An Experimental Object-Lesson in the Ambiguity of Correlation Data," *Organizational Behavior and Human Performance 3:* 440–458.

103. ——, and W. J. Hrapchak, and M. J. Kavanagh (1969) "Consideration and Initiating Structure: An Experimental Investigation of Leadership Traits," *Administrative Scientific Quarterly 14:* 238–253.

104. McCall, M. W., Jr.7)1976) "Leadership Research: Choosing Gods and Devils on the Run," *Journal of Occupational Psychology 49:* 139–153.

105. McClelland, D. C. (1961) *The Achieving Society,* Princeton, N.J.: D. Van Nostrand.

106. Maier, N. R. F. (1949) "Improving Supervision through Training," in A. Kornhauser (ed.), *Psychology of Labor-Management Relations*, Industrial Relations Association.

107. ———. (1063) *Problem Solving, Discussions and Conferences: Leadership Methods and Skills*. New York: McGraw-Hill.

108. ———. (1970) *Problem Solving and Creativity in Individuals and Groups*. Belmont, Calif.: Brooks-Cole.

109. Mann, R. D. (1959) "A Review of the Relationships between Personality and Performance in Small Groups," *Psychological Bulletin 56*, 4: 241–270.

110. Marak, G. E. (1964) "The Evolution of Leadership Structure," *Sociometry 27:* 174–182.

111. Mawhinney, T. C., and J. D. Ford (1977) "The Path Goal Theory of Leader Effectiveness: An Operant Interpretation," *The Academy of Management Review 2:* 398–411.

112. Megargee, E. I. (1969) "Influence of Sex Roles on the Manifestation of Leadership," *Journal of Applied Psychology 53:377–382.*

113. ———, P. Bogart, and B. J. Anderson (1966) "Prediction of Leadership in a Simulated Industrial Task," *Journal of Applied Psychology 50:* 292–295.

114. Meyer, M. W. (1975) "Leadership and Organizational Structure," *American Journal of Sociology 81:* 514–542.

115. Miles, R. H., and M. M. Petty (1977) "Leader Effectiveness in Small Bureaucracies," *Academy of Management Journal 20*, 2: 238–250.

116. Mitchell, T. R. (1970) "Cognitive Complexity and Leadership Style," *Journal of Personality and Social Psychology 16:* 166–173.

117. ———. (1973). "Motivation and Participation: An Integration," *Academy of Management Journal 16:* 160–79.

118. ———, C. R. Smyser, and S. E. Weed (1974) "Locus of Control: Supervision and Work Satisfaction," unpublished, Technical Report No. 74-57, University of Washington.

119. Moos, R. H. (1973) "Conceptualizations of Human Environments," *american Psychologist 28*, 8: 652–665.

120. Mulder, M., J. R. Ritsema van Eck, and R. D. de Jong (1970) "An Organization in Crisis and Non-crisis Situations," *Human Relations 24:* 19–51.

121. ———, and A. Stemerding (1963) "Threat, Attraction to Group, and Need for Strong Leadership," *Human Relations 16:* 317–334.

122. Oberg, W. (1972) "Charisma, Commitment, and Contemporary Organization Theory," *Business Topics 20:* 18–32.

123. Oldham, G. (1976) "The Motivational Strategies Used by Supervisors: Relationships to Effectiveness Indicators," *Organizational Behavior and Human Performance 15:* 66–86.

124. Palmer, G. J. (1962) "Task Ability and Successful and Effective Leadership," *Psychological Reports 11:* 813–816.

125. Pepitone, A. (1964) *Attraction and Hostility*, New York: Atherton Press.

126. Pfeffer, J. (1977) "The Ambiguity of Leadership," *Academy of Management Journal 2:* 104–112.

127. ———, and G. R. Sanancik (1975) "Determinants of Supervisory Behavior: A Role Set Analysis," *Human Relations 28*, 2: 139–153.

128. Reitz, H. J. (1971) "Managerial Attitudes and Perceived Contingencies between Performance and Organizational Response," *Academy of Management Proceedings* 31st Annual Meeting.

129. Rohde, K. J. (1951) "Dominance Composition as a Factor in the Behavior of Small Leaderless Groups," Evanston, Ill.: Northwestern University, Doctoral Dissertation.

130. Ronan, W. W., G. P. Latham, and S. B. Kinne, III. (1973) "Effects of Goal Setting and

Supervision on Worker Behavior in Industrial Situations," *Journal of Applied Psychology 58,* 3: 302–307.

131. Rotter, J. B. (1960) "Generalized Expectancies for Internal versus External Control of Reinforcement," *Psychological Monographs* 80. (1, whole No. 609).

132. Runyon, K. E. (1973) "Some Interactions between Personality Variables and Management Styles." *Journal of Applied Psychology 57:* 288–294.

133. Rush, C. H. (1957) "Leader Behavior and Group Characteristics," in R. M. Stogdill and A. E. Coons, *Leader Behavior: its Description and Measurement,* Columbus: Ohio State University, Bureau of Business Research.

134. Rychlak, J. F. (1963) "Personality Correlates of Leadership among First Level Managers," *Psychological Reports 12:* 43–52.

135. Sabath, G. (1964) "The Effect of Disruption and Individual Status on Person Perception and Group Attraction," *Journal of Social Psychology 64:* 119–130.

136. Salancik, G. R., B. J. Calder, K. M. Rowland, H. Leblebici, and M. Conway (1976) "Leadership as an Outcome of Social Structure and Process," in J. G. Hunt and L. L. Larson (eds.) *Leadership Frontiers,* The Comparative Administrative Research Institute, Kent State University.

137. ———, and J. Pfeffer (1977) "Constraints on Administrator Discretion," *Urban Affairs Quarterly 12:* 474–498.

138. Sales, S. (1972) "Authoritarianism: But as for Me, Give Me Liberty, or Give Me Maybe, a Big, Strong, Leader I Can Honor, Admire, Respect, and Obey," *Psychology Today 8:* 94–143.

139. Sample, J. A., and T. R. Wilson (1965) "Leader Behaviors, Group Productivity, and Rating of Least Preferred Co-worker," *Journal of Personality and Social Psychology 13:* 266–270.

140. Schachter, S., B. Willerman, L. Festinger, and R. Hyman (1961) "Emotional Disruption and Industrial Productivity," *Journal of Applied Psychology 45:* 201–213.

141. Schriesheim, C. S., R. J. House, and S. Kerr (1976) "Leader Initiating Structure: A Reconciliation of Discrepant Research Results and Some Empirical Tests," *Organization Behavior and Human Performance 15:* 197–321.

142. Schuler, R. S. (1973) "A Path-Goal Theory of Leadership: An Empirical Investigation," Michigan State University Doctoral Dissertation.

143. ———. (1976) "Participation with Supervisor and Subordinate Authoritarianism: A Path-Goal Theory Reconciliation," *Administrative Science Quarterly 21:* 320–325.

144. ———. (1977) "Task Design and Organizational Structure as Contingencies for Leader Initiating Structure Effectiveness: A Test of the Path-Goal Theory," unpublished Mimeo. The Ohio State University.

145. Scott, W. C. J. (1976) "Leadership: a functional analysis," in J. G. Hunt and L. L. Larson (eds.) *Leadership: The Cutting Edge,* Carbondale, Ill: Southern Illinois University Press.

146. Shaw, M. E., and J. M. Blum (1966) "Effects of Leadership Style upon Group Performance as a Fucntion of Task Structure," *Journal of Personality and Social Psychology 3:* 238–242.

147. Sheridan, J. E., H. K. Downey, and J. W. Slocum, Jr. (1975) "Testing Causal Relationships of House's Path-Goal Theory of Leadership Effectiveness," in J. G. Hunt and L. L. Larson (eds.), *Leadership Frontiers,* Kent, Ohio: Kent State Unviersity Press.

148. Sherif, M., B. J. White, and O. J. Harvey (1955) "Status in Experimentally Produced Groups," *American Journal of Sociology 60: 370*–379.

149. Shils, E. A. (1965) "Charisma, Order, and Status," *American Sociological Review 30:* 199–213.

150. Sims, H. P., Jr. (1977) "The Leader as a Manager of Reinforcement Contingencies: An Empirical Example and a Model," in J. G. Hunt and L. L. Larson (eds.), *Leadership: The Cutting Edge,* Carbondale, Ill.: Southern Illinois University Press.

151. Sims, H. P., and A. D. Szilagyi (1975) "Leader Reward Behavior and Subordinate Satisfaction and Performance," *Organizational Behavior and Human Performance 14:* 426–438.

152. Smelser, W. T. (1961) "Dominance as a Factor in Achievement and Perception in Cooperative Problem Solving Interactions," *Journal of Abnormal and Social Psychology 62:* 535–542.

153. Stein, R. T., F. L. Geis, and F. Damarin (1973) "Perception of Emergent Leadership Hierarchies in Task Groups," *Journal of Personality and Social Psychology 28,* 1: 77–87.

154. Stinson, J. E., and T. W. Johnson (1974) "The Path-Goal Theory of Leadership: A Partial Test and Suggested Refinement," proceedings of the 17th Annual Conference of the Mid-West Division of the Academy of Management, Kent, Ohio; pp. 18–36. April.

155. Stogdill, R. M. (1948) "Personal Factors Associated with Leadership: A Survey of the Literature," *Journal of Psychology 25:* 35–71.

156. ———. (1974) *Handbook of Leadership: A Survey of Theory and Research.* New York: The Free Press.

157. Szilagyi, A. D., and H. P. Sims (1947) "An Exploration of the Path-Goal Theory of Leadership in a Health Care Environment," *Academy of Management Journal 17:* 622–634.

158. Tannenbaum, A. S., and F. H. Allport (1956) "Personality Structure and Group Literature: An Interpretative Study of Their Relationships through Event-Structure Analysis," *Journal of Abnormal and Social Psychology 53:* 272–280.

159. Tomekovic, T. (1962) "Levels of Knowledge of Requirements as a Motivation Factor in the Work Situation," *Human Relations 15:* 197–216.

160. Torrance, E. P. (1954) "The Behavior of Small Groups under Stress Conditions of Survival," *American Sociological Review 19:* 751–755.

161. Tosi, H. (1970) "A Reexamination of Personality as a Determinant of the Effects of Participation," *Personnel Psychology 23:*191–99.

162. Uleman, J. S. (1972) "The Need for Influence: Development and Validation of a Measure and Comparison with the Need for Power," *Genetic Psychology Monographs 85:* 157–214.

163. Vecchio, R. P. (1977) "An Empirical Examination of the Validity of Fiedler's Model of Leadership Effectiveness," *Organizational Behavior and Human Performance 19:* 180–206.

164. Verba, S. (1961) *Small Groups and Political Behavior: A Study of Leadership,* Princeton, N.J.: Princeton University Press.

165. Vroom, V. H. (1959) "Some Personality Determinants of the Effects of Participation," *Journal of Abnormal and Social Psychology 59:* 322–327.

166. ———. (1976) "Can Leaders Learn to Lead?" *Organizational Dynamics* 17–28.

167. ———, B. D. Grant, and T. S. Cotton (1969) "The Consequences of Social Interaction in Group Problem Solving," *Organizational Behavior and Human Performance 4:* 477–95.

168. ———, and E. W. Yetton (1973) *Leadership and Decision Making,* Pittsburgh: University of Pittsburgh Press.

169. Weber, M. (1947) *the Theory of Social and Economic Organization,* New York: Oxford University Press.

170. Weiss, H. M. (1977) "Subordinate Imitation of Supervisor Behavior: The Role of

Modeling in Organizational Socialization," *Organizational Behavior and Human Performance 19:* 89–105.

171. Wexley, K. N., J. P. Singh, and G. A. Yukl (1973) "Subordinate Personality as a Moderator of the Effects of Participation in Three Types of Appraisal Interviews," *Journal of Applied Psychology 58:* 54–59.

172. Wiggins, J., F. Dill, and R. D. Schwartz (1965) "On Status Liability," *Sociometry 28:* 197–209.

173. Wofford, J. C. (1970) "Factor Analysis of Managerial Behavior Variables," *Journal of Applied Psychology 54:* 169–173.

174. Yukl, G. (1971) "Toward a Behavioral Theory of Leadership," *Organization Behavior and Human Performance 6:* 411–440.

175. Zedp, S. M. (1969) "Intragroup Reinforcement and Its Effects on Leadership Behavior," *Organizational Behavior and Human Performance 4:* 284–298.

176. ———, and W. F. Oakes (1967) "Reinforcement of Leadership Behavior in Group Discussion," *Journal of Experimental Social Psychology 3:* 310–320.

177. Ziller, R. C. (1955) "Leaders Acceptance of Responsibility for Group Action under Conditions of Uncertainty and Risk," *American Psychologist 10:* 475–476.

REFERENCES NOTES

1. Berkowitz, L., and W. Haythorn (1955) "The Relationship of Dominance to Leadership Choice," unpublished Report CRL-LN-55-8, Crew Research Laboratory, AF Personnel & Training Reserve Centre, Randolf AF Base.

2. Fleishman, E. A. (1972) *Manual for the Supervisory Behavior Description Questionnaire,* Washington, D.C.: American Institutes for Research.

3. Halpin, A. W. (1959) *Manual for the Leader Behavior Description Questionnaire,* Columbus: Bureau of Business Research, Ohio State University.

4. Stogdill, R. M. (1973) *Manual for the Leader Behavior Description Questionnaire—Form XII,* Columbus: Bureau of Busienss Research, Ohio State University.

5. Weed, S. E., T. R. Mitchell, and C. R. Smyser (1974) "A Test of House's Path Goal Theory of Leadership in an Organizational Setting," paper presented at the Western Psychological Association Meeting.

6. Cammalleri, J. R., H. W. Hendrick, W. C. Pittman, Jr., H. D. Blout, and D. C. Prather (1972) "Differential Effects of Democratic and Authoritarian Leadership Styles on Group Problem Solving and Processes," unpublished Research Report 72-1, United States Air Force Academy.

7. House, R. J. (1974) "Notes on the Path Goal Theory of Leadership," unpublished Manuscript. University of Toronto.

8. Hemphill, J. K. (1950) *Leader Behavior Description,* Columbus: Personnel Research Ohio State University.

9. Hollander, E. P., J. W. Julian, and F. A. Perry (1960) "Leader Style, Competence and Source of Authority as Determinants of Actual and Perceived Influence," unpublished Technical Report (No. 5), State University of New York at Buffalo.

10. Julian, J. W., and E. P. Hollander (1966) "A Study of Some Role Dimensions of Leader-Follower Relations," unpublished Technical Report, State University of New York at Buffalo.

11. Schmidt, D. E. (1976) "The Least Preferred Coworker (LPC) Measure: A Review and Reinterpretation of the Research," unpublished Manuscript. (Available from Donald E. Schmidt, Department of Psychology, University of Washington, Seattle, Washington, 98195).

12. Schuler, R. S. (1977) "Task Design and Organizational Structure as Contingencies for Leading Initiating Structure Effectiveness: A Test for the Path Goal Theory," unpublished Manuscript.
13. Kerr, S. (1976) "Substitutes for Leadership," unpublished Manuscript, Ohio State University.
14. Vroom, V. H., and A. G. Jago (1976) "On the Validity of the Vroom-Yetton Model," unpublished Manuscript.
15. Field, G. (1977) "Analysis of the Vroom and Yetton Leadership and Decision Making Model," unpublished Manuscript, University of Toronto.
16. Hunt, J. G., and R. S. Schuler (1976) "Leader Reward and Sanctions Behavior Relations with Criteria in a Large Public Utility," unpublished Working Paper, Southern Illinois University.
17. Greene, C. N. (1976) "A Longitudinal Investigation of Performance Reinforcing Behavior and Subordinate Satisfaction and Performance," paper presented at Midwest Academy of Management Meetings.
18. Evans, M. J. (1977) "Point and Counterpoint in OB," paper presented at Academy of Management Meeting.
19. Motowidlo, S. J. (1976) "Needed: New Variables for the Contingency Paradigm of Leadership Effectiveness," paper presented at Canadian Psychological Association, Toront, (June).

PERFORMANCE APPRAISAL EFFECTIVENESS: Its Assessment and Determinants

Jeffrey S. Kane, ADVANCED RESEARCH RESOURCES

ORGANIZATION

Edward E. Lawler III, THE UNIVERSITY OF MICHIGAN

ABSTRACT

This chapter undertakes four basic tasks in response to the increasing demands for more effective performance appraisal procedures than are currently available. First, the meaning of effectiveness as applied to appraisal procedures is examined and methods for measuring its various aspects are set forth. Second, an effort is made to characterize the status of existing models of the determinants of appraisal effectiveness and to demonstrate the need to develop more refined models. Toward this end, the authors next present a programmatic model to guide future theory and research on the determinants of appraisal effectiveness. Finally, a selective review of existing research and theory relevant to such determinants is undertaken with the intention of

Research in Organizational Behavior, Vol. 1, pp. 425–478
ISBN 0-89232-045-1

bringing to the foreground the ideas and findings that more clearly need to be countenanced in any attempt to devise higher level models.

The assessment of how well people have performed their jobs has been the focus of thousands of articles and research reports. In many respects this is not surprising, since performance appraisals play such a key role in most organizations. Appraisals are commonly used in organizations for one or more of four purposes: as a basis for promotion and placement decisions, as a criterion against which selection devices and training programs are validated, as a basis for reward allocations, and as a means of providing development-oriented feedback to individuals.

Despite the attention performance appraisal has received, the effectiveness of appraisal methods in terms of such factors as convergent and discriminant validity, reliability, discriminability, relevance, and freedom from bias remains inadequate for most if not all the purposes that appraisals are intended to serve. The all-too-apparent inadequacies of existing approaches to appraisal have become under increasingly discomforting to most organizations as appraisal systems have come under increasing legal scrutiny. As Edwards (1976) has pointed out, appraisals used for promotion and for virtually any other personnel decision-making purposes must meet the same EEO standards for validity, reliability, freedom from bias, and relevance as are required of employment tests. Moreover, deficiencies in appraisal effectiveness are hampering attempts to validate external selection procedures for EEO purposes. Finally, the weaknesses of existing appraisal methods have always been major problems for those wishing to use merit as a basis for allocating organizational rewards and for those wishing to provide accurate and informative feedback to employees for developmental purposes.

In short, the effectiveness being demanded of appraisal systems has outstripped what the present state of knowledge about appraisal can supply, thereby creating strong pressures to advance our knowledge of the determinants of appraisal effectiveness. With this in mind, we shall focus in this chapter on the developments and issues that we feel are critical to producing significant progress. In doing this, we shall focus on performance appraisal effectiveness from a psychometric standpoint. We shall, therefore, not consider its effectiveness as an instrument of organizational policy or practice, such as for motivating employees. These issues, although important, relate to the effective use of an appraisal system and are logically subordinate to, or downstream from, issues relating to whether an appraisal system has the capacity to be used effectively.

Before appraisal effectiveness can be improved, it must first be understood. The literature at this point is not clear on what constitutes effectiveness. We shall, therefore, begin our discussion by focusing on the definition and assessment of effectiveness. We shall then characterize our present understanding of the determinants of appraisal effectiveness in terms of the contrast between present models of the determinants of appraisal effectiveness and the types of models that are needed. Finally, we shall examine some ideas and findings that suggest how appraisals can be more effective.

CRITERIA OF APPRAISAL EFFECTIVENESS

Five criteria are important to consider in evaluating appraisal effectiveness: validity, reliability, discriminability, freedom from bias, and relevance. Throughout our discussion of these criteria, as well as in all subsequent parts of the chapter, we shall refer to the raters or appraisers as *sources*, to the ratees or appraisees as *objects*, and to the aspects of performance on which people are being appraised as *performance dimensions* or *PDs*. In our discussion of the five effectiveness criteria, we shall lean heavily on the concept of multitrait-multimethod (MT-MM) data (Campbell and Fiske, 1959). Such data consist of ratings by two or more sources for two or more PDs on the same set of objects. By considering sources, PDs, and objects as separate factors in a three-way analysis of variance, many aspects of appraisal effectiveness can be tested for significance. Much of the discussion to follow in this section will focus on how to use the results of this kind of ANOVA to draw conclusions about appraisal effectiveness. Those desiring more background on MT-MM data and the use of ANOVA to evaluate it are referred to Campbell and Fiske (1959), Stanley (1961), Boruch et al. (1970), and Kavanagh et al. (1971).

Validity

Validity is at the top of nearly everyone's list of what constitutes an effective appraisal (see, e.g., Campbell et al., 1970). There also seems to be a consensus that construct validity is the only relevant type of validity to consider, since the other major type—criterion-related validity—requires the availability of a more nearly ultimate measure of job success. If the latter were available, there would be no need for an appraisal, except perhaps to identify specific performance characteristics for developmental purposes.

The conceptualization of construct validity was revolutionized by Campbell and Fiske's (1959) exposition of the multitrait-multimethod (MT-MM) matrix and subsequent methodological developments regarding its analysis (Stanley, 1961; Boruch et al., 1970; Kavanagh et al., 1971). Evidence of the construct validity of an appraisal system can be derived from an analysis of appraisal score variance attributable to source, object, and performance dimension factors, and their interaction. The needed mean square terms for this ANOVA can be obtained either by conducting a standard ANOVA of the raw data, by analyzing the variance-covariance matrix in the manner shown by Stanley (1961), or by combining terms in the (MT-MM) correlation matrix.[1]

Once the mean squares have been computed for all main, interaction, and total effects, we arrive at the question of how to analyze them to assess convergent and discriminant validity. In the case of convergent validity, we encounter some serious problems with the methods that have been proposed for using ANOVA to assess it that have not been addressed in any published sources we know of. In order to establish the nature of these problems, let's first establish how the concept of convergent validity applies to performance appraisals.

Campbell and Fiske (1959) proposed the term convergent validity to refer to the extent to which multiple methods agree in their measurements of the same traits. The degree of convergent validity is reflected by the size of the monotrait-heteromethod correlations, also called the validity diagonals, of the MT-MM matrix. In our terms, each of these coefficients reflects the correlation between sources in the ratings they assign to a set of objects on a PD. The average size of these correlations is inversely proportional to the strength of the source × object interaction and directly proportional to the strength of the object main effect in the analysis of variance of the MT-MM data (Stanley, 1961; Boruch et al., 1970).

At this point a question arises as to whether convergent validity should be viewed from the standpoint of the degree to which it is present or the degree to which it is absent. Kavanagh et al. (1971) advocate evaluating whether it is present to a significant degree, as revealed by a significant object main effect. The problems with this approach are twofold. First, the object main effect could reach significance at useless levels of convergence between sources, especially when the object sample is large. Second, the strength of the object main effect's evidence for the presence of convergent validity may be less or not significantly greater than the strength of the evidence against convergent validity provided by the source × object interaction effect. Thus, the danger exists that one may conclude that convergent validity is present when the evidence for its absence is stronger than the evidence for its presence.

Adopting a disconfirmatory approach to assessing convergen validity by testing whether the null hypothesis of no source × object interaction effect can be rejected is better than the confirmatory approach because it is more conservative. However, it too has some drawbacks. These include its excessive conservatism in larger samples and the fact that it offers no basis for concluding that convergent validity is in fact present.

A way out of this bind is to conceptualize the confirmation of convergent validity as a function of the extent to which evidence for its presence outweights evidence for its absence. This conceptualization suggests a comparison between the object main and source × object interaction effects to determine whether the former is significantly stronger than the latter. This comparison can be made by computing the ratio of the object to the source × object mean squares, which is distributed as the F distribution under the most typically applicable ANOVA model (PDs fixed, sources and objects random) and can be adjusted to meet F distribution assumptions under other models by forming quasi$F$ ratios. This approach will only conclude that convergent validity is present when the confirmatory evidence is significantly greater than the disconfirmatory evidence rather than significantly greater than zero. This guarantees that at least a majority of the total variance relevant to convergent validity (i.e., the sum of the object and source × object mean squares) is due to convergence rather than to nonconvergence if the presence of convergent validity is confirmed. Thus, negligible correlations between sources in their ratings of objects would never be interpreted as indicating convergent validity, regardless of the levels of significance that they reach.

Discriminant validity is revealed by the strength of the object × PD interaction effect, which reflects the degree to which objects are ordered differently on each PD. To the extent that this effect is significant, discriminant validity is present. The *practical* significance of discriminant validity can be assessed by examining the ratio of the object × PD variance component to the total variance (i.e., the omega squared index). This ratio expresses the percent of total variance due to the tendency for objects to be ordered differently on different PDs and has the advantage of being capable of direct comparison with the same ratio obtained in other studies. Previous writers (e.g., Boruch et al., 1970; Kavanagh et al., 1971) have also proposed that the intraclass correlation of the object × PD effect constitutes an expression of the level of discriminant validity. However, this index focuses more on the consistency with which PDs are differentiated than on the differences between PDs, and therefore seems more useful as a reliability estimate than as a discriminant validity index.[2,3]

Reliability

There are three basic forms of reliability: consistency between occasions, consistency between methods, and consistency between items from the same domain. Each of these has relevance to the effectiveness of appraisals under certain conditions, but only the latter two seem to be generally relevant. Consistency between occasions, known generally as test-retest reliability or stability, requires that the source, object, and context of appraisal remain stable across replications. This condition is difficult to create in the laboratory and rarely occurs in real world settings. Stability should therefore be viewed as a generally inappropriate index of appraisal reliability.

To some extent the conception of reliability as consistency between methods, generally known as inter-rater reliability, shares a drawback similar to that of stability when applied to appraisals. It is an entirely appropriate index when used in the laboratory, where the acquaintance of a number of sources with the object can be controlled. However, the assumption of identical acquaintance among sources with members of an object group is unlikely to be met in real world situations, as Borman (1974) and Klimoski and London (1974) have shown. Care should therefore be taken to ascertain that sources are comparable in their acquaintance with objects' previous performance on the PD to be rated before any importance is attached to inter-rater reliability. When this condition prevails, the MT-MM data can be used to obtain an estimate of inter-rater reliability. This estimate is provided by the intraclass correlation for the object effect, which reveals the extent to which there is more similarity in ratings within objects than between objects. Note that it is usually the average agreement of a single source with other sources that is indicative of inter-rater reliability. However, if appraisals are to consist of the collective judgments of multiple sources, the Spearman-Brown prophecy formula should be used to adjust the intraclass correlation accordingly. (Winer, 1971, Tinsley and Weiss, 1975).

The final possibility conceives of reliability as consistency between items from the same domain and is generally known as internal consistency reliability. It is closely related to discriminant validity. Whereas discriminant validity addresses the issue of whether there are sufficient differences between PDs to justify considering them as distinct constructs, internal consistency focuses on whether a single construct is being measured by a purported measure of that construct. Clearly, if the distinctions between PDs are no greater than the error of measurement, no PD measure can be considered to be measuring one PD construct any more than it is measuring any other PD construct. The appraisal, therefore, must be viewed as an unreliable measure of the PDs it purports to measure. It is consequently asserted that the intraclass correlation of the object × PD ANOVA effect constitutes an estimate of the average internal consistency reliability of the ratings of the PDs comprising an appraisal system. This approach makes use of the discriminant validity ANOVA component to express the degree of purity of the constructs that were discriminated. This form of reliability and the method for measuring it holds an advantage over the previous two forms in that it is not as sensitive to differences in source acquaintance with the objects. It holds the disadvantage that if only one construct is reliably measured by several PD measures, it would ascribe zero reliability to the measurement of all PDs.

However, a system that purports to measure multiple PDs but actually measures only one cannot be considered effective anyway, regardless of how reliably that one PD is measured.

Discriminability

The degree to which an appraisal device succeeds in differentiating among objects and thereby imparts variability to the resulting distribution of scores constitutes its discriminability. According to Garner (1960), this is the most fundamental quality that must be possessed by rating methods. Discriminability has two aspects: structural and operational.

Structural Discriminability Structural discriminability refers to the capacity of the appraisal system itself to express the degree of differentiation existing in the object population. This is jointly determined by three characteristics of the system used to elicit and record source judgments (hereafter referred to as the response scale): its score ratio, uncertainty, and profile differentiation. The score ratio of a response scale refers to the number of response alternatives it offers relative to the number of distinct levels of the construct that are actually present in the object population. For example, suppose overall performance was being appraised on a 7-point scale, and the object population was sufficiently large so that twenty-five distinct levels of overall performance existed. In this case seven possible response alternatives are provided by the response scale relative to twenty-five distinguishable levels of performance, yielding a score ratio of 7/25, or 28 percent. Clearly, an appraisal system cannot fully reflect the degree of differentiation in an object population unless its response scales must be characterized by score ratios of at least 100 guishable levels of the construct being appraised. Thus, appraisal system response scales must be characterized by score ratios of at least 100 percent if maximum discriminability is to emerge.

The uncertainty of a response scale refers to a response scale's capacity to impart information. Specifically, uncertainty is the extent to which the number of response alternatives in each region of a scale continuum varies directly with the proportion of a given object population whose performance falls within the respective region. Uncertainty reaches its highest point when the expected proportion of the object population that will be assigned each scale alternative equals $1/K$, where K=the number of alternatives on the response scale. High uncertainty is a desirable characteristic of response scales, because it permits objects to be differentiated in the regions of scales where they tend to cluster. The failure to observe the uncertainty principle in the construction of appraisal rating scales is widespread. For example, the complaint is often heard that 80 percent or more of manager populations are rated at the highest one or two levels of performance. The uncertainty principle indicates that 80 percent of the response alternatives should occur in the region of highest performance in order to maximize discriminability, rather than the 15 to 40 percent typically found.[4]

The third and perhaps the most important aspect of response scales is profile differentiation. Most existing methods of appraisal require sources to characterize each object's performance in terms of a single point on a continuum of goodness (i.e., satisfactoriness). The only appraisal methods that do not fit this description include ranking, forced distribution, and other methods that focus directly on each object's standing in a group. In order to select a single goodness level to represent an object's performance in a PD, the source must make a summary judgment about all the times that the object performed in the PD during the appraisal period. For some PDs this might, of course, involve hundreds of performance instances. These instances of exhibiting a PD can be represented in terms of the proportion of the total number of instances that occurred at each of the levels comprising the goodness continuum. This way of representing the instances of exhibiting a PD will be called a *performance distribution*. The source's task in most rating systems can therefore be conceived as one of attempting to select a single goodness level to represent an entire performance distribution.

Like any other distribution, a performance distribution has a variety of parameters on which its distinguishing features are manifested. These include the distribution's mean, median, mode, variance, kurtosis, skewness, and its definitional equation. Despite this multiplicity of available parameters, the research of human judgment suggests that when forced to characterize a series of events in terms of a single point on an evaluative continuum, people tend to select what they consider to be the average or expected value of the series. Discounting the other features of the frequency distribution of the events in the series is in fact optimal behavior from a statistical perspective: When a distribution is to be represented by a single point on the continuum over which it extends, the error in such a representation is minimized by the use of the mean. Any tendency sources may have to moderate their judgments by considering other features of the object's distribution such as the variability of performance is actually suppressed by anchoring the intervals of the goodness continuum in behavioral terms. Such behavioral anchoring prevents consideration of the overall goodness of the distribution of performance and focuses the source's attention more directly on the object's typical way or outcome of performing.

Ignoring all of the distinctions between people's performance distributions except for their average or expected values can only be rationalized by taking the view that performance is completely determined by stable intra-individual characteristics (i.e., traits). Under such a view, the average of a performance distribution represents an object's true level on the relevant underlying trait. All instances of the object's exhibiting higher or lower goodness levels are ascribed to measurement error, and any deviation of the performance distribution from normality is ascribed to sampling error. However, this view of performance ignores the massive accumulation of evidence that performance, and human behavior generally, is determined at least as much by variable intra- and extra-individual factors as by traits. For example, the effect of variable motivational states is commonly seen in the tendency for periods of high achievement to often be followed by slacking-off periods; and conversely, for periods of "coasting alone" to be followed by compensatory flurries of achievement. Differences in the extent to which people were successful in overcoming the adverse influences of variations in such internal and external factors can therefore be seen as leading to more and less desirable patterns of performance. These differences in the patterns of performance exhibited over a period are reflected in all of the parameters of performance distributions, not just in their means. It follows that differences on the other parameters cannot be viewed as random error but must instead be considered to reflect meaningful distinctions between the ways that people performed—distinctions that are being ignored by virtually all existing appraisal systems.

The way to ensure that appraisals reflect all the distinguishing features of each performance distribution is to elicit ratings in terms of the rates at which each object exhibited the given PD at a series of benchmark goodness levels. The representation of an object's performance distribution that results from this rating process then has to be scored in a way that meaningfully expresses the combined goodness of the distribution on all of its parameters. These specifications can be met through a scoring system that uses profile similarity measures to express the proximity of obtained and ideal distributions along the three parameters of profile variability: elevation, shape, and scatter. This approach to eliciting and scoring ratings of performance over a series of trials will be called *distributional measurement,* and represents what we feel is a new paradigm for the appraisal of performance. In Appendix C we present a detailed elaboration of the distributional measurement paradigm that describes other features of it that are critical to its conceptualization but are not essential to the present discussion. A description of a procedure for operationalizing the approach will also be found in that appendix.

The use of an appraisal method that requires sources to evaluate performance in terms of a single point on a goodness continuum, and that thereby ignores most of the distinguishing features of performance distributions, may not raise serious problems when only gross discriminations are to be made for purposes such as allocating merit pay increases. However, the inadequacy of such methods does raise serious problems when appraisals are used as the bases for promotions and as criteria in the validation of selection and training programs. Range restriction in performance and criterion validity are usually the principal problems to be overcome in such applications of appraisal, and the needed increases in variance and validity can only be achieved through a distributional measurement method that can accurately express all the distinctions among performances.

This extended discussion has been necessary in order to make apparent the important implications of the third aspect of structural discriminability—profile differentiation. This aspect refers to the extent to which a response scale yields measurements that can be scored in distributional measurement terms that express all differences between performance distributions. Nothing approaching maximum discriminability can be achieved unless the response scale can be scored to express more of the differences between performance distributions than merely those between their means.

Operational discriminability   Operational discriminability refers to the extent to which the actual use of an appraisal system succeeds in distinguishing members of a population. The index of this form of discriminability is derived from MT-MM data. It is simply the significance of the object main effect relative to residual error. The magnitude of this effect can be expressed in terms of the omega squared of its variance component, which is useful for comparisons across studies. Note that this thrid use of the object main effect is different from the previous two. Its use in assessing convergent validity involved comparing it to the source × object interaction instead of to residual error. Its use in assessing inter-rater reliability focused on expressing the percent of total object variance that was true variance.

Freedom from Bias

Bias is the systematic tendency for sources to make assessments without regard to actual differences between objects (people), constructs (i.e., performance dimensions), or both. This definition implies three basic forms of bias, each of which can be conceptualized in termsof MT-MM ANOVA effects. Object bias refers to the tendency for some sources to rank certain objects consistently higher or lower than do other sources across all PDs. As such, it is a disregard of actual differences among objects and a bias for or against certain objects. Guilfor (1954) referred to this bias as relative halo, and pointed out that its presence is revealed by a significant source × object interaction effect. In light of our previous discussion of this ANOVA component, we can also view this bias as a major cause of decrements in convergent validity.

Construct bias refers to the tendency for some sources to rank certain performance constructs higher or lower than do other sources. It may therefore be viewed as a disregard of actual differences between PDs in service of a preconceived notion about the relationships between PDs. This bias, which is reflected by a significant source × PD interaction effect, seems to express the influence of different *implicit theories of performance* held by the sources. Such implicit theories about the interrelationships of many other human characteristics have been well-established as having an influence on judgments of such characteristics (e.g., Mulaik, 1964; Passini and Norman, 1966; Rosenberg and Sedlak, 1972). However, the existence of implicit theories of performance has yet to be explored with respect to whether appraisals are reflections of external reality or of the internal belief systems of sources.

Finally, source bias refers to the tendency of some sources to give higher or lower ratings regardless of the object or OD being assessed. When such a condition prevails, the ANOVA will reveal a significant source main effect. This bias has been variously referred to elsewhere as a leniency or severity bias.

Relevance

This final criterion of appraisal effectiveness refers to the extent to which the entire domain of performance, and no extraneous domains, is accurately represented by the PDs on which an appraisal system assesses objects. Brogden and Taylor (1950) posited three dimensions of relevance that have stood the test of time. Slightly changed to orient them toward the primary purpose of appraisal—discrimination between objects—they are as follows:

1 *Deficiency:* The extent to which an appraisal system *fails* to include all of the performance dimensions for which the organization holds recognized standards for satisfactory levels of performance.

2 *Contamination:* The extent to which an appraisal system includes performance dimensions for which an organization holds no recognized standards for satisfactory levels of performance.

3 *Distortion:* The extent to which prevailing standards on the relevant value criterion call for different weights than: a) those being used to differentiate the goodness of the various ways or outcomes of performing on each dimesnion, b) those being used to differentiate the influence of *scores* on different performance dimensions on composite scores, or c) both a and b.

The deficiency dimension of relevance can be assessed through evaluation of the process through which a system was developed. If the process considered for inclusion all performance dimensions on which any party with a formal interest in how well a job is performed recognizes standards, and if it selected PDs for inclusion on the basis of a rigorous assessment of the degree to which standards on them were generally recognized by sources and objects (i.e., the definitiveness of standards), confidence in the system's freedom from deficiency can be high. To the extent that such a course was not followed, the resulting system must be considered deficient until proven otherwise. Post hoc evaluation of a system is possible and requires an assessment of whether the interested parties perceive that any PDs on which they recognize definitive standards were excluded from the system.

Contamination can be quantitatively assessed by collecting data from the salient source and object populations on their perceptions of the standards for satisfactory and unsatisfactory performance on each PD. If the standards for all PDs included in the system are found to be significantly definitive, the system can be considered free from contamination.

Finally, assessing freedom from distortion in the weights attached to the ways or outcomes of performing within each performance dimension, or to the performance dimensions themselves, requires the comparison of the actual weights to consensus weights derived from surveys of formally interested parties. Care should be taken to use effective rather than nominal weights (i.e.g, the weights nominally assigned) as the actual weights in the case of performance dimensions. Each performance dimension's effective weight is computed as the product of its variance and the square of its nominal weight.

THE DETERMINANTS OF APPRAISAL EFFECTIVENESS

Having described the criteria for appraisal effectiveness and the approaches to assessing their attainment, we now turn to the question of what determines the degree to which an appraisal system achieves effectiveness. Our present state of knowledge of the determinants of appraisal effectiveness can be brought into focus by considering where the model currently guiding research on this issue stands in a hierarchy of scientific models. The four basic types of scientific models, in ascending order of specificity, are as follows:

1 *Epigrammatic:* A short statement expressing an outcome as a function of the main and interactive effects of the *classes* of variables thought to determine the outcome (e.g., $B = f(P, E, P \times P, E \times E, P \times E)$, where $B \times$ behavior, $P \times$ characteristics of the person, $E \times$ characteristics of

the environment, $P \times P \times$ interactions between personal characteristics, $E \times E\times$ interactions between environmental characteristics, and $P \times E\times$ interactions between personal and environmental characteristics).

2 *Programmatic:* A statement expressing an outcome as a function of the main and interactive effects of the salient *types or categories* of variables comprising all the classes specified in the epigrammatic model (e.g., $B \times f(K, S, A, T, E_s, E_p, E_o,$ (2-way interactions), (3-way interactions), . . . , (7-way interactions), where $K \times$ knowledge, $S \times$ skill, $A \times$ ability, $T \times$ temperament, $E_s \times$ social environment, $E_p \times$ physical environment, $E_o \times$ organizational environment).

3 *Systemic:* A statement expressing an outcome as a function of the main and interactive effects of the *specific* salient variables comprising all the categories of variables specified in the programmatic model (e.g., $B_i \times f(C, I, U, R, E_1, E_2, E_3,$ (2-way interactions), . . . , (7-way interactions)), where $B_i \times$ voting behavior, $C \times$ knowledge of candidates' positions, $I \times$ skill at gathering information, $U \times$ ability to understand the positions, $R \times$ authoritarianism, $E_1 \times$ social class, $E_2\times$ type of dwelling, and $E_3 \times$ formality of the organizations in which one holds roles).

4 *Causal:* A path diagram, or a series of structural equations, expressing all the causal paths between the variables specified in the systemic model.

The literature on performance appraisal is largely based on a programmatic model that has evolved out of the following epigrammatic model of the determinants of appraisal effectiveness:

$$E = f(M,O,M\times M,O\times O,M\times O)$$

where:
E = appraisal effectiveness

M = appraisal methodology

O = the organizational environment

$M\times M$ = interactions among aspects of appraisal methodology

$O\times O$ = interactions among aspects of the organizational environment

$M\times O$ = interactions between appraisal methodology and organizational environment.

This epigrammatic model envisions main effects attributable to both methodology and environment. In other words, some methodological features may be better in all situations, and some organizational environments may be inherently more or less conducive to high appraisal effectiveness regardless of the methodology used. It further envisions interactions among methodological features, among organizational environment features, and between methodological and organizational environment features.

The process through which the epigrammatic model of appraisal effectiveness has been elaborated into the programmatic model that seems to be guiding current research is fittingly described as one of evolution. The programmatic model did not result from a sudden insight that revealed the salient variable categories, but rather from the gradual accumulation of variable categories that became recognized as salient. It is impossible to say whether this evolutionary process is complete, and therefore the conceptualization of the programmatic model we shall offer is only meant to be exhaustive of the variable categories recognized as salient as of the present time. We propose these salient variable categories to be as follows:

I. Appraisal methodology categories:
   A. Measurement process
   B. Measurement content
   C. Source type
   D. Object type
   E. Administration characteristics
      1. Purpose of appraisal
      2. Nature of feedback
      3. Timing
      4. Due process
      5. Decision process
II. Organizational environment categories
   A. Categories of the objective organization
      1. Characteristics of individual sources
      2. Task characteristics
      3. Social characteristics
      4. Structural characteristics
   B. Categories of phenomenological organization
      1. Characteristics of individual sources
      2. Task characteristics
      3. Social characteristics
      4. Structural characteristics
   C. Objective—phenomenological interface characteristics

The methodological categories specified above are relatively straightforward. The source and object type characteristics categories are meant to refer to the parameters of the source and object roles per se rather than to the characteristics of the individuals in these roles; the administration characteristics category refers to the characteristics of how a system comprised of a given set of measurement process and content, and source and object characteristics is designed, implemented, and used.

The categories subsumed by the organizational environment represent the variety of attributes that have been used to characterize the setting within the boundaries of organizations. This taxonomy of the internal organization can perhaps be represented more meaningfully through the diagram in Figure 1 than through the above list. As shown in Figure 1, organizations are conceived to have in effect a dual existence: one that can be described without relying on the perceptions of its members (i.e., the objective organization, to use the term loosely), and one that is the product of the perceptions of its members (i.e., the phenomenological or experienced organization). The link between the two organizational existences is the individual member, considered either singly or collectively, depending upon the level of collectivity at which the variables being linked exist. The characteristics of the cognitive processes employed in transforming objective cues into an individual or collective version of the phenomenological organization are conceived as comprising a separate category of variables from those subsumed under the objective and phenomenological categories. We conceive of this cognitive process category in terms of Brunswick's Lens model (Brunswick, 1952, 1956; Slovic and Lichtenstein, 1971). The variables it subsumes are therefore the various dimensions used to characterize the operation of that model

Figure 1. A stereoscopic model of organizations.

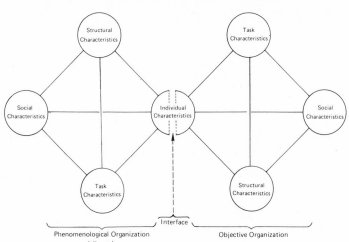

(see the aforementioned references for the specification of these variables).

The selection of the variable categories (i.e., individual, task, social, and structural characteristics) represented in the objective and phenomenological dimensions of the stereoscopic model of the internal organization are based on the conclusions of others who have attacked the taxonomic problem (e.g., Indik, 1971; Udy, 1965; Seiler, 1967; and Campbell et al., 1970). These past efforts seem to converge on these four categories.

It should be noted that our conception of the phenomenological organization is intended to subsume all variables conceived as elements of organizational climate. More precisely, we consider our phenomenological organization to correspond to what James and Jones (1974) call the "perceptual measurement—individual attribute" conception of organizational climate. This latter conception, which is closely associated with the work of Schneider and his colleagues (Schneider, 1972, 1973, 1975; Schneider and Bartlett, 1968, 1970; Schneider and Hall, 1972), holds organizational climate to be the features of the organization as perceived (in descriptive rather than evaluative terms) by its members.

This brings us to the point where we can express the variable categories we have judged to be salient for each of the classes of variables in the epigrammatic model in terms of a programmatic model of appraisal effectiveness:

$$E = f(M_p, M_c, S, O, A, I_o, T_o, G_o, F_o, I_p, T_p, G_p, F_p,$$
$$L, \text{(all 2–14 way interactions))}$$

where: M_p = measurement process

M_c = measurement content

S = source type

O = object type

A = administration characteristics

I_o–F_o = objective individual, task, social, and structural characteristics, respectively

I_p–F_p = phenomenological individual, task, social and structural characteristics, respectively

L = characteristics of objective-phenomenological interface process

Virtually all of the research currently being done on appraisal effectiveness seems to be implicitly guided by aspects of this programmatic model. However, the absence up to the time of this writing of any explicit statement of this programmatic model has apparently served to obscure an important reality: The programmatic model encompasses such a vast array of potentially influential variables that it would take decades to achieve a systemic model through a trial-and-error approach to the identification of the influential variables and their interactions. With the pressure already strong and growing stronger for significant progress in the improvement of appraisal effectiveness, it is clear that what is needed are theories to help us to isolate the key elements of a systemic model more expeditiously.

In an attempt to promote the formulation of the theories needed to expedite progress toward a systemic model, we shall summarize in the brief review that follows some ideas and findings related to the effects of specific variables in each category of the programmatic model of appraisal effectiveness. We shall include the strongest ideas and findings in each category along with those which seem to us to suggest conceptual frameworks potentially capable of integrating an assortment of ideas and findings from multiple categories. Our purpose, therefore, is to be suggestive rather than encyclopedic. The objective-phenomenological interface category will be excluded from this review, since we were unable to find any research on its variables that has been conducted in contexts relevant to appraisal.

Measurement Process

Coombs (1964) has shown that there are logical limits to the number of different ways in which human judges can be asked to furnish data about any given set of objects. These different ways of obtaining human judgments provide a taxonomy of the types of methods that can be used to generate appraisal data. The five basic categories of appraisal methods are principally distinguished by whether they ask the human judge (i.e., the source) to compare performance descriptions for their fit to each object (e.g., forced choice), to compare objects for their fit to a given performance description (e.g., ranking), to compare each object to each performance description (e.g., Likert-type and mixed-standard rating scales), or to compare performance descriptions or objects for their similarity to a standard. These five basic categories of methods can be subdivided further into eighteen more specific *method types* on the basis of whether dominance or proximity judgments are called for, and the specificity demanded in the reporting of judgments. About half of these method types have been tried and found wanting from the standpoint of producing a dramatic breakthrough on the problem of generating effective appraisal data (see Appendix B). Moreover, since the most promising method types have already been tried, it does not seem likely that any great breakthroughs in appraisal effectiveness will result from the application of the method types remaining to be explored.

Recently, most research on appraisal methods has focused on finding better ways of operationalizing the method types that have been in popular use for decades. Most of this attention during the last decade has focused on the evaluation of Smith and Kendall's (1963) technique for minimizing the ambiguity of the descriptive statements used to anchor the intervals of Likert-type rating scales. Unfortunately, the Smith and Kendall technique generally has failed to produce appreciable improvements in the effectiveness of appraisals (e.g., Borman and Dunnette, 1975; Schwab et al., 1975).

Despite this most recent apparent failure to appreciably improve the effectiveness of the conventional rating scale, we feel that the achievement of this goal is still a realistic aspiration in light of the new approach offered by what we have called *distributional measurement*. This approach starts out by asking a more basic question than how rating scales can be better designed. Specifically, it raises the question of just what it is we're trying to measure when we seek to appraise performance. The answer offered is that we should be trying to measure the *distribution* of the rates at which a person exhibited the various levels of performance goodness over multiple trials. This is a radical departure from the conventional view that appraisals need only be concerned with identifying and quantifying the level of performance goodness a person most typically exhibited. This new conceptualization of performance requires the alteration of conventional views as to what method types are appropriate for appraisal purposes and how they should be operationalized. Specifically, only the method type involving the comparison of each object to each performance description is appropriate under the distributional measurement model. This model further specifies that the performance descriptions to which each object is to be compared must be operationalized as the rates at which each level of performance goodness can be exhibited.

No test has been made of whether the distributional measurement approach can improve appraisal effectiveness, because it has never been operationalized. In an attempt to operationalize it, we have developed a new rating approach called Behavioral Discrimination Scales (BDS). We believe that this new technique represents a practical way to operationalize distributional measurement successfully. While it shares the feature of behavioral anchoring with BES and other methods, BDS takes a new approach to reporting and scoring performance on the continuum of behaviors (or outcomes). A complete description of the procedure for developing a BDS system for a job is presented in Appendix C. We shall, therefore, only take enough space here to briefly enumerate the advantages that BDS would seem to offer:

1. It seems to offer a level of discriminability approaching that of ranking techniques without the loss of interval information.

2. Its use of distributional measurement permits entirely new aspects of people's performance to be measured (i.e., the shape and scatter of the profiles of their performance within each performance dimension).

3. It permits people to be directly appraised on factorially complex aspects of their performance (i.e., in terms of behavior and outcomes that represent multiple dimensions).

4. It contains built-in countermeasures against the formation of response sets.

5. It tends to suppress rather than encourage source tendencies to consciously make nonveridical ratings.

6. It directly achieves all the dimensions of relevance through the process of its development.

7. It permits the preparation of feedback reports of high diagnosticity and informativeness.

If the contributions that we foresee BDS making materialize, many of the purely psychometric difficulties with appraisals would be resolved. Gaining control over these difficulties would facilitate more effective research on the other determinants of appraisal effectiveness. Thus, from both an operational and a research standpoint it would seem imperative that the validity of our claims for BDS, and for distributional measurement generally, be evaluated.

Aside from the purely psychometric aspects of the measurement process employed in an appraisal system, there seems to be one other aspect of the measurement process that deserves mention. Appraisals may be oriented toward assessing a person's past performance, present capability, or expected future performance. Only a past performance orientation can be considered to elicit a direct assessment of performance during a prior period. Appraisals oriented toward present capability are attempts to infer capacity to perform from prior periods of performance. Requiring the source to make such an inference leaves the door open to the influence of a variety of judgmental errors and biases. A potentially better approach to assessing capability is to have the source recall the task cycle in which the person performed best during a specified prior period. Performance during the recalled task cycle would then be assessed and the resulting assessment could be interpreted as the person's maximum demonstrated ability to perform. Of course, even the highest level of performance that an object exhibits during a specified prior period would be depressed to an unknown degree from the object's true maximum by motivational factors. However, it is our view that the assessment of the maximum level exhibited is as far as observers can go in assessing another's capability without entering the realm of indefensible speculation.

The remaining alternative among possible time orientations is the future orientation. The use of this orientation is exemplified by the Behavior Expectation Scale approach, which asks for sources' judgments of expected performance. In using this orientation, the source is not only required to infer the object's capability on a construct, but also to predict future motivational levels. It is therefore even more susceptible to judgmental errors and biases than the present capability orientation. Sources are in effect freed from the constraint of having to base their judgments on the levels of performance objects actually exhibited. Expected performance can be specified at any level above or below that typifying prior performance on the basis of expected changes in motivational, skill, or knowledge levels. General impression or friendship is probably the major factor that ends up being assessed when such an orientation is used. This could explain the failure to find appreciable levels of discriminant validity in uses of the BES method. Smith and Kendall (1963) proposed the future orientation as a way of getting around the virtual impossibility of finding a behavior or outcome for each goodness level that describes all persons who performed at the respective level. However, we believe that the future orientation should be abandoned in favor of simply stating the performance descriptions to which objects are compared at the generic level that is widely applicable to objects performing the same jobs.

Measurement Content

In this category, concern is focused on *what* should be assessed by an appraisal system rather than on *how* it should be assessed. There are three basic categories of possible appraisal content: personality traits, behaviors, and outcomes. A major portion of the debate over appraisal content continues to center on whether or not personality traits are appropriate content for performance appraisals. The widely accepted view that a performance appraisal should be a measure of a person's performance rather than of the person per se was well on its way to being equally widely interpreted as excluding traits as content until an article by Kavanagh (1971) attempted to rescue traits from oblivion. He did this by throwing a construct validity lifeline to traits. Specifically, he implied that if traits could be appraised validly and shown to have a place in the "nomological network" underlying an ultimate criterion of performance, they would qualify as legitimate elements of appraisal content.

Identifying traits that hold a place in the nomological network of job performance ignores the important fact that such traits act as *causes or limiters* of performance and do not constitute performance per se. Even though traits may relate in predictable ways to more nearly ultimate criteria of performance, this no more qualifies them as surrogates for such criteria than it qualifies IQ scores as surrogates for school grades. Thus, since traits only constitute characteristics of the person rather than of the person's performance, regardless of the strength of their causal effect on the latter, we contend that traits have no place in the content of appraisal systems.

The exclusion of traits as legitimate appraisal content leaves behaviors and outcomes as the only potentially appropriate types of content. The decision as to which of these types should be used where a choice in fact exists is properly based on the level of *equifinality* characterizing the job in which performance is to be appraised. Equifinality is a characteristic of a job (or of any other goal-directed system) referring to the extent to which the job's objectives or desired outcomes can be reached by two or more equally acceptable paths. If a job lacks equifinality, there is one best way to produce each of a job's desired outcomes and it is therefore appropriate to appraise incumbents on the extent to which they exhibit the optimal behaviors. In contrast, to the extent that equifinality prevails in a job, behaviors are inappropriate as appraisal content, since by definition no one set of behaviors is inherently preferable to the behaviors comprising the other equally acceptable routes. Under such conditions, incumbents should only be appraised on the extent to which they achieve the desired outcomes. Note that we are proposing here that an objective task characteristic (viz., equifinality) interacts with measurement content type to influence appraisal effectiveness.

Source Type

Under consideration here is whether the type of source employed in an appraisal system exerts any impact on the effectiveness of the system. Source types are distinguished by the nature of the relationships they hold with the object population. Seven basic types of sources can be distinguished on this basis, as follows:

1. Immediate supervisor.
2. Second to the Nth most immediate supervisor.
3. Peers.
4. Self.
5. Subordinates.
6. Internal or external clients.
7. Independent trained observers.

Three lines of research yield some insight into considerations that bear on source type. First, appraisal effectiveness has been found to vary directly with the frequency (e.g., Hollander, 1954; Whitla and Tirrel, 1953; Friedson and Rhea, 1965) and relevance to performance (Freeberg, 1969) of the contacts between sources and objects. Second, different types of sources have been found to differ in the specific aspects of any given job's performance that they are able to appraise effectively (e.g., Borman, 1974; Klimoski and London, 1974). Third, no one type of source has consistently been found to be the most effective source; peers, supervisors, and subordinates have all been shown to be good sources of data under certain conditions.

These findings suggest that the conventional use of a single-source type in an appraisal system is suboptimal under most circumstances. No single source type is usually able to gain an accurate view of all aspects of job performance because of each source type's observations being largely restricted to those aspects of object performance that are salient to the formal source-object relationship. When each object's performance is appraised by only one source, it is therefore relatively easy for objects to maximize their performance on only those aspects observable to the type of source appraising them (Lawler and Rhode, 1976). In addition, if objects know that sources are likely to observe their performance at regular intervals, they can easily maximize the source's impression of them by performing their best only on these occasions. Thus, each source type's view of performance is not only limited to those aspects that are visible during the kinds of contacts inherent to its relationship with the object population, but its view is also under the control of the object population to the extent that the occurrence of such contacts is predictable.

The only remedy for the handicaps associated with the use of only one source to appraise each object's performance is to adopt a multiple-source approach to appraisal. In this approach, all aspects of the object's performance that are revealed to each potential source type need to be catalogued. It is then necessary to determine whether each potential source type does in fact have access to significant amounts of information about any important aspects of object job performance. Consideration should also be given to the extent to which the information gained by each source type is under the control of objects. Those source types found to be in a position to contribute useful information are designated to comprise the appraisal system's sources. Each such source type then becomes responsible for completing appraisal reports on those aspects of object performance about which it is directly knowledgeable. An overall performance score is then formed by aggregating the scores from all sources. This approach holds the unique advantage of eliciting information about each performance aspect from the source(s) most knowledgeable about the aspect. Moreover, it refrains from requiring any source type to go beyond its direct knowledge of the past performance of the objects being appraised, which would seem to short-circuit a major source of halo (i.e., object) bias and discriminant validity decrements.

Object Type

Object types vary along two dimensions: level of collectivity and relationship to source, as shown in Figure 2 below. Collectivity level refers to the size of the work unit that is defined to be the object of appraisal. Work units can range in size from one individual worker to the entire organization. Intermediate aggregation approaches might focus on such groups as departments, divisions, or profit centers. Figure 2 shows that there are at least thirty different ways of defining the object of any given appraisal system. This multiplicity of possibilities represents both a problem and an opportunity. It is a problem in that it suggests that a large number of contingencies need to be explored in order to achieve even a working knowledge of the conditions under which each object type is optimal. Yet the very existence of such a range of contingencies offers the opportunity to be able eventually to fine tune appraisal systems to specific circumstances in order to maximize their effectiveness.

Research bearing on the choice of object types is exceedingly sparse, and probably the only well-established findings emerge from the study of prejudice effects. These studies show that appraisals tend to biased against females in traditionally male jobs (e.g., Deaux and Emswiller, 1974; Rosen and Jerdee, 1973, 1974) and against racial minorities (e.g., DeJung and Kaplan, 1962; Flaugher et al., 1969; and Quinn, 1969). Until the day when such biases are eliminated, it may be best to define the object type, where possible, as a level of collectivity that includes both males and females, or minotiries and nonminorities, who are part of the

Figure 2. A two-dimension typology of appraisal objects.

Level of Collectivity

Relationship to source:	Individual	Work group	Dept. or division	Location	Total organization
Immediate subordinate					
Second to n^{th} level subordinate					
Immediate superior					
Peers					
Identity (e.g., self)					
Supplier					

same work group or of some higher-level aggregate. This strategy of shifting to higher levels of collectivity would seem to be a promising way of attacking other forms of object bias as well (e.g., Levine et al., 1974; Leventhal and Whiteside, 1973). Another reason for favoring high levels of collectivity is that in some situations (e.g., when tasks are highly interdependent and autonomous groups or teams are organized) many aspects of performance simply are not visible at the individual level. This strategy, which calls for going to a higher level, sacrifices individual data in return for a greater measure of fairness and validity (Lawler, 1971).

Administration Characteristics

There are five aspects of appraisal administration that appear to be capable of exerting strong effects on appraisal effectiveness and that ought therefore to be of primary concern to researchers focusing on this category.

Purpose of appraisal Appraisal results are commonly used for any one or combination of four purposes: to serve as an internal selection criterion, as the criterion against which other selection devices and training programs are validated, as a basis for reward allocation, and as a means of guiding the way to improved performance. There is no evidence or any other compelling reason to believe that appraisals are inherently more effective for some of these purposes than others. However, there is evidence that in practice it is more difficult to conduct appraisals effectively when important rewards depend on the results (Lawler, 1971).

A number of writers have suggested that using appraisal results for certain combinations of purposes can cause serious problems. For example, it has frequently been suggested that it is not advisable to use appraisals both as development guides and as a basis for reward allocation (see, e.g., Meyer, Kay, and French, 1965). Typically, such appraisals are ineffective as development guides, because feedback of their results generates defensiveness and consequent rejection of the criticism they contain. They also may fail as bases for reward allocation if there is any tendency for sources to balk at making discriminations when there is the prospect of having to confront objects with the appraisal results during feedback sessions.

One way to handle this combination of purposes, which unfortunately is the combination most commonly required of appraisal systems, is to provide sources with frequent opportunities to discuss performance with objects before any consequences actually occur so that there will be no surprises when reward decisions are made and feedback will be separate from the actual pay discussion. This can be done by increasing the frequency of appraisal and feedback during each appraisal period. Pay raises would be based on the aggregate of all appraisals during the period, while each individual appraisal would concentrate on development. We have no proof that this approach would succeed in combining these purposes effectively, but it seems to us to be more plausible than relying on magic formulas for "how to conduct an appraisal interview."

Lawler (1977) has suggested that certain groups are able to handle this problem by openly discussing the performances of members and then reaching a joint pay decision. He cautions, however, that this process is very difficult to develop and that it often requires the help of an outside party. In any case, the development/reward administration conflict is the prime example of what may be a number of incompatible combinations of appraisal purposes.

There are some organizational conditions that are obviously antagonistic to appraisals serving certain purposes effectively, such as their use as development guides in situations where objects neither trust nor respect sources. Unfortunately, there has been little in-depth exploration of either the obvious or the more subtle impacts of organizational properties on the purposes that appraisals can serve effectively. One exception to the general lack of attention to this subject is a treatise by Ghiselli (1969). This work relates to the use of appraisals as internal selection criteria. When used for this purpose, the appraisal's validity in predicting performance at the next highest level for which people are selected is probably the most important basis on which to evaluate the appraisal's effectiveness. We will refer to this type of effectiveness as the appraisal's *promotional validity*.

Ghiselli (1969) derived a model that showed that the promotional validity of appraisals is a function of two categories of organizational properties:

1. Those that maintain the variance of performance at each level despite the fact that each higher level's population constitutes a more thoroughly screened and tested group than the populations of the levels below.
2. Those that increase the relationship between the abilities required at successively higher contiguous levels.

Ghiselli's model implies that the properties that serve to maintain performance variance at successively higher organizational levels are an increasing span of control as the formal hierarchy ascends, and flatness of structure (i.e., a small number of levels). It implies further that an increasingly high relationship between the abilities required at successively higher contiguous levels would be produced by decentralization of authority and decreasing disparity in the nature of tasks performed by successive levels (i.e., increasing delegation).

The theoretical effect of both of the above categories of properties is to increase the correlation between the appraisals received by people at a given organizational level and the appraisals they received in the organizational levels they occupied just previously. However, Ghiselli's model relies on something akin to the Peter Principle, operating to maintain variance in performance at each level. Yet, it may be that as people become more reliant to subordinates as they ascend the organizational hierarchy, opportunities for their personal exceptionality to be evidenced diminish and a leveling effect opposite to Peter Principle expectations occurs. In such cases it would not be the aforementioned properties that would maintain the promotional validity of appraisals, but rather the capacity of appraisals at successively higher levels to accurately discriminate within increasingly narrow ranges.

Feedback The second aspect of appraisal administration that merits mention is feedback. Specifically, when, if ever, should appraisal results be fed back to the object population? We have already seen that feedback may engender reduced discriminability and leniency bias when the appraisal is also being used for the purpose of reward allocation. However, a study by Creswell (1963) suggests that it is not the mere prospect that objects will learn of appraisal results that makes sources timid. It might therefore be inferred that it is the uncomfortable prospect of having to confront objects personally with performance appraisal results in feedback sessions that causes sources to become lenient and nondiscriminating. This inference needs to be investigated empirically, as do the effects of feedback on the effectiveness of appraisals conducted for all other purposes and combinations of purposes.

Timing The third important characteristic of appraisal administration is the timing of appraisals. By timing we mean the length of the time interval between appraisals. There is a noticeable absence of previous theory and research on the issue. It seems to us that Jaques' (1961) Time Span of Discretion (TSD) is a concept that can be usefully applied as a basis for appraisal timing. TSD is the length of time between the point at which the incumbent of a position is assigned a task and the point at which substandard performance could become evident. If the premise is accepted that performance should only be appraised after its effectiveness has had a chance to become apparent, it follows that appraisal periods, (i.e., intervals between appraisals) should consist of at least one TSD.

The actual number of TSDs that should comprise the appraisal period should be determined on the basis of the object population's need for feedback, the number of TSDs needed to make reliable judgments of performance effectiveness, and the availability of time for appraisal completion on the part of the source. The purposes of an appraisal should also influence the optimal length of an appraisal period. Even within a given purpose served by appraisals there may be factors that bear further on what constitutes the optimal appraisal period. This is illustrated by the Levine and Weitz (1971) study of appraisals used for reward allocation purposes. They found that the more difficult the job to be learned, the longer were the intervals needed between employment and first appraisal in order to discriminate among job incumbents. This study also points out that the time interval may need to change as a function of how long the person has been on the job. New employees, for example, seem to need more frequent appraisals than longer-term employees (see, e.g., Van Maanen, 1976). Regardless of how long the optimal appraisal period turns out to be, the point remains that it should be specified in terms of integer multiples of the TSD. Fractions of TSD do not contribute any useful information about performance.

It should be clear from the considerations we have mentioned that the determination of appraisal period length for a given object population requires considerably more care than it is commonly accorded. We have probably only scratched the surface of the domain of factors that should be considered in setting appraisal periods. It nevertheless seems certain, even at the present state of knowledge, that the optimality of appraisal periods exerts an appreciable impact on appraisal effectiveness.

Due process   A fourth administrative characteristic of appraisals that may have an important influence on at least some aspects of appraisal effectiveness is the availability of procedures for appealing appraisals and for having such appeals fairly adjudicated. This due process feature of an appraisal system may act to suppress both conscious biases and careless reporting on the part of sources. Being aware that their appraisal reports will be reviewed both from the organization's and from objects' standpoints should increase the perceived probability that an erroneous report will be exposed. This should tend to motivate the source to do an accurate job of rating. Thus, a review and appeal system may improve all the aspects of appraisal effectiveness.

Decision process   A final aspect of appraisal administration that seems to have an important influence upon appraisal effectiveness is the method by which specific behaviors or outcomes are selected for inclusion in a system. Appraisal content may be developed either with or without the participation of the source and object populations. Borman and Vallon (1974) found that source participation was the principle characteristic of the Smith and Kendall BES method that led to any advantages in reliability and bias reduction it holds over other methods. However, they left unanswered the question of whether the benefits of participative development stem chiefly from the heightened familiarity with appraisal content that results, or from the heightened commitment to making the system succeed that participation tends to engender. Whether the beneficial effects of participation are primarily informational or motivational could have a major impact on the general applicability of participative approaches to selecting appraisal content. While informational effects seem likely to be resistant to moderation by other factors, motivational effects may be nonexistent in situations where participation as a general style of decision making runs counter to prevailing norms and values.

Several studies have suggested that object participation in design may be very desirable. The literature on management by objectives (MBO), for example, argues that this process works best when objectives are jointly set (Lawler, 1971; Porter, Lawler, and Hackman, 1975). Interestingly, this literature also argues for the joint assessment of performance against goals. The presumed advantages of this are very similar to the ones we mentioned for an appeal process. In some ways, a joint assessment process may be viewed as representing the first step, or even the ultimate step, in an appeal process. However, as yet there is only limited evidence for the desirability of object participation in the design and execution of their appraisals, and therefore no definitive conclusions can yet be drawn about either type of participation.

Characteristics of the Organizational Situation

Characteristics of individuals   There is a rather substantial body of literature that demonstrates the importance of individual difference factors in determining the accuracy and nature of performance appraisal ratings. Most studes have involved supervisors rating subordinates on performance, but a few have involved peer ratings, subordinate ratings, and ratings by others. They all point to one quite strong conclusion: Individual differences in rater characteristics make a significant difference in how effectively they rate and appraise the job performance of others.

In a recent study, Borman (1977) showed that 17 percent of the variance in the accuracy of ratings can be attributed to the characteristics of the person doing the rating. He looked at such individual difference variables as personality, interest, ability, and background. According to his study, accurate raters tend to be dependable, stable, good-natured persons who are seldom described in terms such as rebellious, arrogant, careless, headstrong, irresponsible, disorderly, or impulsive. The results of a study by Schneier (1977) are generally consistent with those found by Borman. Schneier reports that more cognitively complex raters provide better BES data than do less cognitively complex individuals.

Other studies have shown that the actual job effectiveness of the rater makes a difference in the kind of ratings that are obtained. Studies by Levy (1960) and by Kirchner and Reisberg (1962) have shown this effect. Kirchner and Reisberg, for example, found that more effective managers tend to value initiative, persistence, broad knowledge, and planning ability. Less effective managers tend to value cooperation, company loyalty, good teamwork, tact, ad consideration. Schneider and Bayroff (1953) and Mandell (1956) have shown that brighter and more successful managers perceive their subordinates in different ways than do less effective managers. Also consistent with the thesis that characteristics òf raters make a difference is some recent evidence that training raters may help to overcome such rating problems as halo and leniency bias and restriction of range. For example, Latham, Wexley, and Purcell (1975) found that a carefully prepared training program for raters can indeed significantly reduce the severity of a number of rating problems.

All these studies seem to point to the conclusion that observers differ considerably in their ability to do an effective job of observing, recalling, and appraising the behaviors and attainments of other persons. These differences are largely attributable to differences in various personal characteristics of observers that influence the kinds of job behaviors and outcomes that they look for and define as effective. A very important implication of this is that any model of performance appraisal effectiveness must take into account the differing perceptions of behavior that are likely to come from raters with different characteristics. In practice, the most effective rating with different characteristics. In practice, the most effective rating system may be one in which greater weight is given to the observations that are provided by people who are more effective raters, or to ask only a select "validated" group of raters for data. Alternatively, it may be that people can be trained to be better raters. Such training is rarely done in organizations and to some extent may be a cure for some of the ineffectiveness of some pople as raters of performance.

There is also evidence that the personal characteristics of the rater interact with the personal characteristics of the object to affect the quality of the appraisal data. Such things as the interpersonal attractiveness to the rater of the person being rated and the degree of similarity or dissimilarity between the rater and the object seem to impact on the accuracy of the ratings that are obtained. Kipnis (1960), for example, has shown that the interpersonal attractiveness of the person being rated is likely to be related to the leniency of the ratings. Attractive people tend to be rated more favorably. There also is a quite large literature on similarity that suggests that people tend to rate people like themselves more favorably. Physical appearance (Carlson, 1967); sex role stereotypes (Deaux and Emswiller, 1974; Rosen and Jerdee, 1973, 1974); family background, kinship, and religious affiliation (Powell, 1963); race, sex, and national origin (Borman, 1974; Hamner, Kim, Bard, and Bigoness, 1974); and attitudinal and personality similarity (Neiva, 1976) all have shown to be related to the favorableness and, presumably, the accuracy of performance appraisal ratings.

Overall, there is a growing body or research that indicates that the characteristics of the rater can interact with the characteristics of the object being rated to influence the validity of the ratings. Apparently, certain combinations of personal and other characteristics tend to produce more accurate ratings than others. Unite the time when training procedures prove capable of counteracting the negative effects of such characteristics, organizations can do little more than recognize these influences and take them into account. This is especially important in the case of race and sex differences, which can leas to particularly unfortunate rating errors. As far as a model of performance appraisal effectiveness is concerned, the evidence suggests that such a model probably needs to contain some fairly complicated interaction effects that take into account both the characteristics of the individual doing the rating and the characteristics of the object being rated.

Task characteristics   There is only a limited amount of evidence on the impact of job characteristics on the effectiveness of performance appraisals. However, there are good reasons to believe that the effects of job characteristics may be substantial. The evidence that does exist shows quite clearly that job certainty and specificity have a strong impact on the accuracy of the performance appraisal ratings that are made by supervisors. Shaw (1972), for example, has shown that the effects of negative stereotyping on ratings are stronger if the position involved is only generally described (e.g., management trainee) than when it is specified in more detail (e.g., industrial engineer). Similarly, Senger (1971) found that the relationship between supervisor-subordinate value similarity and the competence rarings given to subordinates was stronger in general management (low certainty) jobs than in functional management jobs. Nieva (1976) found results very similar to Senger's. She used a measure of job certainty and found that when job certainty was low the similarity-favorability relationship was stronger. All this suggests that when an ambiguity about the task to be performed exists, various kinds of biases and invalid perceptions tend to influence the ratings of job performance.

In addition to job certainty and specificity, there are some other task characteristics that might be expected to influence the effectivenes of performance appraisal ratings. Existing approaches to conceptualizing the psychologically salient dimesnions of tasks—schemes such as those of Hackman and Lawler (1971), Thompson (1967), Perrow (1967), and

Lynch (1974)—suggest a numbe of task attributes that may well moderate or impact upon the effectiveness of performance appraisal ratings. Unfortunately, there is littel research addressed to the nature of such impacts. Nevertheless, some of the more readily conceivable influences of task attributes are worth noting because of their potential importance. One particularly important task attribute that appears in most task classification systems is interdependence. Tasks can range from being highly independent to being highly interdependent. In the former, it would seem that it would be relatively easy to rate the performance of an individual, since the job lends itself to the performance of an individual being highly visible. On the other hand, highly interdependent tasks that require teamwork and a high level of cooperative effort among a group make it very difficult to rate the job performance of the particular individual because his or her peformance cannot be separated from that of others. Thus, one would suspect that these kinds of jobs would be particularly likely to produce biased ratings. Of course, one solution to this problem is to note rate individuals but to rate the performance of the whole interdependent team or group.

Another attribute of tasks that appears in most task classification systems is that of employee autonomy. This dimension concetns the degree to which the employee, through personal effort, can actually control the pace of the work, the nature of the work, and the way in which the work is done. The alternative to a high-autonomy job is a job in which everything is carefully programmed and machine paced. High autonomy jobs would seem to be much easier to rate in terms of the job performance of individuals. For one thing, they open up the possibility for greater variance in actual performance effectiveness and for another, they allow attribution of the different levels of performance directly to the individual rather than to machines or other factors beyond the individual's control. Related to this characteristic is the dimension of task identity that Hackman and Lawler identified. This dimension refers to the degree to which the individual is responsible for a whole piece of work. The prediction here would be that high task identity jobs would lend themselves to better ratings of individuals than would low task identity jobs, since it is easier to discern an individual's contribution when that individual produces an entire piece of work.

Finally, Hackman and Lawler identify feedback as an important task characteristic. By feedback they mean both the kind that comes naturally to the individual as a result of seeing the task progress toward completion (i.e., intrinsic feedback), as well as that which comes from the information and control system of the organization. To the extent that high intrinsic feedback results in the effectiveness of job performance also being more discernible to observers, such observers will be better equipped to assess performance on relevant grounds. If such observational data is strongly supplemented by data from the organization's information and control system, decrements in appraisal effectiveness stemming from information deficiencies seem likely to be minimized. Again, the overriding principle here is that the more objective and plentiful the data the source has, the less need there is to resort to biases and irrelevant data as the basis for appraisals, and the more difficult it is to defend appraisals made on such irrelevant grounds.

Overall, a case can be made that a number of task characteristics need to be considered by any model of performance appraisal effectiveness. Some jobs simply do not lend themselves to the effective assessment of the performance of job holders. Three solutions to this problem suggest themselves: job redesign, assessing performance on the group level, and not assessing performance at all. While the last option speaks for itself, the first and second need a little elaboration. The suggestion that job redesign may produce improvements in performance appraisal effectiveness is certainly not a typical way of looking at performance appraisal, but it is rather consistent with the literature on job redesign (see, e.g., Hackman, 1977). The same characteristics that make a job motivating are those that should make performance in it assessable. Thus, we end up with the interesting conclusion that if you cannot evaluate someone's job performance, the answer may be a job redesign project. The alternative of focusing the appraisal on a large aggregation of individuals follows from the point that the effectiveness of performance is often more discernible at the group level than at the individual level.

Organizational structure   At this point, research has not focused on the relationship between the structural characteristics of organizations, either phenomenological or objective, and performance appraisal effectiveness. There are some reasons, however, to expect that a relationship may exist. A major basis for this belief is the close relationship between job design and organization structure. Organization structure can be looked upon as the overall structure of jobs in an organization. As such, the work and thinking on job structure would seem to have a number of points to suggest about the relationship between organization structure and performance appraisal effectiveness. The more the organization structure is such that individuals end up with clearly identified tasks for which they are responsible and the more these tasks have clear-cut goals and objectives associated with them, the greater the effectiveness that performance appraisal systems should be able to achieve. In organizational terms, this probably means an organization that is characterized by high degrees of decentralization, well-developed profit centers with their own results indicators, and relatively well-defined and structured jobs, but not necessarily small in size or flat in structure. In addition, it means an organization structure that has an information system that collects objective data about the performance of different parts of the organization and of different individuals within those parts (see, e.g., Rhode and Lawler, 1976, and Galbraith, 1973). The argument here is that the more objective data available, the less ambiguity there is about individual performance. Overall, then, although there is little evidence to support it, one would suspect that those kinds of organization structures that lead to tasks on which performance is objectively measured and on which results are clearly visible will lead to more effective performance appraisals.

Social structure There is little evidence that shows how the social structure of an organization influences performance appraisals. Unfortunately, there is no generally accepted taxonomy of the social structural attributes of an organization. When people speak of the social or informal structural properties of an organization, they usually talk about such things as the norms and sanctions that exist, the communication patterns in the organization, and the role and style of leadership that is exerted by the formal and informal leaders.

All of these attributes of organizations may well have a significant impact on the accuracy of the ratings that can be obtained in the performance appraisal process. For example, minimal communication among coworkers and the absence of opportunities for people to observe each other doing work very severely limits the number of people who can conduct appraisals effectively. This becomes an obstacle to achieving overall appraisal effectiveness to the extent that it is desirable to have more than one source involved in the appraisal of each object's performance. In addition, if the norms and sanctions in the organization are such that giving valid data about another's performance is not rewarded, then it can be virtually impossible to obtain valid measures of an individual's performance (see, e.g., Lawler, 1971). This can happen in situations where piece-rate pay plans have developed or where the level of distrust between the organization and its employees becomes extremely high. Lawler (1971) has suggested that climates that are characterized by low trust and openness make performance appraisal particularly difficult to do effectively. In fact, Lawler suggests that unless highly objective data are available on performance, it isn't advisable to assess performance in low trust climates.

Finally, if the management style of the organization prevents participation by subordinates in the setting of performance standards and objectives, then it can be very difficult to appraise performance effectively, because there is likely to be very little agreement on what constitutes good performance and how it can be achieved. Overall, then, there is reason to believe that the informal or social structure of an organization can have an impact on the accuracy of performance appraisals in an organization, but there is very little research that documents the nature of its impact or even identifies the variables that should be of concern.

SUMMARY: DETERMINANTS OF APPRAISAL EFFECTIVENESS

Our review of the factors that influence the effectiveness of appraisals shows that there is a substantial body of literature that argues that a wide array of factors impact upon the effectiveness of the appraisal process. It is interesting, however, that a great deal, indeed most, of the research on performance appraisal has focused on the development of better measurement instruments. So far this effort has not paid off, although there is reason to believe that distributional measurement systems, as exemplified by the BDS method we have proposed, can yield progress from this direction. In any case, one cannot help but conclude that appraisal research has underemphasized the nonmeasurement aspects of appraisal. A great deal of the variance in appraisal effectiveness seems to be due to such things as the administrative characteristics of appraisal systems, the characteristics of appraisers and appraisees, and the organizational conditions that surround the appraisal process and dictate the jobs that individuals have to perform. All this suggests that in their efforts to develop theories and to empirically isolate the key variables influencing appraisal effectiveness, researchers must pay much more attention to contextual issues than has been the case in the past.

APPENDICES

A. Formulas for Variance Components and Intraclass Correlations in a 2-Random, 1-Fixed Factor Unreplicated ANOVA
  B. A Taxonomy of Appraisal Methods
  C. Distributional Measurement and its Operationalization through Behavioral Discrimination Scales

APPENDIX A

At several points in the body of this chapter we referred to the use of variance components and intraclass correlations for the purpose of evaluating appraisal effectiveness criteria. In order to facilitate the use of these statistics for the purposes we recommend, we present in the table below the formulas for computing them for all ANOVA effects in a design in which sources and objects are random and performance dimensions are fixed. While we feel that the latter design is the one most applicable to the typical appraisal research or evaluation situation, it may not fit any number of special circumstances. We therefore urge that these formulas be used only after deciding that their underlying model is relevant in a given situation. Formulas for the variance components appropriate to other designs can be obtained from Vaughan and Corballis (1969), substituting the three-way interaction term for all occurrences of the residual error (Ms_e) term to account for the unreplicated nature of the observations.

Formulas for Variance Components and Intraclass Correlations in a 2-Random, 1-Fixed Factor Unreplicated ANOVA

ANOVA Effect	Variance component (V)	Intraclass correlation
Source (S) (i.e., appraiser)	$\dfrac{MS_S - MS_{S \times O}}{RN}$	$\dfrac{V_S}{V_S + V_E{}^*}$
Object (O) (i.e., appraisee)	$\dfrac{MS_O - MS_{S \times O}}{RC}$	$\dfrac{V_O}{V_O + V_E{}^*}$
Performance Dimension (D)	$(R-1)\,(MS_D - MS_{S \times D} - MS_{O \times D} + MS_{S \times O \times D})/RCN$	$\dfrac{V_D}{V_D + V_E{}^*}$
Source × Object ($S \times O$)	$\dfrac{MS_{S \times O} - MS_{(S \times O \times D)}{}^*}{R}$	$\dfrac{V_{S \times O}}{V_{S \times O} + V_E{}^*}$
Source × Performance Dimension ($S \times D$)	$\dfrac{MS_{S \times D} - MS_{S \times O \times D}}{N}$	$\dfrac{V_{S \times D}}{V_{S \times D} + V_E{}^*}$
Object × Performance Dimension ($O \times D$)	$\dfrac{MS_{O \times D} - MS_{S \times O \times D}}{C}$	$\dfrac{V_{O \times D}}{V_{O \times D} + V_E{}^*}$
Source × Object × Performance Dimension ($S \times O \times D$)	—	—
Error (E)	$MS_{(S \times O \times D)}{}^*$	—
Total	MS_{Total}	—

Notes: 1. R = the number of performance dimensions; C = the number of sources; and N = the number of objects

2. The asterisked (*) occurrences of the three-way interaction term show where it is being substituted for the residual error term under the assumption that any such interaction effect is completely comprised of error. This assumption is both realistic and efficient in the sense of eliminating the necessity to estimate within-cell error through Tukey's procedure (see Winer, 1971, pp. 394–397).

Table B-1. Taxonomy of Appraisal Methods

		Response Format (Searchingness) Alternatives			
Data Type	Source orientation	Single stimuli (Pick $1/n^a$)	Multiple stimuli (Pick k^b/n)	Rank order (Order p^c/n)	Unstructured (Pick any/n)
1. Preferential choice	*Generic:* How do these descriptors compare to the object?				
A. Dominance form	Which descriptor *best* fits the object?	(E.g., forced choice of most descriptive alternative; pair comparison of descriptors)	(E.g., forced choice of k most and/or least descriptive alternatives)		Not applicable
B. Proximity form	Which of these descriptors fit the object, if any?	Not applicable	Not applicable	Not applicable	(E.g., checklists)
2. Single stimulus	*Generic:* How does the object compare to the descriptor *X*?				
A. Dominance form	Is the object higher than, lower than, or equal to descriptor *X*?	(E.g., mixed standard scales)	Not applicable	Not applicable	Not applicable
B. Proximity form	Would you accept descriptor *X* as a description of the object?	(E.g., behaviorally anchored rating scales)	Not applicable	Not applicable	Not applicable
3. Stimulus comparison	*Generic:* How do these objects compare with respect to descriptor *X*?				
A. Dominance form	Which object does descriptor *X* describe *best*?	(E.g., pair comparison of objects)	(E.g., forced distribution; peer nomination)	(E.g., rank order objects)	Not applicable
B. Proximity form	Which of these objects does descriptor *X* describe, if any?	Not applicable	Not applicable	Not applicable	Not applicable

4. Similarities I	*Generic:* How do these descriptors compare to descriptor A in terms of how well they fit the object?				Not applicable
A. Dominance form	Which descriptors fit the object *most* nearly as well as descriptor A does?	Not applicable		Not applicable	
B. Proximity form	Which descriptors fit the object *equally* as well as descriptor A does?		Not applicable?	Not applicable	
5. Similarities II	*Generic:* How do these objects compare to object X in terms of how well the descriptor fits them?				Not applicable
A. Dominance form	Which object does the descriptor fit *most* nearly as well as it fits object X?	Not applicable		Not applicable	
B. Proximity form	Which objects does the descriptor fit *equally* as well as it fits Object X?	Not applicable	Not applicable		

*: No applications known to authors.

a: N = the number of alternatives from which each response is to be selected.

b: $1 \le k \le (n-1)$.

c: $2 \le p \le (n-1)$.

APPENDIX B

Table B-1 represents a taxonomy of appraisal methods based on Coombs' (1964) data type-searchingness structure shceme. Listed vertically at the left of the table are the five basic types of data that human judges can be asked to furnish about any given set of objects. These types of data differ in terms of whether they require the source to compare multiple descriptors (i.e., performance descriptions) for their fit to each object (viz., preferential choice), to compare multiple objects for their fit to each descriptor (viz., stimulus comparison), to compare each descriptor to each object (viz., single stimulus), to compare multiple descriptors for their similarity to a given descriptor in terms of the degree of fit to each object (viz., similarities I), or to compare multiple objects to a given object in terms of the degree to which a descriptor fits them (viz., similarities II). These generic orientations are presented in question form to the right of the respective data types with which they are associated in the table.

All the data types can exist in both dominance and proximity forms. In their dominance form, the source is oreinted toward choosing the best available alternative from among those presented, except in the case of single stimulus data where the task is to decide whether the object has more or less of a characteristic than the amount that the descriptor represents. In the proximity form, the source is oriented toward deciding whether any of the comparisons offered for consideration constitute equalities.

The five types of data can be collected in any of four basic ways, which are described under the Response Format Alternatives heading. These four alternatives, are, respectively, to pick one from the set of possible answers, to pick a specified number of answers from the set of possible answers, to rank the possible answers in the order in which you would pick them, and to pick any number of answers from the set of possible answers.

Of the forty cells in the table, twenty-two are undefined in the sense that the response format contradicts the data type source orientation. Eighteen types of appraisal methods are therefore conceivable under this scheme, of which the authors have been able to supply examples of the application of eight. Most of the methods that to the authors' knowledge haven't been tried fall into the similarities categories.

APPENDIX C

AN INTRODUCTION TO DISTRIBUTIONAL MEASUREMENT

As a first step, let us clarify the terms we shall be using. An *intensity scale* is a continuum comprised of intervals representing different ways or outcomes of performing on a single trial. The intervals of such a scale will be referred to as *intensity levels*. Each intensity level is typically weighted according to the definitiveness of the standards for satisfactory rates of its occurrence, although other weighting criteria may be used. These weights, regardless of the criterion on which they are based, will be called *intensity weights*.

Two other terms need to be defined and distinguished. The *raw occurrence rate* of an intensity level refers to the percentage of the performance trials being considered on which the intensity level was exhibited. A given raw occurrence rate is of different goodness depending on the goodness of the intensity level on which it is exhibited. For example, a *low* raw occurrence rate of a *bad* way or outcome of performing a job is of *high* goodness; a *low* raw occurrence rate of a *good* way or outcome of performing a job is of *low* goodness. We may therefore conceive of *occurrence rate goodness* as the desirability or satisfactoriness of a given range of raw occurrence rates of a given intensity level. Thus, it is possible to conceive of a scale that specifies a goodness value for each raw occurrence rate that may be exhibited on each intensity level. On such a scale, for example, the lowest rates of the worst ways or outcomes of performing a job would fall into the same interval as the highest rates of the best ways of performing the job. A scale of this sort will hereafter be referred to as an *extensity scale*.

Finally, we need to distinguish a performance distribution from a performance profile. A *performance distribution* refers to the plot of a person's raw occurrence rates over the levels of an intensity scale. In contrast, a performance profile will denote the plot of a person's extensity (i.e., occurrence rate goodness) scores over the levels of an intensity scale.

In seeking to measure a person's performance over the levels of an intensity scale in a manner that is sensitive to all the distinguishing features of such performance, we are faced with the choice of measuring the similarity of either performance distributions or performance profiles to their ideals. The use of performance distributions encounters two serious problems which render them inappropriate for the stated purpose. The first of these problems is that the use of performance distributions for this purpose would unjustifiably assume a linear inverse relationship between the magnitude and desirability of the deviations of actual from ideal raw occurrence rates in each intensity level. This relationship seems very likely to be curvilinear rather than linear for a substantial number of ways or outcomes of performing. For example, exhibiting a very bad way or outcome of performing 60 percent of the time instead of 50 percent would probably make much less of a difference to the goodness of a person's performance than would the difference between exhibiting the same kind of performance 10 percent of the time versus never. This kind of distinction would be completely ignored in the comparison of actual to ideal raw occurrence rates that the use of performance distributions in this context would involve.

Secondly, appraising performance in terms of the match between actual and ideal distributions of raw occurrence rates would involve the unjustifiable assumption that a given discrepancy between actual and ideal raw occurrence rates represents the same decrement in goodness for all intensity levels. The fallacy of this assumption is exemplified by an extremely good or bad way or outcome of performing for which a small deviation from its ideal raw occurrence rate would have much stronger positive or negative implications for the goodness of a person's performance than would a deviation of the same size for a marginally good or bad way or outcome of performing.

The joint effect of these two problems with the assessment of performance in terms of the match between actual and ideal performance distributions would be to produce highly inaccurate characterizations of the goodness of object performance. However, both of these problems are explicitly resolved by scaling the raw occurrence rates of each intensity level onto a common extensity (i.e., occurrence rate goodness) continuum. Such an extensity scale directly accounts for any curvilinearity in the relationship between the magnitude and desirability of deviations of actual from ideal raw occurrence rates, and for differences between intensity levels in the relationship between the magnitude and desirability of such deviations. It does this by establishing separate rules of correspondence between the magnitude and desirability of discrepancies between actual and ideal raw occurrence rates for each intensity level. Thus, appropriate measures of the match between actual and ideal performance profiles should yield assessments of the goodness of multiple-trial performance which are both sensitive to all the distinguishing features of each person's performance and much more accurate than those yielded by the same measures applied to performance distributions.

Besides avoiding the problems associated with the use of raw occurrence rates, performance profiles can contribute to appraisal accuracy in a way that performance distributions cannot. The scaling of raw occurrence rates for all intensity levels onto a common extensity scale begins about the following relationship:

The goodness of the average intensity level exhibited by a person will be directly proportional to the average extensity (i.e., occurrence rate goodness) level exhibited by a person over all intensity levels.

In other words, the elevation of a person's performance profile on the extensity (i.e., occurrence rate goodness) scale will be directly proportional to the goodness of the average way or outcome of performing exhibited by the person. An important implication of this relationship is that intensity intervals need not be weighted by their goodness since the goodness of the average way or outcome of performing exhibited by a person is adequately expressed by the performance profile's evaluation of the extensity scale. This permits weights to be assigned to intensity levels in computing actual-to-ideal profile similarity scores that reflect such other values as the extent to which the raw occurrence rates of each intensity interval can be measured reliably and validly, and scored according to definitive standards. Such weighting schemes seem likely to improve discrimination based on the shape and scatter of performance profiles and will not detract from the discriminations based on elevation (i.e., average goodness).

In order to take advantage of all the bases for discrimination furnished by performance profiles, the measure of an obtained profile's similarity to an ideal profile must be sensitive to differences in elevation, shape, and scatter. The ideal profile in all cases will consist of the extensity scale's maximum value over all intensity levels. Of the measures of proximity to this ideal profile that possess the necessary sensitivity (see Carroll and Field, 1974), the city block distance measure seems preferable because of its ease of computation and its amenability to weighting by differential intensity weights to maximize its sensitivity to profile shape and scatter. The best way to express this measure is in terms of the obtained profile's nearness to, rather than its distance from, the ideal profile. This way of expressing the city block distance measure simply involves adding up a person's intensity-weighted extensity scores on all intensity levels used in the assessment of each performance dimension. The resulting totals can then be expressed as percentages of the maximum possible totals for each performance dimension for ease of comprehension.

BEHAVIORAL DISCRIMINATION SCALES: A DISTRIBUTIONAL MEASUREMENT RATING METHOD

The Behavioral Discrimination Scales method (BDS) represents an attempt to achieve the ideal operationalization of the distributional measurement model. This means that the method contains all of the features we previously asserted to be desirable in such an operationalization. However, BDS is not intended to represent the only *useful* way of operationalizing the model; lesser operationalizations will be able to produce the limited improvements in appraisal effectiveness that are all that many situations require. The principal feature for which substitutes will be made in such lesser versions is in fact the key feature of the present version: the scaling of the range of raw occurrence rates for each intensity level of each performance dimension onto a common extensity (i.e., occurrence rate goodness) scale. The procedures for accomplishing this scaling task are the focus of the initial phase of developing a BDS system. The steps comprising this initial phase are described below.

First, a pool of statements describing the full range of satisfactory and unsatisfactory job behaviors and/or outcomes is generated. We recommend that this step be accomplished by having incumbents and their supervisors first list all the job functions carried out daily, periodically (at regular intervals), and occasionally (at irregular intervals). Then the interviewees should be asked to give as many examples as they can of satisfactory and unsatisfactory ways or outcomes of performing each function they list. This way of gathering incidents is essentially a structured Critical Incident generation procedure, which holds the advantage over the conventional unstructured procedure of ensuring a fuller representation of the domain of job functions among the incidents generated.

The resulting pool of incidents is then edited to eliminate duplications, and any of the remaining incidents which exemplify the same generic behavior or outcome are grouped together. A general statement is then written to express the generic behavior or outcome central to each incident or incident grouping, shorn of all description that would tie the statement to specific circumstances. In order to illustrate the type of statement that needs to be abstracted from each incident and incident grouping, consider the following group of incidents that might be collected for the job of printing press operator:

> Had to stop a press run to remove grease from a roller that he or she should have cleaned before the press run started.
> Had to stop a press run to make a paper adjustment that he or she should have made before the press run started.

> Had to stop a press run to fix a mechanical problem that he or she should have discovered in the routine inspection conducted prior to every press run.
> Failed to check the ink reservoir before a press run started, which resulted in a press run stoppage which shouldn't have been necessary.

All of the above incidents are encompassed by the following statement of the generic outcome they share in common:

> Had to stop a press run because of a problem caused by the failure to properly make normal checks and adjustments before the run started.

We shall refer to these statements of generic behaviors and outcomes as *performance specimens*. The reason for deriving these performance specimens instead of using specific incidents as the rating content is that sources are likely to perceive each object to have had a far greater number of opportunities to exhibit such generic behaviors or outcomes than specific incidents. This reduces the number of items necessary to generate ratings on each object. In addition, the greater number of opportunities on which specimen ratings will be based should result in correspondingly higher reliability and score variance.

The performance specimens developed through the process described above are then inserted on a questionnaire administered to a sample of at least twenty job incumbents and their supervisors. Two different forms of the questionnaire, each containing the full pool of specimens are administered, one to each half of the sample. One form asks the following three questions in regard to each specimen.

1. During a normal 6-month period how many times would a person have the opportunity to exhibit this behavior or outcome?

2. It would be *moderately satisfactory* performance to exhibit this behavior or outcome on how many of these occasions?

3. How good or bad is the performance described by this behavior or outcome? (1 = very bad, . . . , 8 = very good)

The other form should contain exactly the same questions except that question 2 should refer to *moderately unsatisfactory* performance.

The results of this survey should be analyzed by first converting question 2 responses to percentages of question 1 responses for each specimen and then computing the *t*-statistic for the difference between the mean percentages of the two subsamples for each specimen. All specimens for which the obtained *t*-value does not reach the .01 *p* level should be eliminated on the ground that definitive standards for their satisfactory and unsatisfactory performance do not exist, and that they therefore cannot be legitimately used to discriminate between satisfactory and unsatisfactory performers. This low p level is needed in order for distinct intervals in the range between satisfactory and unsatisfactory occurrence rates to emerge in the scaling procedure. The elimination of specimens that don't meet this criterion shrinks the specimen pool down to a set of specimens that is capable of legitimately discriminating between more and less satisfactory performers.

Next, each specimen's median occurrence percentage and mean rating on question 3 are computed for the combined sample (i.e., for the satisfactory and unsatisfactory subsamples combined). With these indices along with each specimen's occurrence percentage means and standard deviations in each subsample, the extensity (i.e., occurrence rate goodness) scale value for each specimen can be derived. If the specimen's total sample mean on the survey's goodness scale (i.e., question 3) was above the scale midpoint in the good range, the extensity scale values should be derived according to the specifications in Figure C-1. If the specimen's mean goodness rating was in the bad range, the exact same rules for deriving the extensity scale values as shown in Figure C-1 are used, except that the position of the distributions over the horizontal axis will be reversed, and the extensity intervals will therefore be numbered from 8 to 1, left to right, instead of from 1 to 8 as formerly.

The key elements of the scaling system described above are the standard deviations of the raw occurrence rates (i.e., occurrence percentages)

Figure C-1.   Extensity intervals.

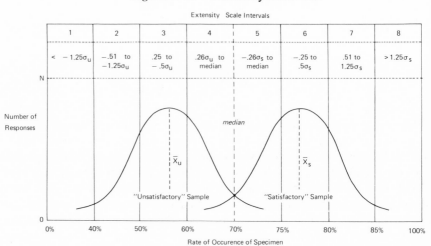

reported in the scaling survey to constitute moderately satisfactory and moderately unsatisfactory performance on each specimen. These standard deviations are used as the basis for determining the range of raw occurrence rates to assign to each extensity interval. This approach is based on the reasoning that the more disagreement there is about what constitutes the benchmark raw occurrence rates, the greater the difference there must be between the midpoints of occurrence rate intervals in order for them to be perceived as differing in goodness. The extensity intervals in the satisfactory and unsatisfactory ranges are separately scaled on the basis of the standard deviation of the raw occurrence rates reported by the subsample surveyed with reference to the respective satisfactoriness range. This permits the extensity intervals to reflect any difference between the two satisfactoriness ranges in the amount of change in the raw occurrence rate needed to bring about a given change in occurrence rate goodness.

The completion of the scaling process signals the end of the initial phase of developing a BDS system. All that remains in the final phase is the construction of the appraisal form and the establishment of scoring procedures. The appraisal form is constructed by listing the specimens in random order at the left side of the form. To the right of each specimen is a column headed by the following question:

> To your personal knowledge, how many times did this person have the opportunity to exhibit this behavior or outcome during the appraisal period? (NOTE: If zero, so indicate and proceed to the next item.)

If the response to this question is greater than zero, the rater is asked to complete the following statement on the same line at the right of the form:

> This person actually exhibited this behavior or outcome on ——————— of these occasions.   (how many?)

The ratings obtained on a BDS appraisal form should be scored in the following manner. First, the frequency assigned to the object on each specimen should be converted to a percentage of his or her opportunities to exhibit the specimen that the source reported knowing about. The extensity scale value corresponding to the resulting percentage for each specimen should then be determined. Each specimen's extensity scale value should then be multiplied by its intensity weight, which can consist of the specimen's t-value from the scaling survey if definitiveness is a valued quality, as we feel it should be. This brings the scoring process to the point where scores on performance dimensions* and on overall performance are ready to be formed.

Performance dimension scores are formed by first multiplying each specimen's extensity-intensity product by any weights (e.g., factor score coefficients) used to reflect the specimen's contribution to each dimension, and then summing the resulting products for the specimens defining each dimension. An overall performance score can be obtained merely by summing or averaging the dimension scores. To permit comparisons between objects and to facilitate comprehension, each object's dimension and overall performance scores should be expressed as percentages of the maximum possible scores the object could have received. Any items excluded in the process of computing an object's raw score should be excluded in computing the object's maximum possible score.

Specimens should be excluded from the computation of an object's observed and maximum possible scores if the object had less than some minimum number of opportunities to exhibit them. Such minimums for each specimen may be set on a variety of bases, but we prefer to use a criterion of one standard deviation below the mean number of opportunities reported for the specimen in the scaling survey sample. The possibility of excluding specimens from the computation of scores also raises the question of what minimum fraction of the specimens defining each performance dimension should be considered sufficient to score a person on the dimension. When dimensions have been defined factor analytically, a reasonable minimum seems to be the number of specimens sufficient to account for half the factor's variance. However, in the case of judgmental dimensions, the minimum must be set more arbitrarily. Probably the most reasonable minimum to set in this case is a number of specimens sufficient to account for half of the sum of the intensity values of all the specimens defining the dimensions.

Costs and Benefits of BDS

It should be apparent that the BDS scoring process is sufficiently complex to require that it be computerized in any operational application of BDS. This requirement, along with the other fairly complex steps needed to construct a BDS system that were described above, make it clear that the development of such a system demands skills exceeding those of

---

*In order to obtain the performance dimensions on which to score an appraisal, the specimen extensity scores obtained in a tryout of the system can be factor analyzed if a large sample of ratings can be obtained. If a large enough sample can't be obtained, dimensions can either be defined judgmentally or through the factor analysis of specimen intercorrelations generated through a procedure the first author has developed for transforming cofrequency estimates by samples as small as $N=1$ to correlations. The details of this latter procedure can be obtained from the first author.

many if not most personnel administrators. This necessarily raises the question of whether a BDS system is worth the effort and expense. Provided that arrangements can be made for computer scoring, our answer is yes, for the reasons described below.

BDS is the first complete operationalization of the distributional measurement model to be proposed, and therefore offers at least the theoretical probability of producing major increases in discriminability over existing rating methods. The only existing method which may approach BDS in its discriminability potential is the Behavior Observation Scales (BOS) methods, recently proposed by Latham and Wexley (1977). However, Latham and Wexley's BOS method departs from distributional measurement in two critical respects which seem likely to lead to its being far less valid and relevant than BDS. First, BOS elicits ratings on all performance 0–19%, 20–39%, . . . , 80–100%), and makes no attempt to express the raw occurrence rate reported for each specimen in terms of the level of goodness it represents on the particular specimen. This approach makes the unjustifiable assumptions that every specimen's occurrence rates are linearly related to their goodness, and that a given deviation from the ideal raw occurrence rate detracts equally from the goodness achieved on every performance specimen (i.e., an equality of slopes assumption). These characteristics of the method can easily lead to inaccurate representations of the proximity of actual to ideal performance. Secondly, in contrast to BDS, BOS can only retain its sensitivity to average performance goodness level as long as the intensity weight attached to each specimen reflects its relative goodness level. This requirement precludes the use of intensity weights reflecting such other useful values as reliability, validity, and definitiveness. Latham and Wexley's disregard of this fact, as evidence by their failure to use goodness values as intensity weights, may be largely responsible for the failure of their BOS scores to explain more than 25 percent of the variance in an objective measure of productivity.

A second advantage of BDS over other appraisal methods is that it may be scored on any one or any combination of the parameters of performance profiles. Thus, it is possible, for example, to examine the validity of selection and training procedures against new aspects of performance on familiar constructs—its shape, elevation, scatter, and their combinations.

Third, BDS shares with Behavior Observation Scales a feature possessed by no other appraisal method: the capacity to appraise factorially complex performance specimens. All other methods systematically exclude specimens which represent performance on more than one dimension. In our view, this seriously curtails the domain of potentially discriminating specimens. Surely, many of the most telling specimens of performance reflect a person's satisfactoriness on more than one dimension. Only be countenancing such specimens can an appraisal system encompass the full range of possibilities for discriminating between objects.

Fourth, BDS prevents the formation of a response set on the part of sources by forcing them to attend to the specific nature of each specimen and its response continuum. It does this by randomly changing both the polarity of the response scale and the opportunity base against which each specimen is rated. Only a few other appraisal methods, including BOS, share the former characteristic, and no other method shares the latter one. Thus, BDS must be considered to have a higher resistance to response sets than any other appraisal method.

Fifth, BDS has greater potential for suppressing source dissembling (i.e., consciously making invalid appraisal ratings) than any other method. We view the tendency to dissemble on the part of sources in the following terms:

$$T_d = (1-P_e)E_d - (P_eN_e)$$

where T_d = the tendency to dissemble

       P_e = the probability of dissembling being exposed

       E_d = the next expected value to the source of the outcomes that could be obtained by dissembling

       N_e = the expected disincentive value to the source of the outcomes resulting from dissembling being exposed

Both E_d and N_e can be decomposed further into their expectancy theoretical components, but we will not take the space to do so here. The point to be made is that holding all other terms in the above model constant, the tendency to dissemble diminishes as the probability of exposure (P_e) increases. This probability necessarily increases when the probability of nonveridical responding being detected increases. The latter probability increases to the extent that one must rate in concrete rather than nebulous terms. Nothing is more concrete than a number, and if a source reports an appreciably wrong number, he or she may be asked to recount the occasions on which the report was based. While one can usually devise justifications for inaccurate evaluative judgments, one would be hard pressed to invent occasions that didn't occur. The recognition of this extreme difficulty of defending inaccurate reports under a BDS system should therefore raise the probability of exposure and reduce the tendency to dissemble.

Sixth, BDS achieves all of the dimensions of relevance in the process of its development. Thus, no further data collection is necessary to conclusively establish its relevance. Deficiency is avoided by involving all formally interested parties in the process of generating performance specimens and retaining all specimens which are found to be significantly definitive. Contamination is avoided by eliminating any specimens on which, by consensus of (the representatives of) all formally interested parties, the standards for satisfactory and unsatisfactory performance are not significantly definitive. Lastly, BDS avoids distortion by using as each specimen's intensity weight an index of the respective specimen's definitiveness derived from the aggregate view of all formally interested parties. The sum of the intensity weights of the specimens associated with each performance dimension may also be viewed as an undistorted weight for use in forming composites of dimension scores.

Finally, the potential in BDS for enhancing the diagnosticity and informativeness of feedback reports appears to be great. Computer-prepared feedback reports could be prepared to convey any or all of the following information:

1. Percentile norms for each performance specimen, dimension, and overall performance based on all persons rated, computed for the present set of appraisals or for any one or combination of prior sets of appraisals.

2. Analyses by performance specimen to reflect the following information:

 a. raw occurrence rate for each specimen reported in the present appraisal,

 b. plots of the raw occurrence rate and extensity scale value achieved on each specimen in all prior appraisals to reflect performance trends,

 c. percentile norms for the occurrence rate of each specimen based on appraisals received by the object population on the present and on all prior appraisals; the object's specimen percentile scores from each prior appraisal may also be plotted over time to discern trends in the object's comparative standing,

3. Plots over time of the object's raw and percentile scores on each dimension and on overall performance.

In summary, it appears that the BDS method has some significant advantages that argue for its use where the resources are available to develop and score it.

FOOTNOTES

1. It should be noted that there may be a problem with the latter approach. The derivations of the formulas expressing the sums of squares components in terms of the MT-MM correlations have been credited to Wolins (Boruch et al., 1970; Kavanagh et al., 1971), but as

far as we can tell neither he nor anyone else has ever published them. Unfortunately, the first publication of the products of these derivations (Boruch et al., 1970), p. 830) contained a serious typographical error, which resulted in the formulas for two sums of squares components being presented as identical. This error was corrected (although not with reference to the error) in an article by Kavanagh et al. that presented further underived formulas for obtaining variance components directly from the MT-MM correlations. All of these formulas have apparently been used by subsequent researchers without comment or effort to replicate the derivations. While the formulas may very well be accurate, the publication and verification of their derivations is long overdue, since they are being used rather frequently.

The standard ANOVA procedure applied to raw data can be used to obtain the needed sums of squares terms until such a time when the Wolins formulas have been properly verified. This approach should not present any problems, since computer programs (e.g., SAS-GLM, BMD5V, SPSS) that will handle any number of objects likely to be encountered are widely available.

2. A word of caution should be mentioned in regard to the use of MT-MM analysis to assess convergent and discriminant validity. The technique relies on the assumption that the multiple sources employed as the method factor are equally capable of assessing each object on each PD, and that there is no dependency between the judgments of different sources. If these assumptions are seriously violated, the technique cannot be used in the form presented above. The assumption most likely to be violated in organizational settings is that sources are equally capable of assessing each object. A possible way of circumventing this problem is to substitute alternate rating techniques for sources as the method factor. The resulting dependency between methods in this case (stemming from the same sources being employed for each method) could be accounted for by adjusting the ANOVA effects through the analysis of covariance, using sources as the (dummy-coded) covariate.

3. Appendix A contains the formulas for constructing various components under the ANOVA model appropriate for the typical appraisal research or evaluation situation (i.e., PDs fixed, sources and objects random).

4. The uncertainty of a response scale is computed by the following formula:

$$U = -\sum_{i=1}^{n} p(x_i)\log_2 p(x_i)$$

where $p(x_i)$ = the probability of response
alternative i being selected

The higher the value of U, the more information a scale is able to convey, and hence the more discriminating it is. Thus, the number of alternatives placed in a region of a response scale should be in direct proportion to the region's probability of being selected, up to the limit of the number of discriminable levels in the region. The use of truncated range response scales in attitude surveys reveals an implicit recognition of this principle. However, little or no such recognition of it exists in the area of appraisal; its explicit recognition as a relevant standard by which to evaluate appraisal systems is long overdue.

REFERENCES

1. Borman, W. C. (1974) "The Rating of Individuals in Organizations: An Alternate Approach, *Organizational Behavior and Human Performance 12:*105–124.
2. ———. (1977) "Some Raters Are Simply Better than Others at Evaluating Performance: Individual Differences Correlates of Rating Accuracy Using Behavioral Scales," paper

presented at the 85th Annual Meeting of the American Psychological Association, San Francisco, California (August).

3. ———, and M. D. Dunnette, (1975) "Behavior Based versus Trait Oriented Ratings: An Empirical Study," *Journal of Applied Psychology 60:* 561–565.

4. Boruch, R. F., J. D. Larkin, L., Wolins, and A. C. and MacKinney (1970) "Alternative Methods of Analysis: Multitrait-Multimethod Data," *Educational and Psychological Measurement 30:* 833–853.

5. Brogden, H. E., and E. K. Taylor, (1952) "The Theory and Classification of Criterion Bias," *Educational and Psychological Measurement 10:* 159–186.

6. Brunswik, E. (1952) *The Conceptual Framework of Psychology,* Chicago: University of Chicago Press.

7. ———. (1956) *Perception and the Representative Design of Experiments,* Berkeley: University of California Press.

8. Campbell, D. T., and D. W. Fiske, (1959) "Convergent an Discriminant Validation by the Multitrait-Multimethod Matrix," *Psychological Bulletin 56:* 81–105.

9. Campbell, J. P., Dunnette, M. D. Dunnette, E. E. Lawler, and K. E. Weick, (1970) *Managerial Behavior, Performance, and Effectiveness,* New York: McGraw-Hill.

10. Carlson, R. E., (1967) "Selection Interview Decisions: The Relative Influence of Appearance and Factual Written Information on an Interviewer's Final Rating," *Journal of Applied Psychology 51:* 461.

11. Carrol, R. M., and J., Field, (1974) "A Comparison of the Classification Accuracy of Profile Similarity Measures," *Multivariate Behavioral Research 9:* 373–380.

12. Coombs, C. H. (1964) *A theory of data,* New York: John Wiley & Sons, Inc.

13. Creswell, M. B. (1963) "Effects of Confidentiality on Performance Ratings of Professional Health Personnel," *Personnel Psychology 16:* 385–393.

14. Deaux, K., and T. Emswiller, (1974) "Explanation of Successful Performance on Sex-linked Tasks: What is Skill for the Male Is Luck for the Female, *Journal of Personality and Social Psychology 29:* 80–85.

15. DeJung, J. E., and J. Kaplan, (1962) "Some Differential Effects of Race of Rater and Ratee on Early Peer Ratings of Combat Aptitude, *Journal of Applied Psychology 46:* 370–374.

16. Edwards, K. J. (1976) "Performance Appraisal and the Law: Legal Requirements and Practical Guidelines," paper presented at the Annual Meeting of the American Psychological Association, Washington, D.C. (September).

17. Flaugher, R. L., J. T. Campbell, and L. W. Pike, (1969) *Prediction of Job Performance for Negro and White Medical Technicians: Ethnic Group Membership as a Moderator of Supervisors' Ratings. (ETS Service Report PR-61-5).* Princeton: Educational Testing Service.

18. Freeberg, N.E. (1969) "Relevance of Rater-Ratee Acquaintance in the Validity and Reliability of Ratings, *Journal of Applied Psychology 53:* 518–524.

19. Friedson, E., and B. Rhea, (1965) "Knowledge and Judgment in Professional Evaluations," *Administrative Science Quarterly 10:* 107–124.

20. Galbraith, J. (1973) *Designing Complex Organizations,* Reading, Mass.: Addison-Wesley.

21. Garner, W. R. (1960) "Rating Scales, Discriminability, and Information Transmission," *Psychological Review 67:* 343–356.

22. Ghiselli, E. E. (1969) "The Efficacy of Advancement on the Basis of Merit in Relation to Structural Properties of the Organization, *Organizational Behavior and Human Performance 4:* 402–413.

23. Guilford, J. P. (1954) *Psychometric Methods, (2nd ed.),* New York: McGraw-Hill.

24. Hackman, J. R. (1977) "Work Design," in J. R. Hackman and J. L. Suttle (eds.),

Improving Life at Work, Santa Monica, Calif.: Goodyear Publishing Co. Inc.

25. ———, and E. E. Lawler, (1971) "Employee Reactions to Job Characteristics," *Journal of applied Psychology 55:* 259–286.

26. Hamner, W. E., J. S. Kim, L. Baird, and W. J. Bigoness, (1974) "Race and Sex as Determinants of Ratings by Potential Employers in a Simulated Work Sampling Task," *Journal of Applied Psychology 59 6:* 705–711.

27. Hollander, E. P. (1954) "Buddy Ratings: Military Research and Industrial Implications." *Personnel Psychology 7:* 385–393.

28. Indik, B. P. (1971) "The Study of Organizational and Relevant Small Group and Individual Dimensions" in W. Ronan and E. P. Prien (eds.), *Perspectives on the Measurement of Human Performance,* New York: Appleton-Century-Crofts.

29. James, L. R., and A. P. Jones, (1974) "Organizational Climate: A Review of Theory and Research," *Psychological Bulletin 81:* 1096–1112.

30. Jaques, E. *Equitable Payment,* New York: John Wiley & Sons, Inc.

31. Kavanagh, M. J. (1971) "The Content Issue in Performance Appraisal: A Review," *Personnel Psychology 24:* 653–668.

32. ———, A. C. MacKinney, and L. Wolins, (1971) "Issues in Managerial Performance: Multitrait-Multimethod Analyses of Ratings, *Psychological Bulletin 75:* 34–49.

33. Kipnis, (1960) D. "Some Determinants of Supervisory Esteem," *Personnel Psychology 13:* 377–391.

34. Kirchner, W. K., and D. J. Reisberg, (1962) "Differences between Better and Less Effective Supervisors in Appraisals of Subordinates," *Personnel Psychology 15:* 295–302.

35. Klimoski, R. J., and M. London, (1974) "Role of the Rater in Performance Appraisal," *Journal of Applied Psychology 59:* 445–451.

36. Latham, G. P., and K. N. Wexley, (1977) "Behavioral Observation Scales for Performance Appraisal Purposes," *Personnel Psychology 30:* 255–268.

37. ———, ———, and E. D. Pursell, (1975) "Training Managers to Minimize Rating Error in the Observation of Behavior," *Journal of Applied Psychology 60 5:* 550–555.

38. Lawler, E. E. (1971) *Pay and Organizational Effectiveness: A Psychological View,* New York: McGraw-Hill.

39. ——— (1977) "Reward Systems," in J. R. Hackman and J. L. Suttle (eds.), *Improving Life at work,* Santa Monica, Calif. Goodyear Publishing Co. Inc.

40. ———, and J. G. Rhode, (1976) *Information and Control in Organizations,* Pacific Palisades, Calif: Goodyear Publishing Co. Inc.

41. Leventhal, G. S. and H. D. Whiteside, (1973) "Equity and Use of Reward to Elicit High Performance," *Journal of Personality and Social Psychology 25:* 75–83.

42. Levine, E. L., and J. Weitz, (1971) "Relationship between Task Difficulty and the Criterion: Should We Measure Early or Late?, *Journal of Applied Psychology 55:* 512–520.

43. Levine, J. M., C. J., Ranelli, and R. S. Valle, (1974) "Self-evaluation and Reaction to a Shifting Other, *Journal of Personality and Social Psychology 29:* 637–643.

44. Levy, S. (1960) "Supervisory Effectiveness and Criterion Relationships," paper presented at the 68th Annual Meeting of the American Psychological Association, Chicago, (May).

45. Lynch, B. P. (1974) "An Empirical Assessment of Perrow's Technology Construct," *Administrative Science Quarterly 19:* 338–356.

46. Mandell, M. M. (1956) "Supervisory Characteristics and Ratings," *Personnel 32:*435–440.

47. Meyer, H. H., E., Kay and R. P. French, Jr. (1965) "Split Roles in Performance Appraisal, *Harvard Business Review, 43* 1:123–129.

48. Mulaik, S.A. (1964) "Are Personality Factors Raters' Conceptual Factors?," *Journal of Consulting PSYCHOLOGY* §–: 506–511.
49. Nieva, V. F. (1976) "Supervisor-subordinate Similarity: A Determinant of Subordinate Ratings and Rewards," University of Michigan, Doctoral Dissertation.
50. Passini, F. T., and W. Norman, (1966) "A Universal Conception of Personality Structure?," *Journal of Personality and Social Psychology 4:* 44–49.
51. Perrow, C. (1967) "A Framework for the Comparative Analysis of Organizations," *American Sociological Review, 32:* 194–208.
52. Porter, L. W., E. E., Lawler, and J. R. Hackman (1975) *Behavior in Organizations,* New York: McGraw-Hill.
53. Quinn, J. L. (1969) "Bias in Performance Appraisal," *Personnel Administration 32:* 40–43.
54. Rosen, B., and T. H. Jerdee (1973) "The Influence of Sex Role Stereotypes on Evaluation of Male and Female Supervisory Behavior," *Journal of Applied Psychology 57:* 44–48.
55. ———, (1974) "The Influence of Sex Role Stereotypes on Personnel Decision," *Journal of Applied Psychology 59:* 9–14.
56. Rosenberg, S., and A. Sedlak, (1972) "Structural Respresentations of Implicit Personality Theory," in L. Berkowitz (ed.), *Advances in Experimental Social Psychology, Vol. 6,* New York: Academic ress, Inc.
57. Schneider, B. (1972) "Organizational Climate: Individual Preferences and Organizational Realities," *Journal of Applied Psychology 56:* 211–217.
58. ———. (1973) "The Perception of Organizational Climate: The Customer's View," *Journal of Applied Psychology 57:* 248– 256.
59. ———. (1975) "Organizational Climate: n Essay," *Personnel Psychology 28:* 447–479.
60. ———, and C.J. Bartlett (1968), "Individual Differences and Organizational Climate, I: The Research Plan and Questionnaire Development," *Personnel Psychology* (1970) "Individual Differences and Organizational Climate, II: Measurement of Organizational Climate by the Multitrait-Multimethod Matrix," *Personnel Psychology 23:* 493–412.
62. ———, and D. T. Hall, (1972) "Toward Specifying the Concept of Work Climate: A Study of Roman Catholic Diocesan Prients, *Journal of Applied Psychology, 56:* 447–455.
63. Schneider, D. E., and A. G., Bayroff (1953), "The Relationship between Rater Characteristics and Validity of Ratings, *Journal of Applied Psychology 37:* 278–280.
64. Schneier, C. C. (1977) "Operational Utility and Psychometric Characteristics of Behavioral Expectation Scales: A Cognitive Reinterpretation, *Journal of Applied Psychology 62:* 541–548.
65. Schwab, D. P., H. Heneman, and T. DeCotiis (1975), "Behaviroally Anchored Rating Scales: A Review of the Literature," *Personnel Psychology 28:* 549–562.
66. Seiler, J. A. (1967) *Systems Analysis in Organizational Behavior,* Homewood, Ill.: Irwin-dorsey.
67. Senger, J. (1971) "Managers' Perceptions of Subordinates' Competences as a Function of Personal Value Orientations, *Academic Management Journal 14,* 3: 415–424.
68. Shaw, E. A. (1972) "Differential Impact of Negative Stereotypes in Employee Selection," *Personnel Psychology, 25:* 333–338.
69. Slovic, P., and S. Lichtenstein, (1971), "Comparison of Bayesian and Regression Approaches to the Study of Information Processing in Judgment," *Organizational Behavior and Human Performance 6:* 649–744.
70. Smith, P. C. and L. M. Kendall, (1963) "Retranslation of Expectations: An Approach to the Construction of Unambiguous Anchors for Rating Scales, *Journal of Applied Psychology, 47:* 149–155.

71. Stanley, J. C. (1961) "Analysis of Unreplicated Three-way Classifications with Applications to Rater Bias and Trait Independence," *Psychometrika 26:* 205–219.
72. Thompson, J. D. (1967) *Organizations in Action* New York: McGraw-Hill.
73. Tinsley, H. A., and D. J. Weiss, (1975) "Interrater Reliability and Agreement of Subjective Judgments," *Journal of Counseling Psychology 22:* 353–376.
74. Udy, S. H. (1965) "The Comparative Analysis of Organizations," in J. G. March (ed.), *Handbook of Organizations,* Chicago: Rand-McNally & Company.
75. Van Maanen, J. (1976) "Breaking In: socialization to Work," in R. Dubin (ed.), *Handbook of Work, Organization, and Society,* Chicago: Rand McNally & Company.
76. Vaughan, G. M., and M. C. Corballis (1969) "Beyond Tests of Significance: Estimating Strength of Effects in Selected ANOVA Designs, *Psychological Bulletin 72:* 204–213.
77. Whitla, D. K., and J. E. Tirrell, (1953) "The Validity of Ratings of Several Levels of Supervisors," *Personnel Psychology 6:* 461–466.
78. Winer, B. J. (1971) *Statistical Principles in Experimental Design (2nd ed.),* New York: McGraw-Hill.

RESEARCH IN CORPORATE SOCIAL PERFORMANCE AND POLICY
An Annual Compilation of Research

Series Editor: **Lee E. Preston, School of Management and Center for Policy Studies, State University of New York, Buffalo**

Volume 1. Published 1978. Cloth 306 pages Institutions: $25.00
ISBN NUMBER: 0-89232-069-9 Individuals: $12.50

CONTENTS: Introduction, Lee E. Preston, **Corporate Social Performance and Policy: A Synthetic Framework for Research and Analysis,** Lee E. Preston, State University of New York, Buffalo. **An Analytical Framework for Making Cross-cultural Comparisons of Business Responses to Social Pressures: The Case of the United States and Japan,** S. Prakash Sethi, University of Texas, Austin. **Research on Patterns of Corporate Response to Social Change,** James E. Post, Boston University. **Organizational Goals and Control Systems: Internal and External Considerations,** Kenneth J. Arrow, Harvard University. **The Corporate Response Process,** Raymond A. Bauer, Harvard University. **Auditing Corporate Social Performance: The Anatomy of a Social Research Project,** William C. Frederick, University of Pittsburgh. **Managerial Motivation and Ideology,** Joseph W. McGuire, University of California, Irvine. **Empirical Studies of Corporate Social Performance and Policy: A Survey of Problems and Results,** Ramon J. Aldag, University of Wisconsin, and Kathryn M. Bartol, Syracuse University. **Social Policy as Business Policy,** George A. Steiner and John F. Steiner, University of California, Los Angeles. **Government Regulation: Process and Substantive Impacts,** Robert Chatov, State University of New York, Buffalo. **Managerial Theory vs. Class Theory of Corporate Capitalism,** Maurice Zeitlin, University of California, Los Angeles. **Appendices A, B, C.**

A 10 percent discount will be granted on all institutional standing orders placed directly with the publisher. Standing orders will be filled automatically upon publication and will continue until cancelled. Please indicate which volume Standing Order is to begin with.

 **JAI PRESS INC.**
P.O. Box 1285
165 West Putnam Avenue
Greenwich, Connecticut 06830

(203) 661-7602 Cable Address: JAIPUBL.

OTHER ANNUAL SERIES OF INTEREST FROM JAI PRESS INC.

Consulting Editor for Economics: Paul Uselding, University of Illinois

ADVANCES IN APPLIED MICRO-ECONOMICS
Series Editor: V. Kerry Smith, Resources for the Future,
  Washington, D.C.

ADVANCES IN ECONOMETRICS
Series Editors: R. L. Basmann, Texas A & M University, and George F.
  Rhodes, Colorado State University

ADVANCES IN ECONOMIC THEORY
Series Editor: David Levhari, The Hebrew University

ADVANCES IN THE ECONOMICS OF ENERGY AND RESOURCES
Series Editor: Robert S. Pindyck, Sloan School of Management,
  Massachusetts Institute of Technology

APPLICATIONS OF MANAGEMENT SCIENCE
Series Editor: Randall L. Schultz, Krannert Graduate School of
  Management, Purdue University

RESEARCH IN AGRICULTURAL ECONOMICS
Series Editor: Earl O. Heady, Director, The Center for Agricultural and
  Rural Development, Iowa State University

RESEARCH IN CORPORATE SOCIAL PERFORMANCE AND POLICY
Series Editor: Lee E. Preston, School of Management and Center for
  Policy Studies, State University of New York, Buffalo

RESEARCH IN ECONOMIC ANTHROPOLOGY
Series Editor: George Dalton, Northwestern University

RESEARCH IN ECONOMIC HISTORY
Series Editor: Paul Uselding, University of Illinois

RESEARCH IN EXPERIMENTAL ECONOMICS
Series Editor: Vernon L. Smith, College of Business and Public
  Administration, University of Arizona

RESEARCH IN FINANCE
Series Editor: Haim Levy, School of Business, The Hebrew University

RESEARCH IN HEALTH ECONOMICS
Series Editor: Richard M. Scheffler, University of North Carolina,
  Chapel Hill and the Institute of Medicine, National Academy of
  Sciences

RESEARCH IN HUMAN CAPITAL AND DEVELOPMENT
Series Editor: Ismail Sirageldin, The Johns Hopkins University

RESEARCH IN INTERNATIONAL BUSINESS AND FINANCE
Series Editor: Robert G. Hawkins, Graduate School of Business
Administration, New York University

RESEARCH IN LABOR ECONOMICS
Series Editor: Ronald G. Ehrenberg, School of Industrial and Labor
Relations, Cornell University

RESEARCH IN LAW AND ECONOMICS
Series Editor: Richard O. Zerbe, Jr., SMT Program, University of
Washington

RESEARCH IN MARKETING
Series Editor: Jagdish N. Sheth, University of Illinois

RESEARCH IN ORGANIZATIONAL BEHAVIOR
Series Editors: Barry M. Staw, Graduate School of Management,
Northwestern University, and Larry L. Cummings, Graduate School of
Business, University of Wisconsin

RESEARCH IN PHILOSOPHY AND TECHNOLOGY
Series Editor: Paul T. Durbin, Center for the Culture of Biomedicine and
Science, University of Delaware

RESEARCH IN POLITICAL ECONOMY
Series Editor: Paul Zarembka, State University of New York, Buffalo

RESEARCH IN POPULATION ECONOMICS
Series Editors: Julian L. Simon, University of Illinois, and Julie DaVanzo,
The Rand Corporation

RESEARCH IN PUBLIC POLICY AND MANAGEMENT
Series Editors: Colin C. Blaydon, Institute of Policy Studies and Public
Affairs, Duke University, and Steven Gilford, Chicago

*ALL VOLUMES IN THESE ANNUAL SERIES ARE AVAILABLE
AT INSTITUTIONAL AND INDIVIDUAL SUBSCRIPTION RATES.
PLEASE WRITE FOR DETAILED BROCHURES ON EACH SERIES*

A 10 percent discount will be granted on all institutional standing orders placed directly
with the publisher. Standing orders will be filled automatically upon publication and will
continue until cancelled. Please indicate which volume Standing Order is to begin with.

JAI PRESS INC.
P.O. Box 1285
165 West Putnam Avenue
Greenwich, Connecticut 06830

(203) 661-7602   Cable Address: JAIPUBL.

RESEARCH IN SOCIAL PROBLEMS AND PUBLIC POLICY
An Annual Compilation of Research
Series Editor: Michael Lewis, Department of Sociology, University of Massachusetts

Research in Social Problems and Public Policy presents original analyses of contemporary social issues and the policy responses they elicit. It is informed by an editorial philosophy which holds that the value of sociology must ultimately be measured by its power to provide an analytic basis for maximizing human serviceability in society. Each contribution will be selected because it breaks new ground and promises to provide fresh premises for policy discourse. On the assumption that the nature of the problem to be studied should determine the method of its study, the papers appearing in this annual series will represent a variety of analytic approaches extant in contemporary society.

Volume 1.   Summer 1979.   Cloth   Approx. 350 pages   Institutions: $25.00
ISBN NUMBER: 0-89232-068-0   Individuals: $12.50

CONTENTS: **Preface,** Michael Lewis. **The Limits of Deinstitutionalization,** Bernard Beck, Northwestern University. **Inequality and Opportunity in the University,** Eve Spangler, Springfield, Mass. **Controlling Ourselves: Deviant Behavior in Social Science Research,** Myron Glazer, Smith College. **Models of the Decision to Conserve,** Robert K. Leik and Anita Sue Kolman, University of Minnesota. **Adoption in America: An Examination of Traditional and Innovative Schemes,** Howard Alstein, University of Maryland, and Rita J. Simon, University of Illinois. **"The Race Relations Industry" as a Sensitizing Concept,** Lewis M. Killiam, University of Massachusetts. **Affluence, Contentment and Resistance to Feminism: The Case of the Corporate Gypsies,** Margaret L. Andersen, University of Delaware. **Drift and Definitional Expansion: Toward the Hypothesis of Race-Specific Etiologies in the Theory of Criminal Deviance,** Michael Lewis and Anthony Harris, University of Massachusetts. **Crowding and Slums: A Statistical Exploration,** Harvey M. Choldin, University of Illinois, Champaign. **Issues in Combining Social Action with Planning: The Case of Advocacy Planning,** Rosalie G. Genovese, Cornell Management Studies Program and St. John Fisher College, Rochester.

> A 10 percent discount will be granted on all institutional standing orders placed directly with the publisher. Standing orders will be filled automatically upon publication and will continue until cancelled. Please indicate which volume Standing Order is to begin with.

 **JAI PRESS INC.**
P.O. Box 1285
165 West Putnam Avenue
Greenwich, Connecticut 06830

(203) 661-7602   Cable Address: JAIPUBL.